T0331758

Infinite Dimensional
and Finite Dimensional
Stochastic Equations
and Applications in Physics

Infinite Dimensional and Finite Dimensional Stochastic Equations and Applications in Physics

Editors

Wilfried Grecksch
Martin-Luther-University, Halle-Wittenberg, Germany

Hannelore Lisei
Babeş-Bolyai University, Cluj-Napoca, Romania

 World Scientific

NEW JERSEY · LONDON · SINGAPORE · BEIJING · SHANGHAI · HONG KONG · TAIPEI · CHENNAI · TOKYO

Published by

World Scientific Publishing Co. Pte. Ltd.

5 Toh Tuck Link, Singapore 596224

USA office: 27 Warren Street, Suite 401-402, Hackensack, NJ 07601

UK office: 57 Shelton Street, Covent Garden, London WC2H 9HE

Library of Congress Control Number: 2020015443

British Library Cataloguing-in-Publication Data
A catalogue record for this book is available from the British Library.

ISBN 978-981-120-978-9 (hardcover)
ISBN 978-981-120-979-6 (ebook for institutions)
ISBN 978-981-120-980-2 (ebook for individuals)

For any available supplementary material, please visit
https://www.worldscientific.com/worldscibooks/10.1142/11538#t=suppl

Printed in Singapore

Preface

Stochastic analysis and its applications belong to the modern fields of mathematical research. Stochastic processes are widely used as mathematical models which describe the evolution in time of random phenomena. They have applications in many disciplines such as chemistry, biology, physics, computer science, engineering fields, economics, finance, etc.

This volume contains survey articles on various aspects of stochastic partial differential equations (SPDEs) and their applications in stochastic control theory and in physics.

The book is concerned with SPDEs viewed as stochastic evolution equations in Hilbert spaces and in Banach spaces. It is necessary to introduce suitable spaces such that the physical models can be studied rigorously by formulating problems with precise boundary conditions, linear or nonlinear perturbations, driven by additive or multiplicative noise terms, involving specific types of stochastic integrals and certain solution concepts for SPDEs. Beside the study of existence and uniqueness of the solutions of SPDEs the volume focuses also on optimal control problems, as well as on the qualitative and long time behavior of the solutions, by investigating attractors and invariant manifolds. The volume contains also a stochastic approach to quantum phenomena; in this sense, quantum Hamilton equations are discussed.

This volume brings together modern techniques from the theory of functional analysis, probability, partial differential equations, optimization, random dynamical systems, numerical approximation, and mathematical physics.

This book is intended not only for graduate students in mathematics or physics, but also for mathematicians, mathematical physicists, theoretical physicists, and science researchers interested in the physical applications of the theory of stochastic processes.

The research reported here has been collaborative, presenting interdisciplinary results between areas of mathematics and physics of research groups

v

in Germany from the Technical University of Munich (TUM), Friedrich Schiller University in Jena, Martin-Luther-University in Halle-Wittenberg, Max Planck Institute for Dynamics of Complex Technical Systems in Magdeburg, University of Münster, and in Romania from the Babeş-Bolyai University in Cluj-Napoca.

The topics presented in this volume are:
• dynamics of stochastic reaction-diffusion equations;
• stochastic Itô-Volterra backward equations in Banach spaces;
• stochastic equations of Schrödinger type;
• optimal control of stochastic Navier-Stokes equations;
• quantum Hamilton equations from stochastic optimal control theory.
Each contribution presents the current state of a research topic and, based on this, presents new results.

In Chapter 1 C. Kuehn and A. Neamţu first present the solution theory for stochastic partial differential equations. The main goal is to provide a bridge between classical Itô theory and recent developments in the context of rough paths and regularity structures. Further, a completely pathwise approach to the construction of stochastic integrals and solutions of several classes of SPDEs is given. These techniques are not restricted to semi-martingales as the Itô calculus and can be applied for instance to the fractional Brownian motion with Hurst index $H \in \left]\frac{1}{3}, \frac{1}{2}\right]$. The generation of a random dynamical system (RDS) from an Itô-type SPDE has been a long-standing open problem, since its solution is defined almost surely. However, once an SPDE can be solved pathwise it is straightforward to obtain an RDS. Beyond the existence theory, a key focus of this chapter is the study of qualitative and quantitative dynamical systems methods for SPDEs. In this regard, the authors survey several results regarding the dynamics of reaction-diffusion SPDEs including steady states, stability, large deviations, attractors, and invariant manifolds.

Stochastic Itô-Volterra backward equations in Banach spaces driven by cylindrical Wiener process are considered by M. Azimi and W. Grecksch in Chapter 2. The existence and uniqueness of so called M-solutions of these equations are proved by using a martingale representation in UMD spaces (UMD is the abbreviation for unconditional martingale difference), an Itô formula in Banach spaces and fixed point theorems. Two duality theorems between linear stochastic forward equations and stochastic backward equations of Itô-Volterra type in Banach spaces are discussed. The results are used to prove an optimality condition of maximum principle type for an optimal control problem with stochastic Itô-Volterra forward equations. A stochastic heat equation is given as example.

In Chapter 3 optimal control problems are discussed by W. Grecksch and H. Lisei for stochastic Schrödinger type equations driven by cylindrical fractional Brownian motions with quadratic objective functions and bilinear controls. First, the state equation has such a structure that a factorization method can be used. An optimality condition of maximum principle type in the form of a sequence of stochastic variational inequalities is proved. One-dimensional stochastic backward equations are especially used. Second, the case of Lipschitz continuous nonlinearities is considered in such a manner that the stochastic controlled nonlinear Schrödinger equation can be transformed into a pathwise equation. An optimality condition of maximum principle type is proved by using stochastic Itô-Volterra backward equations in Hilbert spaces.

In Chapter 4 P. Benner and C. Trautwein study the optimal control of stochastic Navier-Stokes equations driven by Q-Wiener processes. Additive noise and linear multiplicative noise are considered. The solution process of the state equation is defined locally as a mild solution by introducing suitable stopping times also in the three-dimensional case. The solution trajectories are in the domain of a fractional power of the Laplacian. Existence and uniqueness theorems are proved. Using a stochastic maximum principle, the authors derive a necessary optimality condition to obtain an explicit formula that the optimal control has to satisfy. Moreover, they show that the optimal control satisfies a sufficient optimality condition based on higher order Fréchet derivatives of the cost functional of the control problem. By this method control problems of stochastic Navier-Stokes equations especially for two-dimensional and for three-dimensional domains can be solved uniquely.

Köppe and coauthors explain in Chapter 5 how the mathematical theory of optimal control of coupled stochastic differential forward and backward equations for the position of a particle can be used to derive quantum Hamilton equations. The base of the considerations is Nelson's pioneering work, who derived the Schrödinger equation from Newtonian mechanics. The authors establish a stochastic optimal control problem as the equivalent to the deterministic Pontryagin principle in classical mechanics leading to these quantum Hamilton equations of motion. These equations are coupled stochastic differential forward and backward equations for position and momentum of a particle. They typically have to be solved numerically.

Examples for treating problems from quantum physics by using the presented approach are given.

We thank all authors for their interesting contributions, the subjects of their research are an actual and stimulating source of investigation in the domain of stochastic processes and their applications.

W. Grecksch and *H. Lisei*

Contents

Contents

Chapter 1

Dynamics of Stochastic Reaction-Diffusion Equations

Christian Kuehn* and Alexandra Neamțu†

Technical University of Munich,
Faculty of Mathematics,
85748 Garching bei München, Germany

We provide an overview on solution theory and dynamical phenomena for SPDEs, with a strong focus on stochastic reaction-diffusion equations. The main goal of this work is to provide a bridge between the classical theory and the recent developments in the context of rough paths and regularity structures. In this setting, the long-time behavior of SPDEs is discussed.

Contents

---

*E-mail (corresponding author): ckuehn@ma.tum.de
†E-mail: neamtu@ma.tum.de

1. Introduction

Stochastic partial differential equations (SPDEs) represent a very active research field with numerous recent developments and breakthrough results [100, 101]. There are several well-established approaches and methods used to construct solutions for SPDEs, which is always a challenge due to the irregularity of the noise terms that perturb the equation. In applications, such noise terms can quantify the lack of knowledge of certain parameters, finite-size effects, and/or fluctuations occurring due to external perturbations. Since SPDEs have become a key modelling tool in applications, there has been a growing interest in studying their dynamical phenomena [18, 42, 44, 63, 105, 144].

The main goal of this work is to provide a survey on different approaches to solution theory and dynamical properties for SPDEs, which is accessible for a wide community interested in modern methods in stochastic analysis, dynamics and applications.

This work is structured as follows: Section 2 is devoted to the solution theory of SPDEs driven by general additive and multiplicative noise. The aim is to illustrate the classical Itô-theory, the random field approach and also recent developments in the context of rough paths and regularity structures. The first three subsections 2.1-2.3 provide a concise survey of several broadly used classical solution concepts for semi-linear SPDEs such as mild, weak, strong, martingale and variational solutions [51, 119, 151]. In Section 2.4 we investigate more complicated quasilinear SPDEs and their pathwise mild solutions [124]. Section 2.5 presents an alternative to the Itô calculus via the random field approach [52, 53] based on Walsh integration theory [165]. In contrast to the first five subsections of this chapter, Section 2.6 starts by presenting one possible pathwise approach to the construction of stochastic integrals and solutions of SPDEs. Pathwise techniques are not restricted to semi-martingales as the Itô calculus or Walsh theory and apply, for instance, to certain types of fractional Brownian motion [85, 95, 107]. Finally, we conclude the part on solution theory in Section 2 with singular SPDEs. Due to the high irregularity of the noise, such SPDEs become ill-posed, as also justified in Section 2.5. Therefore, new tools such as regularity structures [100] are required in order to give an appropriate meaning of the solution.

In Section 3 the emphasis is on the long-time qualitative and quantitative dynamical behavior of SPDEs. Section 3.1 collects concepts from the theory of random dynamical systems [4] which will be employed in Sections 3.3 and 3.4 to construct random invariant manifolds [24, 64, 84, 125, 167] and attractors [42, 45, 55, 88–90, 157] for SPDEs. We provide examples for SPDEs which fit into the theory of random dynamical systems and discuss the difficulties that occur in this sense for general SPDEs driven by multiplicative noise. To overcome such obstacles it is helpful to rely on a pathwise approach. Based on this we provide a center manifold theory for PDEs with irregular forcing in Section 3.3. Section 3.2 deals with various stability concepts for SPDEs such as stability in probability, almost sure exponential stability and moment exponential stability [38, 59, 84, 138]. We point out the main techniques required in this setting and illustrate them on several SPDE examples. We do not focus only on asymptotic stability, but also discuss metastability, i.e. dynamics which looks stable on very long time scales [15, 17, 18]. Here we explain, how large deviation principles [44, 75, 111] can be helpful in order to investigate such a behavior. We conclude Section 3 mentioning further dynamical phenomena for SPDEs, which were not discussed in this work due to space limitations, and we provide a brief outlook.

2. Solution Theory for SPDEs

We start by fixing the main notation used throughout this chapter. Let $(\mathcal{H}, \|\cdot\|_{\mathcal{H}})$ be an infinite-dimensional separable Hilbert space and let $\mathcal{L}(\mathcal{H})$ denote the space of linear operators on $\mathcal{H}$ and $\mathcal{L}_2(\mathcal{H})$ the space of Hilbert-Schmidt operators. The scalar product on a Hilbert space will be denoted by $[\cdot, \cdot]$ and $\langle \cdot, \cdot \rangle$ stands for the duality pairing. The dual space of a Banach space $\mathcal{V}$ will be denoted by $\mathcal{V}^*$. For a linear operator A, we write A^* for its adjoint and $\mathrm{Dom}(A)$ for its domain. Furthermore, $(\Omega, \mathcal{F}, \mathbb{P})$ denotes a complete probability space, $(\mathcal{F}_t)_{t \geq 0}$ is a filtration, $\mathcal{B}(\cdot)$ signifies the Borel-sigma algebra on a certain space, $\otimes$ indicates the product σ-algebra as well as product measures, and $\mathbb{E}$ stands for the mathematical expectation. We use the abbreviations $\mathbb{P}$-a.s. (almost surely) and a.e. (almost everywhere). For $d \geq 1$ and a domain $D \subseteq \mathbb{R}^d$ and $1 < p < \infty$, $L^p(D, \mathcal{H})$ denotes the space of p-integrable (over D) $\mathcal{H}$-valued functions, whereas $L^p(\Omega, \mathcal{F}, \mathbb{P}; \mathcal{H})$ refers to the space of $\mathcal{H}$-valued random variables defined on the probability space $(\Omega, \mathcal{F}, \mathbb{P})$ having finite moments of order p. For $k \in \mathbb{R}$ and $1 < p < \infty$, $W^{k,p}(D, \mathcal{H})$ denote the usual Sobolev spaces (i.e. contain functions that belong to $L^p(D, \mathcal{H})$ and their weak derivatives up to order k are again in $L^p(D, \mathcal{H})$). Whenever we incorporate Dirichlet / Neumann boundary conditions, we write $W_D^{k,p}$ / $W_N^{k,p}$. For zero Dirichlet boundary conditions we write for simplicity $W_0^{k,p}$. Finally we fix a time horizon $T > 0$ and set $\Delta_T := \{(s,t) \in [0,T] \times [0,T] : t \geq s\}$.

In order to deal with stochastic evolution equations in infinite-dimensional spaces we firstly introduce the following Hilbert space-valued processes. Further details can be looked up in [51, Chapter 4], [160, Chapter 6] and [119, 151].

Definition 1. An $\mathcal{H}$-**isonormal process** on Ω is a mapping $\mathcal{W} : \mathcal{H} \to L^2(\Omega)$ such that the following two conditions hold:

- for every $h \in \mathcal{H}$ the random variable $\mathcal{W}h$ is centered Gaussian, more precisely $\mathcal{W}h \sim N(0, \|h\|_{\mathcal{H}}^2)$;
- $\mathbb{E}(\mathcal{W}h_1 \cdot \mathcal{W}h_2) = [h_1, h_2]_{\mathcal{H}}$ for every $h_1, h_2 \in \mathcal{H}$.

There are several classical examples for the choice of $\mathcal{H}$. For instance, if $\mathcal{H} := L^2(0,T)$ then $W(t) := \mathcal{W}1_{[0,t]}$ defines a (real-valued) Brownian motion on $[0,T]$. Furthermore, if $\mathcal{H} := L^2(D)$ where $D \subseteq \mathbb{R}^d$ for $d \geq 1$ one obtains *white noise* on D.

Definition 2. An $L^2(0,T;\mathcal{H})$-isonormal process is called $\mathcal{H}$-**cylindrical Brownian motion** on $[0,T]$.

An $\mathcal{H}$-cylindrical Brownian motion will be denoted by $W_{\mathcal{H}}$. From the definition one immediately has that:

- $(W_{\mathcal{H}}(t)h)_{t \in [0,T]}$ is a real-valued Brownian motion for every $h \in \mathcal{H}$;
- $\mathbb{E}(W_{\mathcal{H}}(t)h \cdot W_{\mathcal{H}}(s)g) = \min\{s,t\}[g,h]_{\mathcal{H}}$ for every $g, h \in \mathcal{H}$ and $s,t \in [0,T]$.

Setting $\mathcal{H} := L^2(D)$ in the Definition 2 we remark that a $L^2(D)$-cylindrical Brownian motion provides the mathematical model for **space-time white noise** on $[0,T] \times D$. This explains why $\mathcal{H}$-cylindrical Brownian motions appear naturally in the context of stochastic partial differential equations. As in the finite-dimensional case, one intuitively expects to represent $W_{\mathcal{H}}$ as a series given by

$$W_{\mathcal{H}}(t) = \sum_{n=1}^{\infty} w_n(t)e_n, \tag{1}$$

where $(w_n(\cdot))_{n \geq 1}$ are one-dimensional independent standard Brownian motions and $(e_n)_{n \geq 1}$ stands for an orthonormal basis in $\mathcal{H}$. However, such a series does not always describe a genuine $\mathcal{H}$-valued process, since it may not always converge in $L^2(\Omega, \mathcal{F}, \mathbb{P}; \mathcal{H})$. Therefore, one is interested in necessary conditions that ensure the convergence of the series (1) in $L^2(\Omega, \mathcal{F}, \mathbb{P}; \mathcal{H})$. One can show that there always exists a larger space $\overline{\mathcal{H}}$ that contains $\mathcal{H}$ such that (1) converges in $L^2(\Omega, \mathcal{F}, \mathbb{P}; \overline{\mathcal{H}})$, according to Proposition 2.5.2 in [51] and Proposition 2.3.4 in [151]. Namely, if the embedding operator $\mathcal{H} \hookrightarrow \overline{\mathcal{H}}$ is Hilbert-Schmidt, i.e. the map $\mathcal{H} \ni x \mapsto x \in \overline{\mathcal{H}}$ is Hilbert-Schmidt, then (1) converges in $L^2(\Omega, \mathcal{F}, \mathbb{P}; \overline{\mathcal{H}})$. Conditions that ensure the convergence of (1) in $\mathcal{H}$ can be expressed in terms of trace-class operators on $\mathcal{H}$. In this context we recall that a non-negative self-adjoint operator $Q \in \mathcal{L}(\mathcal{H})$ is called **trace-class** if

$$\mathrm{Tr}\, Q = \sum_{n=1}^{\infty} [Q e_n, e_n]_{\mathcal{H}} = \sum_{n=1}^{\infty} \lambda_n < \infty, \tag{2}$$

where the sequence of eigenvectors $(e_n)_{n \geq 1}$ of Q forms an orthonormal basis of $\mathcal{H}$ and $(\lambda_n)_{n \geq 1}$ are the corresponding eigenvalues. This leads to the following definition / characterization of an $\mathcal{H}$-valued Wiener process. For more details, see Proposition 4.1 in [51].

Definition 3. An $\mathcal{H}$-valued stochastic process is called Q-**Wiener process** (notation $(W^Q(t))_{t \in [0,T]}$) if the following two conditions hold:

- $(W^Q(t))_{t \in [0,T]}$ is a Gaussian process on $\mathcal{H}$ with mean zero and covariance operator tQ for $t \geq 0$;
- For any $t \geq 0$ the process $(W^Q(t))_{t \in [0,T]}$ has the following representation

$$W^Q(t) = \sum_{n=1}^{\infty} \sqrt{\lambda_n} w_n(t) e_n, \tag{3}$$

where $w_n(t) := \frac{1}{\sqrt{\lambda_n}}(W^Q(t), e_n)$ for $n \in \mathbb{N}$ are independent real-valued standard Brownian motions on $(\Omega, \mathcal{F}, \mathbb{P})$ and the series (3) converges in $L^2(\Omega, \mathcal{F}, \mathbb{P}; \mathcal{H})$. As previously introduced, $(e_n)_{n \geq 1}$ is the sequence of eigenvectors of Q with corresponding eigenvalues $(\lambda_n)_{n \geq 1}$.

The requirement that the trace of Q must be finite can immediately be derived computing the variance of W^Q as given by (3) as follows:

$$\mathbb{E}\|W^Q(t)\|_{\mathcal{H}}^2 = \mathbb{E}\left(\sum_{n=1}^{\infty} \lambda_n w_n^2(t)[e_n, e_n]_{\mathcal{H}} \right) = \sum_{n=1}^{\infty} \lambda_n \mathbb{E} w_n^2(t) = t\, \mathrm{Tr}\, Q. \tag{4}$$

Of course, the covariance of W^Q is given by

$$\mathbb{E}(W^Q(t) \cdot W^Q(s)) = \min\{s,t\} \operatorname{Tr} Q, \quad s,t \in [0,T].$$

The construction of the stochastic integral of Hilbert-space valued processes with respect to a Q-Wiener process (more general cylindrical Wiener processes) can be seen as the natural generalization of the finite-dimensional Itô-integral to Hilbert spaces. Further details can be looked up in [51, Section 4.2], [131, 151]. As in the finite-dimensional one has an analogue square-integrability condition together with an Itô-isometry.

Theorem 1. *Let $\mathcal{U}$ denote a further separable Hilbert space and let ϕ : $[0,T] \times \Omega \to \mathcal{L}_2(\mathcal{U},\mathcal{H})$ be $\mathcal{B}([0,T]) \otimes \mathcal{F}; \mathcal{B}(\mathcal{L}_2(\mathcal{U},\mathcal{H}))$-measurable such that*

$$\mathbb{P}\left(\int_0^T \|\phi(s)\|_{\mathcal{L}_2(\mathcal{U},\mathcal{H})}^2 \, \mathrm{d}s < \infty \right) = 1.$$

*Then ϕ is stochastically integrable with respect to an $\mathcal{U}$-cylindrical Brownian motion $W_{\mathcal{U}}$ and the following **Itô-isometry** holds true:*

$$\mathbb{E}\left\| \int_0^t \phi(s) \, \mathrm{d}W_{\mathcal{U}}(s) \right\|_{\mathcal{H}}^2 = \mathbb{E} \int_0^t \|\phi(s)\|_{\mathcal{L}_2(\mathcal{U},\mathcal{H})}^2 \, \mathrm{d}s, \quad \text{for } t \in [0,T]. \quad (5)$$

After introducing the preliminary notions on infinite-dimensional stochastic processes and integrals, we can now consider the SPDE

$$\begin{cases} \mathrm{d}u(t) = [Au(t) + f(t,u(t))] \, \mathrm{d}t + g(t,u(t)) \, \mathrm{d}W_{\mathcal{U}}(t), & t \in [0,T] \\ u(0) = u_0 \in \mathcal{H}, \end{cases} \quad (6)$$

where the assumptions on the coefficients are precisely specified in each section. As a typical example, the reader may already think of $u = (u(t))_{t \in [0,T]} = (u(t,x))_{t \in [0,T], x \in D}$ (where the space-dependency is usually dropped whenever there is no risk of confusion), the Laplacian $A = \Delta$ and sufficiently regular maps f, g. The aim of the remaining part of this section is to provide an overview and discuss various solution concepts for (6).

2.1. Mild, Weak and Strong Solutions

Throughout this section the stochastic basis $(\Omega, \mathcal{F}, (\mathcal{F}_t)_{t \in [0,T]}, \mathbb{P})$ is fixed and we make the following assumptions on the coefficients of (6). The initial condition u_0 is $\mathcal{F}_0$-measurable, the linear operator A generates a C_0-semigroup $(S(t))_{t \geq 0}$ on $\mathcal{H}$, the nonlinear terms $f : [0,T] \times \Omega \times \mathcal{H} \to \mathcal{H}$,

$g : [0,T] \times \Omega \times \mathcal{H} \to \mathcal{L}_2(\mathcal{U}, \mathcal{H})$ are $(\mathcal{B}([0,t]) \otimes \mathcal{F}_t \otimes \mathcal{B}(\mathcal{H}); \mathcal{B}(\mathcal{H}))$-measurable, respectively $(\mathcal{B}([0,t]) \otimes \mathcal{F}_t \otimes \mathcal{B}(\mathcal{H}); \mathcal{B}(\mathcal{L}_2(\mathcal{U}, \mathcal{H})))$-measurable for every $t \in [0,T]$ and $(W_{\mathcal{U}}(t))_{t \in [0,T]}$ is an $\mathcal{U}$-cylindrical Brownian motion. For notational simplicity, the ω-dependence of f and g has been dropped.

Similar to the deterministic PDE case, we recall the following solution concepts for (6).

Definition 4. An $\mathcal{H}$-valued adapted process $(u(t))_{t \in [0,T]}$ having $\mathbb{P}$-a.s. Bochner integrable trajectories is called a **strong solution** for (6) if for all $t \in [0,T]$

$$\int_0^t u(s) \, \mathrm{d}s \in \mathrm{Dom}(A), \quad \mathbb{P}-a.s.,$$

and for all $t \in [0,T]$:

$$u(t) = u_0 + \int_0^t [Au(s) + f(s, u(s))] \, \mathrm{d}s + \int_0^t g(s, u(s)) \, \mathrm{d}W_{\mathcal{U}}(s), \quad \mathbb{P}-a.s.$$

Definition 5. An $\mathcal{H}$-valued adapted process $(u(t))_{t \in [0,T]}$ having $\mathbb{P}$-a.s. Bochner integrable trajectories is called a **weak solution** for (6) if for every test function $\zeta \in \mathrm{Dom}(A^*)$ and $t \in [0,T]$ we have:

$$\langle u(t), \zeta \rangle = \langle u(t), u_0 \rangle + \int_0^t [\langle u(s), A^*\zeta \rangle + \langle f(s, u(s)), \zeta \rangle] \, \mathrm{d}s$$

$$+ \int_0^t \langle g(s, u(s)), \zeta \rangle \, \mathrm{d}W_{\mathcal{U}}(s), \quad \mathbb{P}-a.s.$$

Clearly, a strong solution is also a weak one.

Definition 6. An $\mathcal{H}$-valued adapted process $(u(t))_{t \in [0,T]}$ is called **mild solution** for (6) if

$$\mathbb{P}\left(\int_0^T \|u(s)\|_{\mathcal{H}}^2 \, \mathrm{d}s < \infty \right) = 1 \tag{7}$$

and for $t \in [0,T]$ the **variation of constants** (or **Duhamel's**) **formula** holds true:

$$u(t) = S(t)u_0 + \int_0^t S(t-s)f(s, u(s)) \, \mathrm{d}s \tag{8}$$

$$+ \int\limits_0^t S(t-s)g(s,u(s)) \; \mathrm{d}W_{\mathcal{U}}(s), \quad \mathbb{P} - a.s. \tag{9}$$

The first integral in (8) will be referred to as **deterministic convolution** and the second one will be called **stochastic convolution**. This is well-defined due to the assumption (7). If additionally the C_0-semigroup $(S(t))_{t\geq 0}$ is analytic ([149, Chapter 3]), then one can derive optimal regularity results for the stochastic convolution [51, Sections 5.4 and 6.4], [162, 163]. Using a classical fixed-point argument, one can prove the following existence result of mild solutions for (6), see Theorem 7.2 in [51].

Theorem 2. *Let f and g additionally satisfy the following Lipschitz and growth boundedness assumptions:*

1) *there exists a constant $L > 0$ such that for all $u, v \in \mathcal{H}$, $t \in [0,T]$ and almost all $\omega \in \Omega$ we have:*

$$\|f(t,\omega,u) - f(t,\omega,v)\|_{\mathcal{H}} + \|g(t,\omega,u) - g(t,\omega,v)\|_{\mathcal{L}_2(\mathcal{U},\mathcal{H})} \leq L\|u - v\|_{\mathcal{H}}; \tag{10}$$

2) *there exists a constant $l > 0$ such that for all $u \in \mathcal{H}$, $t \in [0,T]$ and almost all $\omega \in \Omega$ we have:*

$$\|f(t,\omega,u)\|_{\mathcal{H}}^2 + \|g(t,\omega,u)\|_{\mathcal{L}_2(\mathcal{U},\mathcal{H})}^2 \leq l^2(1 + \|u\|_{\mathcal{H}}^2). \tag{11}$$

There exists a unique (up to equivalence) mild solution of (6).

Naturally one can obtain similar assertions under weaker assumptions on the coefficients (e.g. local Lipschitz continuity, dissipativity, local monotonicity etc.) using for instance cut-off and localization techniques, see [51, 131, 151].

Remark 1.

1) Under the assumptions on the coefficients stated above, weak and mild solutions for (6) are equivalent, see Theorem 5.4 [51] and [147, Theorem 12]. A meaningful example where strong, weak and mild solutions are equivalent is given by the linear stochastic heat equation perturbed by a trace-class Brownian motion, [51, Theorem 5.14] and [160, Theorem 8.10].

2) The solution concepts discussed in this subsection are **strong in the probabilistic sense**, meaning that the probability space together with the processes defined on them have been fixed.

We point out that this existence result of mild solutions for SPDEs carries over with suitable modifications to the Banach space-valued case, see [160, 162] and the references specified therein. First of all, one considers Banach spaces which have certain geometric properties (which are satisfied by all relevant reflexive spaces appearing in the PDE theory such as Sobolev, Besov, Bessel potential spaces) and replaces the Hilbert-Schmidt operators by γ-radonifying ones (Definition 5.8 [160]). Furthermore, one has to impose a stronger Lipschitz assumption in (10), a so-called "randomized Lipschitz condition" ([162, Section 5]). This can be interpreted as a Gaussian version of Lipschitz continuity. This is mainly necessary due to the fact that if a Banach space $\mathcal{V}$ has the property that $g(u)$ is stochastically integrable (with respect to a Brownian motion) for every $\mathcal{V}$-valued function u and Lipschitz continuous function $g : \mathcal{V} \to \mathcal{V}$, then $\mathcal{V}$ is isomorphic to a Hilbert space. Therefore, it is highly non-trivial to find an appropriate Banach space in order to set up a fixed-point argument and conclude existence results for Banach space-valued evolution equations in the Itô-setting. In Section 2.6 we present a pathwise approach, where such technical difficulties are not encountered.

2.2. *Martingale Solutions*

The notion of martingale solution for an SPDE is the same as weak solution for a finite-dimensional SDE. In this context, since one also has weak in the PDE sense, recall Definition 5 one refers to weak solutions in the probabilistic case as *martingale solutions*.

Definition 7. If for given data $\mathcal{H}$, Q, $u_0 \in \mathcal{H}$ and coefficients A, f, g there exists a probability space $(\Omega, \mathcal{F}, \mathbb{P})$ together with a filtration $(\mathcal{F}_t)_{t \geq 0}$ and a Q-Wiener process such that u is the mild solution of the SPDE

$$
\begin{cases}
du = [Au + f(u)]\, dt + g(u)\, dW^Q(t), & t \in [0, T] \\
u(0) = u_0 \in \mathcal{H},
\end{cases}
\tag{12}
$$

then $(\Omega, \mathcal{F}, \mathbb{P}, (\mathcal{F}_t)_{t \geq 0}, W^Q, u)$ is called a **martingale solution** for (12).

This means that for an initial configuration one seeks a probability space together with a Q-Wiener process which is defined on it, such that u is a mild solution for (12). This means that the probability space and W^Q are part of the solution. The general strategy for existence of martingale solutions is based on compactness arguments, [51, Section 8.3] or on the

Girsanov theorem, see [51, Section 10.3]. We shortly describe the first situation which additionally assumes that the semigroup $(S(t))_{t\geq0}$ is *compact* in order to apply Arzela Ascoli's theorem. For example, this holds true if A is the realization of a uniformly elliptic operator on a bounded domain and it will also be explored in Chapter 3.

Theorem 3. *Let* $(S(t))_{t\geq0}$ *be compact,* $f : \mathcal{H} \to \mathcal{H}$, $g : H \to \mathcal{L}_2(\mathcal{H})$ *be continuous mappings and* W^Q *a* Q*-Wiener process on* $\mathcal{H}$. *If additionally* f *and* g *satisfy the linear growth condition*

$$\|f(u)\|_{\mathcal{H}} + \|g(u)\|_{\mathcal{L}_2(\mathcal{H})} \leq c(1 + \|u\|_{\mathcal{H}}), \quad x \in \mathcal{H}, \tag{13}$$

then the SPDE (12) *has a martingale solution.*

The proof of this theorem relies on the following steps.

1) One constructs solutions u^n for regular coefficients f^n and g^n on a probability space $(\Omega, \mathcal{F}, \mathbb{P})$ with a filtration $(\mathcal{F}_t)_{t\geq0}$ and a Q-Wiener process W^Q. Here $f^n := P_n(f((P_n u)))$ and $g^n := P_n(g((P_n u)))$ for $u \in \mathcal{H}$ where P_n denotes the projection on the finite-dimensional subspace spanned by $\{e_1, \ldots e_n\}$. Again $(e_n)_{n\geq1}$ stands for an orthonormal basis in $\mathcal{H}$. The main idea is to approximate f^n and g^n by Lipschitz continuous mappings that satisfy (13) (with possibly another constant c independent on n).

2) One shows that the sequence of laws $\{\text{Law}(u^n)\}_{n\geq1}$ converges weakly in $C([0,T], \mathcal{H})$ to a measure μ.

3) One constructs the solution u of (12) with the law μ on a new probability space $(\tilde{\Omega}, \tilde{\mathcal{F}}, \tilde{\mathbb{P}})$ for a new filtration $(\tilde{\mathcal{F}}_t)_{t\geq0}$ with respect to a new Q-Wiener process $\tilde{W}^Q$. The main tools required to perform this step are Skorohod's embedding and martingale representation theorems.

For further details on this topic, see [51, Chapter 8] and [34].

2.3. *Variational Solutions*

In contrast to the previous sections, we deal now with *nonlinear operators* A satisfying suitable monotonicity, coercivity and growth conditions and discuss the variational approach of [12, 13, 131, 151, 153]. Further solution concepts for nonlinear operators will be presented in Section 2.4. Similar to (6) we consider

$$\begin{cases} \mathrm{d}u(t) = A(t, u(t)) \, \mathrm{d}t + g(t, u(t)) \, \mathrm{d}W_\mathcal{U}(t), & t \in [0, T] \\ u(0) = u_0 \in \mathcal{H}. \end{cases} \tag{14}$$

Let $\mathcal{V}$ stand for a reflexive Banach space such that the embeddings $\mathcal{V} \hookrightarrow \mathcal{H} \hookrightarrow \mathcal{V}^*$ are continuous and dense. Then $(\mathcal{V}, \mathcal{H}, \mathcal{V}^*)$ is a **Gelfand triple**. The coefficients

$$A : [0, T] \times \mathcal{V} \times \Omega \to \mathcal{V}^* \text{ and } g : [0, T] \times \mathcal{V} \times \Omega \to \mathcal{L}_2(\mathcal{U}, \mathcal{H})$$

satisfy suitable measurability conditions and the following assumptions:

1) (Hemicontinuity) For all $u, v, x \in \mathcal{V}$, $\omega \in \Omega$ and $t \in [0, T]$ the map

$$\mathbb{R} \ni \lambda \mapsto \ _{\mathcal{V}^*}\langle A(t, u + \lambda v, \omega), x \rangle_{\mathcal{V}}$$

   is continuous.

2) (Local monotonicity) There exists $c \in \mathbb{R}$ such that for all $u, v \in \mathcal{V}$

$$2 \ _{\mathcal{V}^*}\langle A(\cdot, u) - A(\cdot, v), u - v \rangle_{\mathcal{V}} + \|g(\cdot, u) - g(\cdot, v)\|^2_{\mathcal{L}_2(\mathcal{U}, \mathcal{H})}$$
$$\leq c \|u - v\|^2_{\mathcal{H}} \text{ on } [0, T] \times \Omega.$$

3) (Coercivity) There exists $\alpha \in (1, \infty)$ and constants $c_1 \in \mathbb{R}$ and $c_2 \in (0, \infty)$ and an $(\mathcal{F}_t)$-adapted process $\overline{f} \in L^1([0, T] \times \Omega, dt \otimes \mathbb{P})$ such that for all $v \in \mathcal{V}$ and $t \in [0, T]$

$$2 \ _{\mathcal{V}^*}\langle A(t, v), v \rangle_{\mathcal{V}} + \|B(t, v)\|^2_{\mathcal{L}_2(\mathcal{U}, \mathcal{H})} \leq c_1 \|v\|^2_{\mathcal{H}} - c_2 \|v\|^p_{\mathcal{V}} + \overline{f}(t), \text{ on } \Omega. \tag{15}$$

4) (Boundedness) There exist $c_3 \in [0, \infty)$ and an $(\mathcal{F}_t)$-adapted process $h \in L^{\frac{p}{p-1}}([0, T] \times \Omega, dt \otimes \mathbb{P})$ such that for all $v \in \mathcal{V}$ and $t \in [0, T]$

$$\|A(t, v)\|_{\mathcal{V}^*} \leq h(t) + c_3 \|v\|^{p-1}_{\mathcal{V}}, \text{ on } \Omega,$$

   where p is the constant from assumption 3).

Example 1. Let $D \subset \mathbb{R}$ be an open bounded domain and consider the Gelfand triple

$$\mathcal{V} := W_0^{1,2}(D) \hookrightarrow L^2(D) \hookrightarrow W^{-1,2}(D).$$

We assume for simplicity that $g : \mathcal{V} \to \mathcal{L}_2(L^2(D))$ is Lipschitz continuous and point out the following operators:

- $Au := \Delta u + f(u)\nabla u$, where $f : \mathbb{R} \to \mathbb{R}$ is a bounded Lipschitz function,
- and $Au := \Delta u - |u|^{m-2}u + \eta u$, for $\eta < 0$ and $1 \leq m < 2$,

which both satisfy the properties introduced above. ◆

For further applications, see [131, Section 3], [90, Section 3], [12, 13, 151] and the references specified therein. Additional examples are also provided in Section 3.2.

Definition 8. A continuous $\mathcal{H}$-valued process u is called a **variational solution** for (14) if for its $dt \otimes \mathbb{P}$-equivalent class $\overline{u}$ we have the following:

- $\overline{u} \in L^p([0,T] \times \Omega, dt \otimes \mathbb{P}, \mathcal{V}) \cap L^2([0,T] \times \Omega, dt \otimes \mathbb{P}, \mathcal{H})$, where p is the constant from assumption 3).

- $u(t) = u_0 + \int\limits_0^t A(s, \overline{u}(s)) \, ds + \int\limits_0^t g(s, \overline{u}(s)) \, dW_{\mathcal{U}}(s), \quad \mathbb{P}-a.s., t \in [0,T].$

The main existence and uniqueness result in this setting is given in [131, Theorem 1.1].

Theorem 4. *Under the above assumptions, for any given initial condition $u_0 \in L^2(\Omega, \mathcal{F}_0, \mathbb{P})$ there exists a unique variational solution of (14).*

The proof of this statement relies on a Galerkin approximation combined with suitable a-priori estimates of the solution. The variational approach is also applicable if one further perturbs (14) by a general additive fractional noise as in [153]. Results regarding Lévy-type noise can be looked up in [90]. Results regarding the equivalence of mild and variational solutions can be found in [164, Section 7.5].

2.4. *The Quasilinear Case*

Using similar notations as in Section 2.3 we are now interested in *semigroup methods* and *mild solutions* for SPDEs constituted by

$$\begin{cases} du(t) = A(u(t))u(t) \, dt + g(u(t)) \, dW_{\mathcal{U}}(t), & t \in [0,T] \\ u(0) = u_0 \in \mathcal{H}. \end{cases} \tag{16}$$

To this aim we let for simplicity $g : \mathcal{H} \to \mathcal{L}_2(\mathcal{U}, \mathcal{H})$ be Lipschitz continuous. Furthermore we denote by $\mathcal{P}_T$ the set of all $\mathcal{H}$-valued continuous $(\mathcal{F}_t)$-adapted processes on $[0,T]$ and impose the following restrictions on A:

1) For every $u \in \mathcal{P}_T$, $A(u)$ is a **sectorial operator** of angle $0 < \phi < \frac{\pi}{2}$, namely

$$\sigma(A(u)) \subset \Sigma_\phi := \{\lambda \in \mathbb{C} : |\arg \lambda| < \phi\}, \text{ for } u \in \mathcal{P}_T,$$

where $\sigma(A(u))$ denotes the spectrum of $A(u)$.

2) For $u \in \mathcal{P}_T$ the resolvent operator $(\lambda\mathrm{Id} - A(u))^{-1}$ satisfies the **Hille-Yosida estimate**, i.e., there exists $\widetilde{M} \geq 1$ such that

$$||(\lambda\mathrm{Id} - A(u))^{-1}||_{\mathcal{L}(\mathcal{H})} \leq \frac{\widetilde{M}}{|\lambda| + 1}, \text{ for } \lambda \notin \Sigma_\phi \text{ and } u \in \mathcal{P}_T.$$

3) Let $0 < \nu \leq 1$ be fixed and $L \geq 1$. Then

$$||A^\nu(u)(A(u)^{-1} - A(v)^{-1})||_{\mathcal{L}(\mathcal{H})} \leq L||u-v||_{\mathcal{H}}, \text{ for } u, v \in \mathcal{P}_T. \quad (17)$$

The first two assumptions guarantee that for $u \in \mathcal{P}_T$, $A(u)$ generates a **parabolic evolution system** U^u, which is a family of linear operators depending on two-time parameters and having similar properties with analytic C_0-semigroups, consult [1, Section II.2] and [149, Chapter 3]. Naturally, since u is a stochastic process, U^u is obviously also ω-dependent. One has for every $\omega \in \Omega$ that

$$U^u(t, t, \omega) = \mathrm{Id}, \quad \text{for all } t \in [0, T];$$

$$U^u(t, s, \omega)U^u(s, r, \omega) = U^u(t, r, \omega), \quad \text{for all } r \leq s \leq t.$$

Assumption 3) represents a classical Lipschitz continuity, which can be weakened according to [124]. One can easily incorporate a locally Lipschitz drift term. Keeping this in mind, one would expect to obtain by fixed-point methods a mild solution for (16). This should be given by the variation of constants formula

$$u(t) = U^u(t, 0, \omega)u_0 + \int_0^t U^u(t, s, \omega)g(u(s)) \, \mathrm{d}W_\mathcal{U}(s). \quad (18)$$

However, this cannot hold true, since the terms appearing in the stochastic convolution in (18) do not satisfy the necessary properties required in order to define the Itô-integral. More precisely, one can show that [164, Proposition 2.4] the mapping $\omega \mapsto U^u(t, s, \omega)$ is only $\mathcal{F}_t$-measurable but *not $\mathcal{F}_s$-measurable*. A possible ansatz to overcome this fact is to use the Skorohod integral for non-adapted integrands and Malliavin calculus [146]. Another possibility is to exploit the integration by parts formula in order to define (18) in a meaningful way. This concept of solution is referred to *pathwise mild* and was developed in [152] for *nonautonomous linear random operators* A.

Definition 9. A continuous $\mathcal{H}$-valued adapted process $(u(t))_{t \in [0,T]}$ is called a **pathwise mild solution** for (16) if it satisfies

$$u(t) = U^u(t, 0, \omega)u_0 + U^u(t, 0, \omega) \int_0^t g(u(s)) \, \mathrm{d}W_\mathcal{U}(s)$$

$$- \int\limits_0^t A(u(s))U^u(t,s,\omega) \int\limits_s^t g(u(r)) \; \mathrm{d}W_{\mathcal{U}}(r) \; \mathrm{d}s, \quad \mathbb{P} - a.s. \qquad (19)$$

The last two terms in the previous formula form a *generalized stochastic convolution*. The term (19) is well-defined due to the $\mathbb{P}$-a.s. Hölder continuity of the integral $\int\limits_0^{\cdot} g(u(s)) \; \mathrm{d}W_{\mathcal{U}}(s)$ which is necessary in order to compensate the fact that $\|A(u(s))U^u(t,s,\omega)\|_{\mathcal{L}(\mathcal{H})} \leq \dot{C}(\omega)(t-s)^{-1}$ for $0 \leq s < t \leq T$. In the nonautonomous case, one can also use the *forward integral* of Russo-Vallois [154] to define the convolution (18) and to construct a solution to the corresponding SPDE which coincides with the pathwise mild solution as argued in [152, Section 4.5].

Theorem 5. *Under the previous assumptions, the quasilinear problem* (16) *has a unique local-in-time pathwise mild solution.*

The statement can be proved using fixed-point arguments as in [124]. This can be achieved in two steps. First of all, fixing a process $v \in \mathcal{P}_T$ and setting $A_v(t) := A(v(t,\omega))$, we obtain a Cauchy problem with time-dependent random drift

$$\mathrm{d}u = A_v(t)u(t) \; \mathrm{d}t + g(u(t)) \; \mathrm{d}W_{\mathcal{U}}(t), \quad t \in [0,T]. \qquad (20)$$

Using [152, Theorem 5.3] we infer that (20) has a pathwise mild solution given by

$$u(t) = U^v(t,0)u_0 + U^v(t,0) \int\limits_0^t g(u(s)) \; \mathrm{d}W_{\mathcal{U}}(s)$$

$$- \int\limits_0^t A_v(s)U^v(t,s)A_v(s) \int\limits_s^t g(u(r)) \; \mathrm{d}W_{\mathcal{U}}(r) \; \mathrm{d}s, \quad \mathbb{P} - a.s.,$$

where $U^v(t,s)$ is the random parabolic evolution operator generated by A_v. For the sake of brevity we dropped the ω-dependence in the previous formula. The second step is to prove that the mapping

$$\Phi(v) := u, \text{ for } v \in \mathcal{P}_T$$

maps $\mathcal{P}_T$ into itself and is a contraction if the time-horizon T is chosen sufficiently small. As applications we mention the stochastic Shigesada-Kawasaki-Teramoto model from mathematical biology, which was introduced to analyze population segregation by induced cross-diffusion, see [124, Section 4]. Written in divergence form, this is given by

$$\mathrm{d}u = \mathrm{div}(\widetilde{\mathcal{A}}(u)\nabla u) \; \mathrm{d}t + f(u) \; \mathrm{d}t + g(u) \; \mathrm{d}W_{\mathcal{U}}(t),$$

where for $u := (u_1, u_2)^T$, the matrix $\widetilde{\mathcal{A}}$ is constituted by

$$\widetilde{\mathcal{A}}(u) := \begin{pmatrix} k_1 + 2cu_1 + au_2 & au_1 \\ bu_2 & k_2 + 2du_2 + bu_1, \end{pmatrix}$$

and the nonlinear term is the same as in the classical Lotka-Volterra model, i.e.

$$f(u) := \begin{pmatrix} \delta_{11}u_1 - \gamma_{11}u_1^2 - \gamma_{12}u_1u_2 \\ \delta_{21}u_2 - \delta_{21}u_1u_2 - \gamma_{22}u_2^2 \end{pmatrix}.$$

The constants above are assumed to be positive and chosen such that the matrix $\widetilde{\mathcal{A}}(u)$ is positive definite.

We conclude this subsection pointing out that there are numerous other solution concepts for quasilinear equations such as martingale solutions [60], strong (in the probabilistic sense) and weak (in the PDE sense) solutions [110], where the coefficients are approximated with locally monotone ones, as discussed in Section 2.3, kinetic solutions [56, 69], entropy solutions [54], solution concepts for quasilinear SPDEs via rough paths theory [148], paracontrolled calculus [7, 79] and regularity structures [87].

2.5. *Random Field Approach*

In this section we briefly present the main ideas of an alternative method to [51] (Section 2.1) to construct solutions for SDPEs, see [52, 53] for further details. As described at the beginning of Section 2, the framework of [51] relies on defining stochastic integrals with respect to Hilbert space-valued processes, whereas the random field approach is based on Walsh integration theory [165] with respect to martingale measures. This approach will also be explored in Section 3 in order to compute moments of the solutions and Lyapunov-exponents. As a reminder, the concept **random field** refers to a family of random variables indexed by several parameters, usually space and time. We now fix $d \geq 1$ and consider the heat equation

$$\frac{\partial u(t,x)}{\partial t} = \Delta u(t,x) + u(t,x)\dot{W}(t,x), \quad t > 0 \text{ and } x \in \mathbb{R}^d, \quad (21)$$

where $\dot{W} = \partial_t W$ and the noise is white in time and (possibly) correlated in space, i.e.

$$\mathbb{E}[\dot{W}(t,x)\dot{W}(s,y)] = \delta_0(t-s)f(x-y), \quad s,t > 0 \text{ and } x,y \in \mathbb{R}^d.$$

Here δ_0 is the Dirac distribution. We further assume that the initial condition u_0 is constant and f is a positive definite function having a **spectral measure** μ, i.e., $\mathcal{F}\mu = f$, where $\mathcal{F}$ denotes the Fourier-transform. There are several examples for f such as:

1) $f = \delta_0$. In this case one obtains **space-time white-noise** (on $\mathbb{R}_+ \times \mathbb{R}^d$), recall Definition 2. The spectral measure is $\mu(\mathrm{d}\xi) = (2\pi)^{-d/2}\mathrm{d}\xi$.

2) (**Riesz kernels**) $f(x) = |x|^{-\eta}$ for $0 < \eta < d$. Here the spectral measure is $\mu(\mathrm{d}\xi) = c_{d,\eta}|\xi|^{-(d-\eta)}\mathrm{d}\xi$, where the constant $c_{d,\eta}$ depends only on the dimension d and on η.

3) (**Fractional kernels**) For $d = 1$ and $H \in (0,1)$ let $f(x) = c_H|x|^{2H-2}$ with $\mu(\mathrm{d}\xi) = c_H|\xi|^{1-2H}\mathrm{d}\xi$. We notice that if $H \leq 1/2$, f is not a measure anymore.

Let $\overline{\mathcal{U}}$ stand for a separable Hilbert space which will be described below. For $G : [0,T] \times \Omega \to \overline{\mathcal{U}}$ (satisfying suitable measurability assumptions), we introduce the notation

$$\int_0^T G(s)\,\mathrm{d}W(s) := \int_0^T \int_{\mathbb{R}^d} G(s,y)W(\mathrm{d}s,\mathrm{d}y) \qquad (22)$$

and emphasize that the stochastic integral (22) is defined in the sense of Walsh [165]. More precisely, letting $\overline{\mathcal{U}}$ be the Hilbert space obtained by the completion of $C_0^\infty(\mathbb{R}^d)$ under

$$\|\phi\|_{\overline{\mathcal{U}}}^2 = (2\pi)^{d/2} \int_{\mathbb{R}^d} \mu(\mathrm{d}\xi)\,|\mathcal{F}(\phi)(\xi)|^2,$$

one has for an adapted process $G \in L^2([0,T] \times \Omega, \overline{\mathcal{U}})$ the *isometry*

$$\mathbb{E}\left(\int_0^T G(s)\,\mathrm{d}W(s)\right)^2 = \mathbb{E}\|G\|_{L^2(0,T,\overline{\mathcal{U}})}^2 = \mathbb{E}\int_0^T \mathrm{d}t \int_{\mathbb{R}^d} \mu(\mathrm{d}\xi)\,|\mathcal{F}(G(t))(\xi)|^2.$$

$$(23)$$

Definition 10. We call a stochastic process $\{u(t,x) : t > 0, x \in \mathbb{R}^d\}$ satisfying

$$u(t,x) = \Gamma_t u_0(x) + \int_0^t \int_{\mathbb{R}^d} \Gamma_{t-s}(y-x)u(s,y)W(\mathrm{d}s,\mathrm{d}y), \quad \mathbb{P} - a.s., \qquad (24)$$

for all $t > 0$ and $x \in \mathbb{R}^d$, a **mild random field solution** of (21).

Here Γ is the fundamental solution of the heat equation / heat kernel, i.e.

$$\Gamma_t(x) = (4\pi t)^{-d/2}\exp\left(\frac{-|x|^2}{4t}\right)$$

and the convolution appearing in (24) is constructed via Walsh's theory. In this case, one can easily compute the Fourier-transform of the heat kernel, i.e.

$$\mathcal{F}(\Gamma_t)(\xi) = \exp(-4\pi^2 t |\xi|^2), \quad \xi \in \mathbb{R}^d$$

and observe that

$$\int_0^T \exp(-4\pi^2 t |\xi|^2) \, \mathrm{d}t = \frac{1}{4\pi^2 |\xi|^2}(1 - \exp(-4\pi^2 T |\xi|^2)).$$

Recalling (23), in order that

$$\int_0^T \mathrm{d}t \int_{\mathbb{R}^d} \mu(\mathrm{d}\xi) \, |\mathcal{F}(\Gamma_t)(\xi)|^2 < \infty,$$

one has to impose that the spectral measure satisfies

$$\int_{\mathbb{R}^d} \frac{\mu(\mathrm{d}\xi)}{1 + |\xi|^2} < \infty. \tag{25}$$

Condition (25) is referred to as **Dalang's condition**. For space-time white-noise, recall $\mu(\mathrm{d}\xi) = (2\pi)^{-d/2}\mathrm{d}\xi$, it is immediately clear that (25) holds true only in $d = 1$. Consequently, in higher space-dimensions, the problem (21) becomes ill-posed. This aspect was investigated in [100] and will also be addressed in Section 2.7.

Existence of mild random field solutions for (21) was obtained in [52]. In this framework, one has an analogue statement to Theorem 2. Namely, if one incorporates measurable drift and diffusion coefficients f and g in (21) (using the same notations as in Theorem 2) satisfying Lipschitz and growth boundedness conditions, one obtains a mild random field solution $\{u(t, x) : t > 0, x \in \mathbb{R}^d\}$ such that for every $p \geq 1$

$$\sup_{t \geq 0} \sup_{x \in \mathbb{R}^d} \mathbb{E}|u(t, x)|^p < \infty.$$

The proof uses Picard iteration, see [52], [53, Theorem 4.3].

2.6. *Mild Solutions for Rough SPDEs*

In this section we would like to deal with a random forcing, which unlike Brownian motion is no longer a semi-martingale and its trajectories are not independent. A meaningful example in this sense is given by the **fractional Brownian motion** (fBm), originally introduced by B. Mandelbrot and J. van Ness in [137]. This is also a Gaussian process, but its covariance function additionally depends on a Hurst index / parameter $H \in (0,1)$. If $H = 1/2$ one obtains Brownian motion, if $H \neq 1/2$, this process exhibits different behavior from Brownian motion. The key feature which is required to comprehend the next steps is the Hölder continuity of its trajectories for an exponent $\alpha < H$. The entire solution theory relies only on this regularity result and is developed within the framework of the *rough path approach* [77, 95, 99, 135]. Similar to Skorohod integrals [146], rough path techniques also go beyond the semi-martingale case, where Itô calculus and Walsh theory no longer apply. The major difference is that these provide a completely pathwise construction of stochastic integrals and implicitly of the solution of SPDEs. However, many solution concepts and techniques that are available in the context of deterministic PDEs have not been completely developed for rough SPDEs (e.g. weak solutions, variational inequalities, etc). Therefore, this has been a very active research field in the last few years with numerous recent developments, see for instance [57, 58, 86, 95, 98, 109] and the references specified therein.

In the finite-dimensional case, the theory is well-understood [77, 78, 135] and rough path techniques have been successfully applied to investigate dynamical aspects of solutions, see [9, 91, 125].

In this section, we provide the main intuition of the construction of a mild solution for a SPDE similar to (6), but driven by multiplicative fractional noise. For further details regarding the following approach, see [107]. We fix the **Hurst parameter** $H \in (1/3, 1/2]$ and consider the non-linear SPDE (12), driven by a trace-class fBm B^H, more precisely

$$\begin{cases} du = [Au + f(u)] \, dt + g(u) \, dB^H(t), & t \in [0,T] \\ u(0) = u_0 \in \mathcal{H}. \end{cases} \tag{26}$$

Here, B^H is a $\mathcal{U}$-valued fBm and is defined as in (3) taking a sequence of independent standard real-valued fBm-s $(w_n)_{n \in \mathbb{N}}$ having the same Hurst index $H \in (1/3, 1/2]$. Furthermore, we assume that A generates an *analytic C_0-semigroup* and for $\beta \in \mathbb{R}$ we denote by $\mathcal{H}_\beta := \text{Dom}((-A)^\beta)$ the domains of its fractional powers. These can be identified with Sobolev spaces for certain ranges of β. For a detailed description of these spaces, see [149, Chapter 3]. The diffusion coefficient $g : \mathcal{H} \to \mathcal{L}(\mathcal{U}, \mathcal{H})$ is supposed to be three times continuously differentiable with bounded derivatives and also Lipschitz continuous as a mapping $g : \mathcal{H} \to \mathcal{L}(\mathcal{U}, \mathcal{H}_\beta)$. Typical examples of such operators are given by integral operators with smooth kernels, see [57, 85, 107]. Naturally, for finite-dimensional noise one can consider polynomials with smooth coefficients, see [57, 86]. For simplicity we set $f = 0$, since this term does not require any additional arguments. Similar to (6) we call a mild solution for (26) a process u that satisfies the variation of constants formula

$$u(t) = S(t)u_0 + \int_0^t S(t - s)g(u(s)) \, \mathrm{d}B^H(s). \tag{27}$$

However, additional measurability and adaptedness conditions are not required. We now provide a purely pathwise definition of the stochastic convolution

$$\int_0^t S(t - s)g(u(s)) \, \mathrm{d}B^H(s). \tag{28}$$

The strategy to define (28) relies on an approximation procedure. We firstly consider a smooth path B^H and a (Hölder) continuous trajectory u. The general argument eventually follows considering smooth approximations of B^H. Thereafter, the passage to the limit entails a suitable construction/interpretation of all the expressions above according to [107, Section 5].

Throughout this section we use the standard rough path notation, i.e. the value at time t of a function y is given by y_t instead of $y(t)$. For $y : [0, T] \to \mathcal{H}$, $y_{st} := y_t - y_s$ denotes an **increment**. We use the same notation to indicate the evaluation at two time points s and t of an element $y : \Delta_T \to \overline{\mathcal{H}}$, where $\overline{\mathcal{H}}$ stands for an arbitrary separable Hilbert space. For the semigroup S we keep the notation $S(\cdot)$.

Our aim is to define (28) using Riemann-Stieltjes sums and a Taylor expansion for g. In the following $\mathcal{P}$ stands for a partition of the interval $[0, t]$. By a *formal computation*, this ansatz reads as

$$\int_0^t S(t - r)g(u_r) \, \mathrm{d}B_r^H = \sum_{[v_1, v_2] \in \mathcal{P}} S(t - v_2) \int_{v_1}^{v_2} S(v_2 - r)g(u_r) \, \mathrm{d}B_r^H$$

$$\approx \sum_{[v_1, v_2] \in \mathcal{P}} S(t - v_2) \Big[\int_{v_1}^{v_2} S(v_2 - r)g(u_{v_1}) \, \mathrm{d}B_r^H$$

$$+ \int_{v_1}^{v_2} S(v_2 - r)\mathrm{D}g(u_{v_1})(u_r - u_{v_1}) \, \mathrm{d}B_r^H \Big]$$

$$=: \sum_{[v_1,v_2]\in\mathcal{P}} S(t-v_2)\Big[\omega^S_{v_1v_2}(g(u_{v_1})) + z_{v_1v_2}(\mathrm{D}g(u_{v_1}))\Big].$$

(29)

Here we introduced the notation

$$\omega^S_{v_1v_2}(G(y_{v_1})) := \int_{v_1}^{v_2} S(v_2-r)g(u_{v_1})\,\mathrm{d}B^H_r,$$

(30)

respectively

$$z_{v_1v_2}(\mathrm{D}g(y_{v_1})) := \int_{u}^{v} S(v_2-r)\mathrm{D}g(u_{v_1})(u_r-u_{v_1})\,\mathrm{d}B^H_r.$$

(31)

By a classical integration by parts formula, see Theorem 3.5 in [149] one can argue that the term ω^S can be defined for a rough input B^H. This is no longer the case for z. By a classical rough path ansatz, where one *plugs in the definition of the solution itself in the construction of the rough integral*, one obtains the following approximation for z:

$$z_{st}(E) \approx \sum_{[v_1,v_2]\in\mathcal{P}} S(t-v_2)\Big[b_{v_1v_2}(E,g(u_{v_1})) + a_{v_1v_2}(E,u_{v_1})\Big] - \omega^S_{st}(Eu_s),$$

where E is a placeholder that stands for $\mathrm{D}g$ and

$$b_{v_1v_2}(E,g(u_{v_1})) := \int_{v_1}^{v_2} S(v_2-r)E\int_{v_1}^{r} S(r-q)g(u_{v_1})\,\mathrm{d}B^H_q\,\mathrm{d}B^H_r,$$

respectively

$$a_{v_1v_2}(E,u_{v_1}) := \int_{v_1}^{v_2} S(v_2-r)ES(r-v_1)u_{v_1}\,\mathrm{d}B^H_r.$$

This indicates that we have to define a, b and ω^S in order to fully characterize z. As already emphasized, the previous formal computation has been conducted under the assumption that B^H is a smooth path. In this case a, b and ω^S are well-defined. However, the main challenge is to construct these processes for rough inputs B^H and to show that (29) is indeed the right way to define (27). These arguments contain suitable approximation techniques ([107, Section 5]) combined with the additional necessary assumption $g : \mathcal{H} \to \mathcal{L}(\mathcal{U},\mathcal{H}_\beta)$, since it is not possible to define the *iterated integral b* for arbitrary operators [57, Remark 4.3]. Another essential tool is represented by the Sewing Lemma, which ensures the existence of the rough integral as a limit of Riemann-Stieltjes sums, see [107, Theorem 4.1] and the references specified therein. Putting all these together, one concludes that the *pathwise* solution of (26) is given by a pair (u, z), where

$$u_t = S(t)u_0 + \sum_{[v_1, v_2] \in \mathcal{P}} S(t - v_2) \Big[\omega_{v_1 v_2}^S(g(u_{v_1})) + z_{v_1 v_2}(Dg(u_{v_1})) \Big]$$

(32)

$$z_{st}(Dg) = \sum_{[v_1, v_2] \in \mathcal{P}} S(t - v_2) \Big[b_{v_1 v_2}(Dg(u_{v_1}), g(u_{v_1})) + a_{v_1 v_2}(Dg(u_{v_1}), u_{v_1}) \Big]$$
$$- \omega_{st}^S(Dg(u_s)).$$

(33)

Here u is referred to the *path* component of the solution and z is the *area* term.

Theorem 6. *The rough SPDE* (26) *has a unique local-in-time solution* (u, z), *where the two components* u *and* z *are given by* (32) *and* (33).

The existence of such a solution is obtained by a fixed-point argument in a suitable function space, which is chosen in order to compensate the time-singularity in zero of $(S(t))_{t \geq 0}$. This is required since the terms appearing in the Riemann-Stieltjes approximations must exhibit a certain Hölder regularity to make the Sewing Lemma applicable. However, using estimates for analytic semigroups, one has for a fixed $\gamma \in (0, 1]$ that

$$\|S(t)u_0 - S(s)u_0\|_{\mathcal{H}} = \|(S(t - s) - \mathrm{Id})S(s)u_0\|_{\mathcal{H}}$$
$$\leq C(t - s)^\gamma \|u_0\|_{\mathcal{H}_\gamma}$$
$$\leq C(t - s)^\gamma s^{-\gamma} \|u_0\|_{\mathcal{H}}.$$

(34)

Keeping (34) in mind, one is motivated to introduce for $\gamma_1, \gamma_2 \in (0, 1]$ the function space

$$C^{\gamma_1, \gamma_2}([0, T], \mathcal{H}) := \left\{ y \in C([0, T], \mathcal{H}) \ : \ \sup_{0 < s < t \leq T} s^{\gamma_1} \frac{\|y_t - y_s\|_{\mathcal{H}}}{(t - s)^{\gamma_2}} < \infty \right\}.$$

Regarding the area component, we also have

$$C^{\gamma_1, \gamma_2}(\Delta_T, \overline{\mathcal{H}}) := \left\{ y \in C(\Delta_T, \overline{\mathcal{H}}) \ : \ \sup_{0 < s < t \leq T} s^{\gamma_1} \frac{\|y_{st}\|_{\overline{\mathcal{H}}}}{(t - s)^{\gamma_2}} < \infty \right\},$$

for a certain Hilbert space $\overline{\mathcal{H}}$. Using (34) one has that $S(\cdot)u_0 \notin C^\gamma([0, T], \mathcal{H})$ but $S(\cdot)u_0 \in C^{\gamma, \gamma}([0, T]; \mathcal{H})$. Regarding this aspect together with further necessary regularity conditions required for the Sewing Lemma, the appropriate space for the fixed-point argument turns out to be

$$\bigg\{(u, z) \; : \; u \in C^{\beta, \beta}([0, T], \mathcal{H}) \text{ and}$$

$$z \in C^{\alpha}(\Delta_T, \mathcal{L}(\mathcal{L}(\mathcal{H} \otimes \mathcal{U}, \mathcal{H}), \mathcal{H})) \times C^{\alpha + \beta, \beta}(\Delta_T, \mathcal{L}(\mathcal{L}(\mathcal{H} \otimes \mathcal{U}, \mathcal{H}), \mathcal{H}))\bigg\}.$$

Here $\alpha < H$ stands for the Hölder regularity of B^H and β is chosen such that $\alpha + 2\beta > 1$ and gives the time-regularity of the solution of (26).

Due to the Taylor expansion employed in (29) one obtains certain quadratic estimates of the solution, therefore the global-in-time existence is not straightforward. This follows e.g. under an additional boundedness assumption on g, using regularizing properties of analytic semigroups, as argued in [57, 108].

Before pointing out some concluding remarks, we emphasize that for finite-dimensional noise, the solution theory for (26) simplifies. In this case, one can show that the solution (u, z) is given by

$$(u, z) = \bigg(S(\cdot)u_0 + \int\limits_0^\cdot S(\cdot - r)g(u_r) \, \mathrm{d}B_r^H, g(u)\bigg), \qquad (35)$$

similar to the ODE case [77, Chapter 8]. This can be achieved by means of a fixed-point argument in a function space that incorporates additional *space regularity* in order to compensate the missing time-regularity in (34). More precisely, one can prove that the following approximation

$$\int\limits_0^t S(t - r)g(u_r) \, \mathrm{d}B_r^H = \lim_{|\mathcal{P}| \to 0} \bigg(\sum_{[v_1, v_2] \in \mathcal{P}} S(t - v_1)g(u_{v_1})B_{v_1 v_2}^H$$

$$+ \sum_{[v_1, v_2] \in \mathcal{P}} S(t - v_1)\mathrm{D}g(u_{v_1})g(u_{v_1}) \int\limits_{v_1}^{v_2} (B_r^H - B_{v_1}^H) \, \mathrm{d}B_r^H \bigg)$$

can be used to define the stochastic convolution. For further details regarding this topic, see [86].

Remark 2.

1) The techniques are applicable also in the Banach space-valued setting. For finite-dimensional noise, the solution of (26) is given by (35). For infinite-dimensional noise, one needs to ensure that the series (3) defines a Banach space-valued fractional Brownian motion, see [35, Section 3] and [160, Proposition 8.8].

2) We emphasize that rough paths techniques are not restricted to the case of fBm, but apply to a wider class of *Gaussian processes*, their covariance functions satisfy certain criteria, as specified in [77, Chapter 10] and [78, Chapter 15].

3) If the trajectories of the fBm are more regular, i.e. $H \in (1/2, 1)$, then (27) can be defined using the Young integral, see [98], as

$$\int_0^t S(t - r)g(u_r) \, \mathrm{d}B_r^H = \lim_{|\mathcal{P}| \to 0} \sum_{[v_1, v_2] \in \mathcal{P}} S(t - v_2)\omega_{v_1 v_2}^S (g(u_{v_1})).$$

4) Another possibility to introduce mild solutions for such SPDEs is to decompose the integral in a series of one-dimensional integrals and define these using fractional calculus, see [85, 140]. However, such an approach leads to restrictions on the coefficients and on the trace of the noise. Using this ansatz, one needs to impose as in [140] that $\mathrm{Tr}\, Q^{1/2} = \sum_{n=1}^{\infty} \sqrt{\lambda_n} < \infty$, compare (2). Another choice would be to take $g : \mathcal{H} \to \mathcal{L}_2(\mathcal{U}, \mathcal{H})$ as in [85].

5) If one considers (26) driven by additive fractional noise, then (28) can easily be defined as a suitable Itô-type integral, see [66, 139].

2.7. *Renormalized Solutions*

Here we review briefly an approach for singular SPDEs using regularity structures [100, 101]. To illustrate the main issue, let us consider the stochastic bistable **Nagumo equation** on the torus with space-time white noise, i.e.,

$$\partial_t u = \Delta u + f(u) + \xi, \qquad f(u) := u(1 - u)(u - p), \tag{36}$$

where $u = u(t, x)$, $x \in \mathbb{R}^d/\mathbb{Z}^d =: \mathbb{T}^d$ for $d = 2$ or $d = 3$, $\xi = \xi(t, x) = \partial_t W^{\mathrm{Id}}(t, x)$ is space-time white noise, Δ is the Laplacian, and $p \in \mathbb{R}$ is a parameter; we remark that the case of a cubic nonlinearity also occurs in the **Allen-Cahn**/**Chaffee-Infante**/Φ^4/**Real-Ginzburg-Landau**/**Schlögl** models, which share similar features regarding existence theory of solutions. The fundamental issue of (36) is that the roughness of the space-time white noise forcing ξ interplays with the nonlinearity to prevent the application of more classical solution concepts. To understand this, let us introduce the **parabolic scaling** $\mathfrak{s} := (2, 1, 1)$ for $d = 2$ and $\mathfrak{s} := (2, 1, 1, 1)$ for $d = 3$, which can be used to define a norm

$$\|(t, x)\|_{\mathfrak{s}} := |t|^{1/2} + \sum_{j=1}^{d} |x_j|,$$

which is better adapted to the natural scaling of the operator $L := \partial_t - \Delta$. Let $C_{\mathfrak{s}}^\alpha$ denote the space of Hölder continuous functions $\phi : \mathbb{R} \times \mathbb{R}^d \to \mathbb{R}$ with the scaled norm, which is straightforward to define for $\alpha \geq 0$. For $\alpha < 0$, fix $r = -\lfloor \alpha \rfloor$ and let $\mathcal{B}_{\mathfrak{s},0}^r$ denote the space of all C^r-functions supported on $\{z \in \mathbb{R}^{d+1} : \|z\|_{\mathfrak{s}} \leq 1\}$. A Schwartz distribution $v \in \mathcal{S}(\mathbb{R}^{d+1})$ belongs to $C_{\mathfrak{s}}^\alpha$ if it belongs to the dual space of C_0^r and for every compact set $\mathfrak{K} \subset \mathbb{R}^{d+1}$ there exists a constant $C > 0$ such that

$$\langle v, \mathcal{S}_{\mathfrak{s},z}^\delta \eta \rangle \leq C\delta^\alpha, \quad \forall \delta \in (0,1], \ \forall z \in \mathfrak{K}, \ \forall \eta \in \mathcal{B}_{\mathfrak{s},0}^r \text{ with } \|\eta\|_{C^r} \leq 1,$$

where $(\mathcal{S}_{\mathfrak{s},z}^\delta \eta)(\bar{t}, \bar{z}) := \delta^{-(d+2)}\eta(\delta^{-2}(t - \bar{t}), \delta^{-1}(x - \bar{x}))$. Plainly, $C_{\mathfrak{s}}^\alpha$ generalized α-Hölder functions do not blow up worse near each point than a power divergence with exponent α. One can prove that space-time white noise satisfies

$$\xi \in C_{\mathfrak{s}}^\alpha \quad \text{for } \alpha = -\frac{d+2}{2} - \kappa \text{ and any } \kappa > 0. \tag{37}$$

Suppose we would want to solve (36) by re-formulating it as a fixed-point of an iterated map M

$$u^{(k+1)} = L^{-1}(f(u^{(k)}) + \xi) =: M(u^{(k)}). \tag{38}$$

Suppose we start with $u^{(0)} = 0$, then we get $M(0) = L^{-1}(0 + \xi) = L^{-1}(\xi) = u^{(1)}$. Now $u^{(1)}$ has higher regularity in comparison to ξ since applying L^{-1} corresponds to convolution with the heat kernel. Classical Schauder theory tells us that $u^{(1)} \in C_{\mathfrak{s}}^{\alpha+2}$ so we gained two orders of regularity. For $d = 2$ we get $\alpha + 2 = -2 - \kappa + 2 = -\kappa < 0$ while for $d = 3$ we have $\alpha + 2 = -\frac{5}{2} - \kappa + 2 = -\frac{1}{2} - \kappa < 0$. So in both cases, we get that $u^{(1)}$ is just a Schwartz distribution with negative Hölder regularity. However, if we now want to compute $u^{(2)}$, then we have to make sense of $f(u^{(1)})$, which is just not possible within classical distribution/generalized function theory as we can only multiply distributions, which have non-negative sums of their Hölder regularity exponents. A natural attempt to still use a fixed point approach is to first smooth the noise, e.g., via a mollifier

$$\rho_\varepsilon(t, x) := \varepsilon^{-(d+2)}\rho(t\varepsilon^{-2}, x\varepsilon^{-1}), \qquad \xi_\varepsilon := \rho_\varepsilon * \xi,$$

where ρ is a compactly supported function of integral 1, and $*$ denotes the space-time convolution. Then it is classical that

$$\partial_t u_\varepsilon = \Delta u_\varepsilon + f(u_\varepsilon) + \xi_\varepsilon, \tag{39}$$

has smooth, even global-in-time, solutions for sufficiently regular initial. However, it turns out that directly taking a limit $\varepsilon \to 0$ leads to divergence so the SPDE is singular. Classical considerations in phase transitions / bifurcation theory suggest that the divergence can be prevented if the equation is **renormalized** to

$$\partial_t u_\varepsilon = \Delta u_\varepsilon + f(u_\varepsilon) + C(\varepsilon) u_\varepsilon + \xi_\varepsilon, \tag{40}$$

where $C(\varepsilon) = \mathcal{O}(\ln \varepsilon)$ for $d = 2$ and $C(\varepsilon) = \mathcal{O}(\varepsilon^{-1})$ for $d = 3$ are computable diverging functions as $\varepsilon \to 0$. If the initial condition $u(0, \cdot)$ belongs to a sufficiently regular Hölder space, e.g., continuity is sufficient, then one can prove [100] that there exists a sequence of local-in-time solutions u_ε such that

$$u_\varepsilon \to u_0 \quad \text{as } \varepsilon \to 0 \text{ in probability independently of } \rho.$$

The last statement is one example application of the theory of regularity structures, i.e., it establishes the existence of a **renormalized solution** u_0. To prove the convergence, a very elaborate construction is necessary. The basic steps are (I) building a regularity structure and a model space adapted to the SPDE, (II) solving the fixed point problem in the abstract space of modeled distribution, (III) proving that the renormalized modeled distributions converge and one may reconstruct an actual distribution as a solution. Basically, this procedure has been worked out as an abstract result applicable to many singular SPDEs [31, 32, 41, 100]. We only illustrate part of the first step to discuss, to which classes of singular SPDEs, we may apply the theory of regularity structures.

The first step is to build a **regularity structure** $(\mathcal{A}, \mathcal{T}, \mathcal{G})$, where $\mathcal{A} \subset \mathbb{R}$ is an index set bounded below without accumulation points in $\mathbb{R}$, $\mathcal{T} = \oplus_{\alpha \in \mathcal{A}} \mathcal{T}_\alpha$ is the **model space** consisting of a graded sum of Banach spaces with $\mathcal{T}_0$ isomorphic to $\mathbb{R}$, and $\mathcal{G}$ is the structure group. We shall not discuss the structure group but try to illustrate, how $\mathcal{A}$ and $\mathcal{T}$ are constructed. The idea is to construct an abstract jet space of symbols capable of representing suitable regularity classes of functions. For $\mathcal{T}_\alpha$ with $\alpha \in \mathbb{N}$, we simply take the space of homogeneous polynomials of degree α in $(d+1)$ variables $X_0, X_1, \ldots, X_d$ where X_0 represents the time variable, and X_j is to be interpreted as a formal symbol. We count **homogeneity** $|\cdot|_{\mathfrak{s}}$ of a polynomial in the parabolic scaling, i.e.,

$$|X^k|_{\mathfrak{s}} = |k|_s := 2k_0 + \sum_{j=1}^{d} k_j, \quad X^k = X_0^{k_0} \cdots X_d^{k_d}.$$

We also set $|\mathbf{1}|_{\mathfrak{s}} = 0$ with the unit element $\mathbf{1}$ spanning $\mathcal{T}_0$. Furthermore, we let Ξ be a symbol representing the noise with $|\Xi|_{\mathfrak{s}} = \alpha_0 := -(d+2)/2 - \kappa$ for any fixed (small) $\kappa > 0$. The convolution/integration against the heat kernel is represented by a map $\mathcal{I} : \mathcal{T} \to \mathcal{T}$ and $|\mathcal{I}(\tau)|_{\mathfrak{s}} = |\tau|_{\mathfrak{s}} + 2$ for any $\tau \in \mathcal{T}$. Consider the abstract iterated map

$$U^{(k+1)} = \mathcal{I}(f(U^{(k)}) + \Xi),$$

where we hope that iteration yields a fixed point. If we iterate the map, we get new symbols, e.g., with $U^{(0)} = 0$ we get $U^{(1)} = \mathcal{I}(\Xi)$ with $|\mathcal{I}(\Xi)|_{\mathfrak{s}} = -\alpha_0 + 2$. Now one can iterate again, which yields second and third powers $\mathcal{I}(\Xi)^2$ and $\mathcal{I}(\Xi)^3$ as well as $\mathcal{I}(\mathcal{I}(\Xi))$ and so on including various combinations involving polynomials if they are represented in the initial condition. Now suppose we continue this procedure for all possible iterates and all possible symbols with non-negative homogeneity as initial condition. Then we check homogeneities and only keep symbols with negative homogeneity. For $d = 2$ this yields

$$\Xi, \ \mathcal{I}(\Xi)^3, \ \mathcal{I}(\Xi)^2, \ \mathcal{I}(\Xi),$$

with homogeneities $-2 - \kappa$, -3κ, -2κ and $-\kappa$ respectively. Carrying out the same calculation for $d = 3$ is a good exercise yielding more symbols. For both cases, the procedure terminates only giving a finite number of negatively homogeneous symbols, so we can set $\mathcal{A}$ as only containing all the negative homogeneities and use the negatively homogenous symbols to span $\mathcal{T}_\alpha$ for $\alpha < 0$. The condition that $\mathcal{A}$ is bounded from below effectively means that the SPDE is **locally subcritical** and the theory of regularity structures is designed for this class of singular SPDEs. Roughly speaking, for the class of locally subcritical singular SPDEs, the theory of regularity structures works and provides a local-in-time renormalized solution. The full procedure can be found in the works [31, 32, 41, 100], while some particular examples can be found in [19, 20, 86, 102, 103].

Another breakthrough in the theory of singular SPDEs is constituted by the **paracontrolled calculus** developed in [96]. This was successfully applied to the KPZ equation [97], the Φ_3^4 model [40] and singular quasilinear problems [7, 79]. A comparison between the two methods can be looked up in [8].

3. Dynamics: Concepts & Results

The aim of this section is to provide an overview and discuss recent developments regarding the long-time behavior of the solutions of SPDEs, again with a strong focus on stochastic reaction-diffusion equations. Similar to the deterministic setting [123, 159, 161], one is interested to investigate the stability of steady states, predict if changes of stability (bifurcations) occur, analyze if trajectories become (exponentially) close to each other or look for sets where trajectories accumulate. There are several ways to describe such kind of phenomena in stochastic dynamics. We focus on the random dynamical system theory [4], which is used to define invariant sets such as manifolds or attractors for the flow generated by an SPDE. We also indicate several other possibilities to quantify the long-time behavior of SPDEs, such as invariant measures, large deviations or sample path approaches, depending also on the different noise terms driving these SPDEs, as discussed in Section 2.

3.1. Random Dynamical Systems

As seen in Section 2, there are several different ways to perturb a PDE by a random input, such as trace-class (fractional) Brownian motion, space-time white noise, noise which is white in time and correlated in space, etc. Therefore, we introduce the next concept, which indicates a model of the noise driving the PDE. Recall that $\mathcal{H}$ denotes an arbitrary separable Hilbert space and $(\Omega, \mathcal{F}, \mathbb{P})$ stands for a probability space, which will further be specified later on.

Definition 11. (Metric dynamical system (MDS)) Let $\theta : \mathbb{R} \times \Omega \to \Omega$ be a family of $\mathbb{P}$-preserving transformations (meaning that $\theta_t \mathbb{P} = \mathbb{P}$ for all $t \in \mathbb{R}$) with the following properties:

(1) the mapping $(t, \omega) \mapsto \theta_t \omega$ is $(\mathcal{B}(\mathbb{R}) \otimes \mathcal{F}, \mathcal{F})$-measurable;
(2) $\theta_0 = \mathrm{Id}_\Omega$;
(3) $\theta_{t+s} = \theta_t \circ \theta_s$ for all $t, s, \in \mathbb{R}$.

Then the quadruple $(\Omega, \mathcal{F}, \mathbb{P}, (\theta_t)_{t \in \mathbb{R}})$ is called a **metric dynamical system**.

In order to simplify the notation we write $\theta_t \omega$ for $\theta(t, \omega)$. We always assume that $\mathbb{P}$ is **ergodic** with respect to $(\theta_t)_{t \in \mathbb{R}}$, which means that any invariant subset has zero or full measure. Next, we introduce now the MDS corresponding to a genuine $\mathcal{H}$-valued process having stationary increments. As examples, one can consider a trace-class (fractional) Brownian motion as given in (3).

Example 2. Let $C_0(\mathbb{R}, \mathcal{H})$ denote the set of continuous $\mathcal{H}$-valued functions, which are zero at zero and are equipped with the compact open topology.

Furthermore, by taking $\mathbb{P}$ as the Wiener measure on $\mathcal{B}(C_0(\mathbb{R}, \mathcal{H}))$ having a trace-class covariance operator Q on $\mathcal{H}$ and applying Kolmogorov's theorem about the existence of a continuous version yields the canonical probability space $\Omega := (C_0(\mathbb{R}, \mathcal{H}), \mathcal{B}(C_0(\mathbb{R}, \mathcal{H})), \mathbb{P})$. To obtain an ergodic metric dynamical system we introduce the **Wiener shift**, which is defined by

$$\theta_t \omega(\cdot) = \omega(t + \cdot) - \omega(t), \text{ for } \omega \in C_0(\mathbb{R}, \mathcal{H}). \qquad (41)$$

Whenever we work with the RDS approach, i.e. Sections 3.3 and 3.4, we only consider the subset of all $\omega \in C_0(\mathbb{R}, \mathcal{H})$ with subexponential growth and work with this new metric dynamical system using the same notations $(\Omega, \mathcal{F}, (\theta_t)_{t \in \mathbb{R}}, \mathbb{P})$. ♦

Remark 3. Let $\overline{W}_{\mathcal{H}}$ stand for a process with stationary increments having a trace-class covariance operator Q on $\mathcal{H}$. Then one can embed $\overline{W}_H$ into a canonical probability space, as given in Example 2 and identify

$$\overline{W}_{\mathcal{H}}(t, \omega) = \omega(t), \text{ for } \omega \in C_0(\mathbb{R}, \mathcal{H}).$$

We will use this notation in Sections 3.4 and 3.3.

In this framework, one often has to deal with random constants whose values have to be controlled along the orbits of θ. Therefore, the following concept plays a key role. For more details, see Proposition 4.1.3 in [4].

Definition 12. A positive real-valued random variable Y on a MDS $(\Omega, \mathcal{F}, \mathbb{P}, (\theta_t)_{t \in \mathbb{R}})$ is called **tempered** if there exists a $(\theta_t)_{t \in \mathbb{R}}$-invariant set of full measure such that

$$\lim_{t \to \pm\infty} \frac{\log^+ Y(\theta_t \omega)}{|t|} = 0.$$

Remark 4. Note that temperedness is equivalent to **subexponential growth**. The only alternative to this situation in the ergodic case is that

$$\limsup_{t \to \pm\infty} \frac{\log^+ Y(\theta_t \omega)}{|t|} = \infty.$$

In the deterministic theory of dynamical systems, one obtains under certain assumptions on the coefficients the flow property of the solution operator. Since our PDE is now perturbed by a non-autonomous, irregular forcing, it is natural to expect that the time-evolution of the noise must play a role in this flow property. We describe the precise mathematical formalism of this fact [4].

Definition 13. (Random dynamical system (RDS)) A continuous **random dynamical system** on $\mathcal{H}$ over a metric dynamical system

$(\Omega, \mathcal{F}, \mathbb{P}, (\theta_t)_{t \in \mathbb{R}})$ is a mapping

$$\varphi : \mathbb{R}^+ \times \Omega \times \mathcal{H} \to \mathcal{H}, \quad (t, \omega, x) \mapsto \varphi(t, \omega, x),$$

which is $(\mathcal{B}(\mathbb{R}^+) \otimes \mathcal{F} \otimes \mathcal{B}(\mathcal{H}), \mathcal{B}(\mathcal{H}))$-measurable and satisfies:

1) $\varphi(0, \omega, \cdot) = \mathrm{Id}_{\mathcal{H}}$ for all $\omega \in \Omega$;
2) $\varphi(t + \tau, \omega, x) = \varphi(t, \theta_\tau \omega, \varphi(\tau, \omega, x))$, $x \in \mathcal{H}$, $t, \tau \in \mathbb{R}^+$ and $\omega \in \Omega$;
3) $\varphi(t, \omega, \cdot) : \mathcal{H} \to \mathcal{H}$ is continuous for all $t \in \mathbb{R}^+$.

The property 2) is referred to as **cocycle property**. If one drops the ω-dependence, one obtains exactly the flow property from the deterministic case. Here we see that the shift $\theta_t \omega$ describes the evolution of the noise. Keeping this mind, we observe that RDS can be interpreted as the generalization of non-autonomous deterministic dynamical systems. Referring to [4], it is well-known that an Itô-type stochastic ordinary differential equation generates a random dynamical system under natural assumptions on the coefficients. This fact is based on the flow property, see [128, 155], which can be obtained by Kolmogorov's theorem about the existence of a (Hölder)-continuous random field with finite-dimensional parameter range, i.e. the parameters of this random field are the time and the non-random initial data. However, the generation of a RDS from an SPDE (as considered in Sections 2.1–2.5) has been a long-standing open problem, since Kolmogorov's theorem breaks down for random fields parametrized by infinite-dimensional Hilbert spaces, see [143]. As a consequence it is not trivial how to obtain a RDS from an SPDE, since its solution is defined almost surely, which contradicts Definition 13, where all properties must hold for all $\omega \in \Omega$. Particularly, this means that there are exceptional sets which depend on certain parameters of the SPDE, and it is not clear how to define a RDS if more than countably many exceptional sets occur. This problem was fully solved only under very restrictive assumptions on the structure of the noise driving the SPDE. More precisely, for *additive* and *linear multiplicative* noise, one can perform certain *Doss-Sussmann-type transformations* [61, 158] and reduce the corresponding SPDE to a PDE with random coefficients. Such a PDE can be solved pathwise for every realization of the noise. We provide here two examples of such transformations. Based on these we analyze attractors and invariant manifolds for SPDEs in Sections 3.3 and 3.4.

Example 3. (SPDE with additive noise) Consider the SPDE on $\mathcal{H}$

$$\mathrm{d}u(t) = [Au(t) + f(u(t))] \, \mathrm{d}t + \mathrm{d}W^Q(t), \tag{42}$$

where W^Q is a trace-class Brownian motion in $\mathcal{H}$. We further assume that the semigroup generated by A is exponentially stable, i.e. there exist constants $\widetilde{M} \geq 1$ and $\overline{\mu} > 0$ such that $\|S(t)\|_{\mathcal{H}} \leq \widetilde{M}e^{-\overline{\mu}t}$ for $t > 0$. The unique *stationary* solution of the *linear* SPDE

$$du(t) = Au(t)\, dt + dW^Q(t) \tag{43}$$

is given by the **Ornstein-Uhlenbeck process**, which can be represented by the convolution

$$\int\limits_{-\infty}^{t} S(t-s)\, dW^Q(s).$$

Taking the canonical probability space corresponding to W^Q, as introduced in Example 2, and identifying $W^Q(t, \omega) = \omega(t)$, for $\omega \in \Omega$ we introduce the process $(t, \omega) \mapsto Z(\theta_t \omega)$ as

$$Z(\theta_t \omega) := \int\limits_{-\infty}^{t} S(t-s)\, d\omega(s) = \int\limits_{-\infty}^{0} S(-s)\, d\theta_t\omega(s). \tag{44}$$

Subtracting (44) from (42) one obtains

$$du(t) = [Au(t) + f(u(t) + Z(\theta_t\omega))]\, dt, \tag{45}$$

which is a PDE with random nonautonomous coefficients. ♦

Example 4. (SPDE with linear multiplicative noise) Let W stand for a one-dimensional standard Brownian motion. Consider the SPDE on $\mathcal{H}$ with Stratonovich noise

$$du(t) = [Au(t) + f(u(t))]\, dt + u \circ dW(t). \tag{46}$$

Performing the transformation $u^* = ue^{-z(\omega)}$, where z is the one-dimensional Ornstein-Uhlenbeck process and dropping the $*$-notation, one obtains

$$du(t) = [Au(t) + z(\theta_t\omega)u(t)]\, dt + \overline{f}(\theta_t\omega, u(t))\, dt, \tag{47}$$

where $\overline{f}(\omega, u) := e^{-z(\omega)}f(e^{z(\omega)}u)$.

In both of the previous cases, one can show that the solution operators of (45) and (47) generate random dynamical systems. Furthermore, the dynamical systems generated by the original SPDEs are equivalent with those generated by (45) and (47). This means that it is enough to work with the transformed equations and transfer all the results such as fixed-points, manifolds or attractors to the initial SPDEs. However, for general nonlinear multiplicative noise, this technique is obviously no longer applicable. As a consequence of this issue, dynamical aspects for SPDEs such as stability, Lyapunov exponents, multiplicative ergodic theorems, random attractors, random invariant manifolds have not been investigated in their full generality. We discuss situations in the following sections and point out how the techniques presented in Section 2.6 can be employed to investigate the dynamics of SPDEs driven by multiplicative noise. Further applications in this sense can be looked up in [9, 59, 69, 80, 84, 125] and the references specified therein.

3.2. *Stability*

In this section we discuss several stability concepts for SPDEs such as

- stability in probability;
- almost sure exponential stability;
- moment exponential stability;
- metastability.

Common tools used to investigate such concepts rely on the existence of Lyapunov functionals [112, 138], random dynamical systems methods, ergodic theory and Lyapunov exponents [3, 4, 37, 83, 84, 130, 132], or large deviation theory [18, 74, 75]. There are numerous works dealing with stability statements for Itô-type SPDEs. Arguments via Lyapunov functionals heavily use the Markov property of the solution and semi-martingale techniques. Therefore, it is a challenging open problem to get optimal stability results for SPDEs driven by fractional Brownian motion, such as (26). Progress in this direction was made in [59, 83, 84].

In this section we only refer to SPDEs driven by Brownian motion, where the precise assumptions are stated below. We write $u(t, u_0)$ in order to refer to a solution u of such an SPDE at time $t > 0$ having $u_0 \in \mathcal{H}$ as initial condition. Again, the probability space $(\Omega, \mathcal{F}, \mathbb{P})$ is fixed. Let $q \geq 1$.

Definition 14. We call a *global-in-time* solution u of an SPDE:

1) **stochastically stable** or **stable in probability** if for every pair $\varepsilon \in (0, 1)$ and $r > 0$, there exists a $\delta = \delta(\varepsilon, r) > 0$ such that
$$\mathbb{P}(\|u(t, u_0)\|_{\mathcal{H}} < r \text{ for all } t \geq 0) \geq 1 - \varepsilon,$$
for $\|u_0\|_{\mathcal{H}} < \delta$;

2) **almost sure (global) exponentially stable** if for all $u_0 \in \mathcal{H}$

$$\limsup_{t \to \infty} \frac{1}{t} \log(\|u(t, u_0)\|_{\mathcal{H}}) < 0 \quad \text{a.s.;} \tag{48}$$

3) **q-th moment exponentially stable** if for all $u_0 \in \mathcal{H}$

$$\limsup_{t \to \infty} \frac{1}{t} \log(\mathbb{E}\|u(t, u_0)\|_{\mathcal{H}}^q) < 0.$$

In general moment and almost sure exponential stability do not imply each other, see [112, 138] for further details. For linear SPDEs, the quantity on the left-hand side of (48) is often called **Lyapunov exponent**, see [112]. For a related, yet different, definition of Lyapunov exponents in random dynamical system theory, see [4, Section 3.2] and [3, 37, 71, 130].

Remark 5. Note that one can also investigate **local exponential stability**, meaning that there exists a *random neighbourhood* $\mathcal{N}(\omega)$ of zero such that for $u_0 \in \mathcal{N}(\omega)$, the relation (48) is satisfied, compare Definition 11 in [83].

We now illustrate the concepts introduced in Definition 14 for general SPDEs satisfying the assumptions formulated in Section 2.3. We first provide abstract conditions which ensure the stability of solutions of SPDEs and present thereafter concise examples. Using the same notations as in Section 2.3 we recall that $(\mathcal{V}, \mathcal{H}, \mathcal{V}^*)$ stands for a Gelfand triple and let $\overline{\beta} > 0$ such that $\|x\|_{\mathcal{H}} \le \overline{\beta}\|x\|_{\mathcal{V}}$. We further impose the following *coercivity* condition, compare (15). There exist constants $\overline{\alpha} > 0$, $\overline{\mu} > 0$, $\overline{\lambda} \in \mathbb{R}$ and a nonnegative continuous function h such that

$$2\,_{\mathcal{V}^*}\langle A(t, v), v\rangle_{\mathcal{V}} + \|g(t, v)\|_{\mathcal{L}_2(\mathcal{U}, \mathcal{H})}^2 \le -\overline{\alpha}\|v\|_{\mathcal{V}}^p + \overline{\lambda}\|v\|_{\mathcal{H}}^2 + h(t)e^{-\overline{\mu}t}, \quad v \in \mathcal{V}, \tag{49}$$

where $p > 1$ and for arbitrary $\delta > 0$ we have $\lim_{t \to \infty} \frac{h(t)}{e^{\delta t}} = 0$. Then the following results ensure the mean square and almost sure exponential stability for the solutions of SPDEs, compare [38, Theorems 2.2,2.3].

Theorem 7. *Let assumptions 1), 2), 4) in Section 2.3 and (49) hold true. Then, if u is a global solution of (14), there exists constants $\varepsilon > 0$, $C > 0$ such that*

$$\mathbb{E}\|u(t)\|_{\mathcal{H}}^2 \le Ce^{-\varepsilon t}, \quad t \ge 0,$$

if either one of the following hypotheses are satisfied:

a) $\overline{\lambda} < 0$ *(for every $p > 1$);*

b) $\overline{\lambda} \cdot \overline{\beta}^2 - \overline{\alpha} < 0$ *(for p = 2).*

Theorem 8. *Under the assumptions of Theorem 7, there exist positive constants* $\widetilde{M}$, ε *and a subset* $N_0 \subset \Omega$ *with* $\mathbb{P}(N_0) = 0$ *such that for each* $\omega \notin N_0$, *there exists a positive random number* $\widetilde{T}(\omega)$ *such that*

$$\|u(t)\|_{\mathcal{H}}^2 \leq \widetilde{M} e^{-\varepsilon t}, \quad t \geq \widetilde{T}(\omega).$$

The proofs of the previous theorems rely on the Itô-formula ([151, Theorem 4.2.5]), martingale arguments and the Gronwall Lemma. For a better comprehension we now provide examples where the previous abstract stability criteria are applicable.

Example 5. Let $a \in \mathbb{R}$, $b : \mathbb{R} \to \mathbb{R}$ be a Lipschitz continuous function with $b(0) = 0$ and let W stand for a one-dimensional standard Brownian motion. We consider a stochastic heat equation on $\mathcal{H} := L^2(0, \pi)$ with Dirichlet boundary conditions given by

$$\begin{cases} \mathrm{d}u = [\Delta u + au] \, \mathrm{d}t + b(u) \, \mathrm{d}W(t), \quad t > 0, \quad x \in (0, \pi) \\ u(t, 0) = u(t, \pi) = 0, t > 0 \\ u(0, x) = u_0(x). \end{cases} \tag{50}$$

Here $\mathcal{V} := W_0^{1,2}(0, \pi)$, $A(t, u) := \Delta u + au$ and $g(t, u) := b(u)$. One can verify that for $u \in \mathcal{V}$:

$$2 \, {}_{\mathcal{V}^*}\langle A(t, u), u \rangle_{\mathcal{V}} + \|g(t, u)\|_{\mathcal{L}_2(\mathbb{R}, \mathcal{H})}^2 \leq -2\|u\|_{\mathcal{V}}^2 + (2a + l^2)\|u\|_{\mathcal{H}},$$

where l is the Lipschitz constant of b. Therefore (49) holds for $p = 2$ and in order to apply Theorem 7 with $\overline{\alpha} = 2$ and $\overline{\lambda} = 2a + l^2$, we have to choose $\overline{\beta} > 0$ such that $(2a + l^2)\overline{\beta}^2 - 2 < 0$, which can be achieved e.g. setting $\overline{\beta} := \frac{\pi}{\sqrt{2}}$. ♦

For a nonlinear operator A, we provide the following example.

Example 6. Let $a > 0$ and $b : \mathbb{R} \to \mathbb{R}$ be a Lipschitz continuous with Lipschitz constant $l > 0$ and assume $b(0) = 0$. Furthermore, W stands for a one-dimensional standard Brownian motion. We let $2 < p < \infty$ and consider the SPDE on $\mathcal{H} := L^2(0, 1)$

$$\begin{cases} \mathrm{d}u(t, x) = \left[\frac{\partial}{\partial x} \left(\left| \frac{\partial u(t,x)}{\partial x} \right|^{p-2} \frac{\partial u(t,x)}{\partial x} \right) - au(t, x) \right] \mathrm{d}t + b(u(t, x)) \, \mathrm{d}W(t), \\ u(t, 0) = u(t, 1) = 0, \quad t > 0 \\ u(0, x) = u_0(x), \qquad x \in [0, 1]. \end{cases}$$

Here, $\mathcal{V} := W_0^{1,p}(0,1)$, $g(t,u) = b(u)$ and the nonlinear monotone operator $A : \mathcal{V} \to \mathcal{V}^*$ is defined as

$$\mathcal{V}^* \langle Au, v \rangle_{\mathcal{V}} = \int_0^1 \left| \frac{\partial u(x)}{\partial x} \right|^{p-2} \frac{\partial u}{\partial x} \frac{\partial v}{\partial x} \, \mathrm{d}x - a \int_0^1 u(x)v(x) \, \mathrm{d}x, \quad u, v \in \mathcal{V}.$$

One can check the (49) holds true for $h \equiv 0$, $p > 2$ $\overline{\alpha} = 2$ and $\overline{\lambda} = -\kappa$, where $\kappa > 0$ is chosen such that $l^2 < 2a - \kappa$, see [38, Example 3.2]. In this case one obtains again due to Theorems 7 and 8 the mean square and almost sure exponential stability of the solution. ♦

A very simple example, where almost all sample paths of the solution *do not* tend exponentially to zero is constituted by the following SDE.

Example 7. Let a, $b > 0$ be two constants and W stand for a one-dimensional standard Brownian motion. We consider the SDE

$$\begin{cases} \mathrm{d}u = -au(t) \, \mathrm{d}t + (1+t)^{-b} \, \mathrm{d}W(t), \quad t > 0 \\ u(0) = 0. \end{cases} \tag{51}$$

Setting $A(t,x) := -ax$, $g(t,x) := (1+t)^{-b}$ and letting $\langle \cdot, \cdot \rangle$ denote the usual scalar product in $\mathbb{R}$, we easily have that

$$2 \langle A(t,x), x \rangle + \|g(t,x)\|^2 = -2ax^2 + (1+t)^{-2b},$$

where the last term obviously does not exponentially tend to zero. Therefore (49) is not satisfied. Here, using the law of iterated logarithm, one can immediately verify that for the solution of (51)

$$u(t) = \mathrm{e}^{-at} \int_0^t \mathrm{e}^{as}(1+s)^{-b} \, \mathrm{d}W(s),$$

it holds that

$$\limsup_{t \to \infty} \frac{1}{t} \log |u(t)| = 0, \quad \text{a.s.}$$

Taking another diffusion coefficient, e.g. $g(t,x) := \mathrm{e}^{-bt}$, the solution of the (51) will become almost sure exponentially stable. ♦

Using the random field approach described in Section 2.5 and considering (21) driven by space-time white-noise, i.e. $f = \delta_0$, one can show the following assertion regarding the moments of the mild random field solution

$\{u(t,x) : t > 0, x \in \mathbb{R}\}$. According to [73, 116], for $k \geq 1$ and $C > 0$, the moments grow exponentially, which means that

$$\mathbb{E}|u(t,x)|^k \sim \exp(Ck^3 t),$$

consequently

$$\tilde{\lambda}(k) := \limsup_{t \to \infty} \frac{1}{t} \mathbb{E} \log |u(t,x)|^k \sim Ck^3.$$

Therefore, one observes

$$\tilde{\lambda}(1) < \frac{\tilde{\lambda}(2)}{2} < \dots \frac{\tilde{\lambda}(k)}{k} < \dots.$$

This phenomenon is called **full intermittency**, which means that the random field solution develops high peaks concentrated on small sets for large time values. For further details on this topic, see [10, 11, 73, 116]. Results regarding Lyapunov-exponents for parabolic SPDEs on *bounded domains* are available within the RDS approach using **Oseledets' multiplicative ergodic theorem** for *compact operators* in [37, 71, 130, 132, 143]. Beyond the Lyapunov spectrum, one is interested in further spectral properties for S(P)DEs, such as *dichotomy spectrum*. To our best knowledge this was dealt with only in the finite-dimensional setting [36].

If one is not interested in asymptotic stability as $t \to +\infty$, one may also study **metastability**, i.e., dynamics which looks stable on very long time scales. For example, consider the Allen-Cahn SPDE [17]

$$du = \left[\partial_x^2 u + u - u^3\right] dt + \sqrt{2\varepsilon} \, dW^{\mathrm{Id}}, u(0,x) = u_0(x), \qquad (52)$$

where $u = u(t,x)$, $x \in \mathbb{R}/\mathbb{Z} = \mathbb{S}^1$, and we assume small noise $0 < \varepsilon \ll 1$. With the tools from Section 2.1, one may show that (52) has a global-in-time solution, e.g., a mild solution $u_{u_0}^\varepsilon(t)$, where the subscript reminds us of the dependence on the initial condition and the superscript of the parameter dependence. Furthermore, (52) is a stochastic **gradient system**

$$du = -\nabla_{\mathcal{H}} V(u) \, dt + \sqrt{2\varepsilon} \, dW^{\mathrm{Id}},$$

$$\text{where } V(\zeta) := \int_{\mathbb{S}^1} \frac{1}{2} \|\nabla \zeta(x)\|_{\mathcal{H}}^2 + \frac{1}{4}(\zeta(x)^2 - 1)^2 \, dx,$$

where V is called a **potential** and where we use the space $\mathcal{H} = L^2(\mathbb{S}^1)$. Let $P_t(h)(u_0) := \mathbb{E}[h(u_{u_0}^\varepsilon(t))]$, for a test function $h : \mathcal{H} \to \mathbb{R}$, denote the **transition semigroup** associated to the solution process. In general, a measure μ is called an **invariant measure** for P_t if

$$P_t^* \mu = \mu \qquad \forall t \geq 0.$$

One may prove [51] that quite a number of dissipative reaction-diffusion equations have an invariant measure using a tightness argument. For (52), one may write the invariant measure more explicitly as

$$\mu(d\zeta) = \frac{1}{Z_0} e^{-V(\zeta)/\varepsilon} \, d\zeta, \quad Z_0 := \int_{\mathcal{H}} e^{-V(\zeta)/\varepsilon} \, d\zeta, \qquad (53)$$

which is also known as the **Gibbs measure** in this context; technically, the notation in (53) is to be understood more precisely using the Gaussian free field to get rid of the problem that there is no Lebesgue measure on $L^2(\mathcal{H})$. Although the invariant measure provides good insight regarding the bimodal stationary density of the gradient system (52), it does not provide direct information regarding finite-time dynamics for some fixed $T > 0$ such as transition times between neighborhoods of deterministically stable steady states, e.g., between $u \equiv \pm 1$. It is expected that the solution of (52) approaches one of these equilibria and stays in a neighbourhood of these equilibria for a long time. Eventually, the system will transition from one of these neighborhoods to the other one. Such a behavior has been analyzed using large deviations in [18]. The main ideas can be summarized as follows. We assume that $u_{u_0}^{\varepsilon}(t)$ converges to a deterministic quantity $\overline{u}_{u_0}(t)$ as $\varepsilon \to 0$, i.e., for any fixed $\delta > 0$:

$$\lim_{\varepsilon \to 0} \mathbb{P} \left(\sup_{t \in [0,T]} \|u_{u_0}^{\varepsilon}(t) - \overline{u}_{u_0}(t)\|_{\mathcal{H}} > \delta \right) = 0.$$

Furthermore, u^* stands for an asymptotically stable equilibrium of the deterministic equation (52), $\mathcal{D}_0 := \{u_0 \in \mathcal{H} : \lim_{t \to \infty} \|\overline{u}_{u_0}(t) - u^*\|_{\mathcal{H}} = 0\}$ and

$$\tau_{u_0}^{\varepsilon} := \inf\{t > 0 : u_{u_0}^{\varepsilon}(t) \notin \mathcal{D}_0\}$$

is the **first-exit time**. As $\varepsilon \to 0$, the time required for this rare event to occur grows exponentially. Therefore, one investigates quantities like

$$\lim_{\varepsilon \to 0} \varepsilon \log \mathbb{E}\tau_{u_0}^{\varepsilon}, \quad \lim_{\varepsilon \to 0} \varepsilon \log \tau_{u_0}^{\varepsilon}, \quad \lim_{\varepsilon \to 0} u_{u_0}^{\varepsilon}(\tau_{u_0}^{\varepsilon}),$$

which provide the expected exit-time, exit-time and exit location/shape. For instance, one straightforward possibility to estimate the expected exit-time is

$$\mathbb{E}(\tau_{u_0}^{\varepsilon}) \leq T \sum_{k=0}^{\infty} \mathbb{P}(\tau_{u_0}^{\varepsilon} \geq kT). \qquad (54)$$

Due to the Markov property, one can show that for $k \in \mathbb{N}$:

$$\sup_{u_0 \in \mathcal{D}_0} \mathbb{P}(\tau_{u_0}^{\varepsilon} \geq kT) \leq \left(\sup_{u_0 \in \mathcal{D}_0} \mathbb{P}(\tau_{u_0}^{\varepsilon} \geq T) \right)^k.$$

Plugging this in (54) leads to

$$\mathbb{E}(\tau^{\varepsilon}_{u_0}) \leq T\left(1 - \sup_{u_0 \in D} \mathbb{P}(\tau^{\varepsilon}_{u_0} \geq T)\right)^{-1}$$

$$\leq T\left(\inf_{u_0 \in \mathcal{D}_0} \mathbb{P}(\tau^{\varepsilon}_{u_0} < T)\right)^{-1}.$$

This entails

$$\limsup_{\varepsilon \to 0} \varepsilon \log \mathbb{E}(\tau^{\varepsilon}_{u_0}) \leq -\liminf_{\varepsilon \to 0} \inf_{u_0 \in \mathcal{D}_0} \varepsilon \log \mathbb{P}(u^{\varepsilon}_{u_0} < T).$$

In order to bound the right-hand side of the previous inequality, it is helpful to prove a **large deviation principle (LDP)** [18, 44, 49, 51, 68, 74, 75, 111]. The key object in this case is the **good rate function**

$$\mathcal{F}_{[0,T]}(\gamma) = \frac{1}{2}\int_0^T \int_{\mathbb{S}^1} |\partial_t \gamma(t,x)|^2 \, dx \, dt,$$

where γ is a sufficiently regular path and we set $\mathcal{F}_{[0,T]}(\gamma) = +\infty$ if the last integral does not exist. One may prove $\mathcal{F}_{[0,T]}$ is lower semi-continuous and has compact level sets. Furthermore, it helps us to estimate the probability of the solution process $(u^{\varepsilon}_{u_0}(t))_{t \in [0,T]}$ of (52) as it satisfies an LDP

$$\liminf_{\varepsilon \to 0} 2\varepsilon \ln \mathbb{P}\left((u^{\varepsilon}_{u_0}(t))_{t \in [0,T]} \in S_o\right) \geq -\inf_{\gamma \in S_o} \mathcal{F}_{[0,T]}(\gamma),$$

$$\liminf_{\varepsilon \to 0} 2\varepsilon \ln \mathbb{P}\left((u^{\varepsilon}_{u_0}(t))_{t \in [0,T]} \in S_c\right) \leq -\inf_{\gamma \in S_c} \mathcal{F}_{[0,T]}(\gamma),$$

where S_o and S_c denote the sets of all open and closed subsets of the space of continuous paths respectively. The idea is that since we are able via an LDP to control the probability that certain sample paths appear in the small noise regime, we can, e.g., analyze first-exit times [44, 111] from a given sufficiently nice subset $\mathcal{D}_0 \subset L^2(\mathcal{H})$. For example, the LDP implies for a subset Γ of continuous paths that

$$\mathbb{P}\left((u^{\varepsilon}_{u_0}(t))_{t \in [0,T]} \in \Gamma\right) \approx e^{-\inf_{\Gamma} \mathcal{F}_{[0,T]}/(2\varepsilon)}$$

to be understood asymptotically, and more precisely as logarithmic equivalence, as $\varepsilon \to 0$. It is evident that the previous formulas indicate that **metastability** occurs as the probabilities to exit from a deterministically stable steady state become exponentially small in the limit $\varepsilon \to 0$. Using small noise limits and LDP estimates carries one quite far. The results range from sharp analytical exit-time asymptotics for reaction-diffusion systems [15, 18] to very broadly applicable numerical algorithms to capture metastability [94, 121].

3.3. Invariant Manifolds

The aim of this section is to analyze sets that contain the trajectories of an SPDE that converge to an equilibrium in forward / backward time or remain bounded for large time. For *deterministic* dynamical systems, these sets are called stable / unstable and center manifolds. Moreover, for initial conditions belonging to those sets, the corresponding solution must also evolve within the set. This property is called **invariance**. For deterministic systems, this can often be verified quite readily. For stochastic systems it is not a-priori clear, what a meaningful analogue of this concept is. We describe this fact within the RDS approach. We firstly recall that $(\Omega, \mathcal{F}, (\theta_t)_{t \in \mathbb{R}}, \mathbb{P})$ stands now for the *metric dynamical system*, constructed in Example 2. The elements of this quadruple will be denoted by ω and $\theta_t \omega$ represents the shift introduced in (41).

We illustrate the theory of invariant sets for stochastic evolution equations of the form (47) and indicate later on how this can be extended to more complicated SPDEs such as (26). We focus on *center manifolds*, due to their numerous properties and applications, i.e. center manifold theory allows a reduction to finite dimensional dynamics [62]. Therefore *stochastic center manifolds* have been intensively investigated in the literature, see [4, 47, 48, 62, 125, 167, 169] and the references specified therein.

We impose the following restrictions on the coefficients of (46). The drift $f : \mathcal{H} \to \mathcal{H}$ is assumed to be a locally Lipschitz nonlinear term with $f(0) = f'(0) = 0$. The spectrum of the linear operator A is supposed to contain eigenvalues with zero and strictly negative real parts, i.e. $\sigma(A) = \sigma^c(A) \cup \sigma^s(A)$, where $\sigma^c(A) = \{\lambda \in \sigma(A) : \text{Re}(\lambda) = 0\}$ and $\sigma^s(A) = \{\lambda \in \sigma(A) : \text{Re}(\lambda) < 0\}$. The subspaces generated by the eigenvectors corresponding to these eigenvalues are denoted by $\mathcal{H}^c$ respectively $\mathcal{H}^s$ and are referred to as *center* and *stable* subspace. These subspaces provide an invariant splitting of $\mathcal{H} = \mathcal{H}^c \oplus \mathcal{H}^s$. We denote the restrictions of A on $\mathcal{H}^c$ and $\mathcal{H}^s$ by $A_c := A|_{\mathcal{H}^c}$ and $A_s := A|_{\mathcal{H}^s}$. Since $\mathcal{H}$ is finite-dimensional we obtain that $S^c(t) := e^{tA_c}$ is a *group* of linear operators on $\mathcal{H}^c$. Moreover, there exist projections P^c and P^s such that $P^c + P^s = \text{Id}_{\mathcal{H}}$ and $A_c = A|_{\mathcal{R}(P^c)}$ and $A_s = A|_{\mathcal{R}(P^s)}$, where $\mathcal{R}$ denotes the range of the corresponding projection. Additionally, we impose the following dichotomy condition on the semigroup. We assume that there exist two exponents $\tilde{\gamma}$ and $\tilde{\beta}$ with $-\tilde{\beta} < 0 \le \tilde{\gamma} < \tilde{\beta}$ and constants $M_c, M_s \ge 1$, such that

$$\|S^c(t)x\|_{\mathcal{H}} \le M_c e^{\tilde{\gamma}t}\|x\|_{\mathcal{H}}, \quad \text{for } t \le 0 \text{ and } x \in \mathcal{H}; \tag{55}$$

$$\|S^s(t)x\|_{\mathcal{H}} \le M_s e^{-\tilde{\beta}t}\|x\|_{\mathcal{H}}, \quad \text{for } t \ge 0 \text{ and } x \in \mathcal{H}. \tag{56}$$

There are numerous operators that satisfy the spectral conditions imposed above. For instance let $\mathcal{H} := L^2(0, \pi)$ and set

$$Au := \Delta u + u$$

with domain $\text{Dom}(A) = H^2(0, \pi) \cap H_0^1(0, \pi)$. Its spectrum is given by $\{1 - n^2 : n \geq 1\}$ with corresponding eigenvectors $\{\sin(nx) : n \geq 1\}$. The eigenvectors give us the center subspace $\mathcal{H}^c = \text{span}\{\sin x\}$ and the stable one $\mathcal{H}^s = \text{span}\{\sin(nx) : n \geq 2\}$.

We now investigate center manifolds for the SPDE (46), under the above assumptions. Let φ denote the RDS generated by (46).

Definition 15. (Random center manifold) We call a set $\mathcal{M}^c(\omega)$ a **random center manifold** if

- $\mathcal{M}^c(\omega)$ contains all trajectories $\varphi(t, \cdot, \cdot)$ which are bounded in forward and backward time;
- $\mathcal{M}^c(\omega)$ has a graph structure. This means that there exists a function $h^c(\omega, \cdot) : \mathcal{H}^c \to \mathcal{H}^s$ with $h(\omega, 0) = 0$ such that

$$\mathcal{M}^c(\omega) = \{x + h^c(\omega, x) : x \in \mathcal{H}^c\}; \qquad (57)$$

- the **tangency condition** $Dh^c(\omega, 0) = 0$ holds true;
- $h^c(\cdot, x) : \Omega \to \mathcal{H}^s$ is measurable for every $x \in \mathcal{H}^c$;
- $h^c(\omega, \cdot) : \mathcal{H}^c \to \mathcal{H}^s$ is Lipschitz / smooth;
- the following **invariance** property holds true: if $x \in \mathcal{M}^c(\omega)$ then $\varphi(t, \omega, x) \in \mathcal{M}^c(\theta_t \omega)$ for $t \in \mathbb{R}^+$.

If (57) is satisfied for $x \in \mathcal{H}^c \cap \mathcal{B}(0, r(\omega))$, then $\mathcal{M}^c$ is called a *local* center manifold. Here $\mathcal{B}(0, r(\omega))$ denotes a random neighbourhood of the origin.

Remark 6. In the RDS approach the suitable concept for invariance of a random set [4, 64] is that each orbit starting inside this random set, evolves and remains there omega-wise modulo the changes that occur due to the noise. These changes can be characterized by a suitable shift of the fiber of the noise. Another concept is constituted by the almost sure invariance of a deterministic set under stochastic influences, more precisely this means that each orbit starting inside this deterministic set remains there almost surely, see [169] and the references specified therein.

We continue our deliberations regarding the existence of invariant sets for (47). Since there are no stochastic differentials / integrals one can prove the existence of random center manifolds similar to the deterministic case. There one uses the Lyapunov-Perron method, which seeks that the trajectories of (47) that remain close to the center subspace under the dynamics. This can be equivalently formulated as a fixed-point problem in a suitable function space. More precisely, one introduces the continuous-time **Lyapunov-Perron map / transform** for (47) is given by

$$J(\omega, u, u(0))[t] := S^c(t)e^{\int_0^t z(\theta_\tau\omega)\,d\tau} P^c u(0)$$

$$+ \int_0^t S^c(t-r)e^{\int_r^t z(\theta_\tau\omega)\,d\tau} P^c \overline{f}(\theta_r\omega, u(r))\,dr$$

$$+ \int_{-\infty}^t S^s(t-r)e^{\int_r^t z(\theta_\tau\omega)\,d\tau} P^s \overline{f}(\theta_r\omega, u(r))\,dr. \qquad (58)$$

Further details regarding this operator can be found in [167], [65, Section 6.2.2], [47, Chapter 4] and the references specified therein. The next natural step is to show that (58) possesses a fixed-point in a certain function space. One possible choice turns out to be $BC^{\tilde{\eta},z}(\mathbb{R}^-, \mathcal{H})$, see [65, p. 156]. This space is defined as

$$BC^{\tilde{\eta},z}(\mathbb{R}^-, \mathcal{H}) := \left\{ u : \mathbb{R}^- \to \mathcal{H}, \ \sup_{t \le 0} e^{-\tilde{\eta}t - \int_0^t z(\theta_\tau\omega)\,d\tau} \|u(t)\|_\mathcal{H} < \infty \right\}$$

and is endowed with the norm

$$\|u\|_{BC^{\tilde{\eta},z}} := \sup_{t \le 0} e^{-\tilde{\eta}t - \int_0^t z(\theta_\tau\omega)\,d\tau} \|u(t)\|_\mathcal{H}. \qquad (59)$$

Here $\tilde{\eta}$ is determined from (55) and (56), namely one has $-\tilde{\beta} < \tilde{\eta} < 0$. Note that the previous expressions are well-defined since

$$\lim_{t \to \pm\infty} \frac{|z(\theta_t\omega)|}{|t|} = 0,$$

according to [64, Lemma 2.1] and the references specified therein. Under a suitable smallness assumption on the Lipschitz constant of f (gap condition) one can show that J possesses a fixed-point $\Gamma(\cdot, \omega, u(0))$ for $u(0) \in \mathcal{H}^c$. Since such growth conditions on f can be quite restrictive in applications, one usually introduces a cut-off function to truncate the nonlinearity outside a random ball around the origin. This fixed-point characterizes the random center manifold $\mathcal{M}^c(\omega)$ for (47). More precisely, one can show that

$\mathcal{M}^c(\omega)$ can be represented locally by the graph of a function $h^c(\omega, \cdot)$, where $h^c(\omega, u(0)) = P^s\Gamma(0, \omega, u(0))$ for $u(0) \in \mathcal{H}^c \cap \mathcal{B}(0, r(\omega))$, i.e.

$$h^c(\omega, u(0)) = \int_{-\infty}^{0} S^s(-\tau) e^{\int_{\tau}^{0} z(\theta_r \omega)\,\mathrm{d}r} P^s \overline{f}(\theta_\tau \omega, \Gamma(\tau, \omega, u(0)))\,\mathrm{d}\tau. \qquad (60)$$

Here $\mathcal{B}(0, r(\omega))$ denotes a random neighbourhood of the origin, i.e. the radius $r(\omega)$ depends on the intensity/magnitude of the noise.

Example 8. Let $a > 0$, $\sigma > 0$ and W stand for a one-dimensional Brownian motion. For the SPDE

$$\begin{cases} du = (\Delta u + u - au^3)\,dt + \sigma u \circ dW(t) \\ u(0, t) = u(\pi, t) = 0, \text{ for } t \geq 0 \\ u(x, 0) = u_0(x), \qquad \text{for } x \in [0, \pi], \end{cases} \qquad (61)$$

the transformation into a PDE with random coefficients leads to

$$\frac{\partial u}{\partial t} = \Delta u + u + \sigma z(\theta_t \omega)u - ae^{2\sigma z(\theta_t \omega)}u^3, \qquad (62)$$

as discussed in Example 4. Regarding the discussion above, one can infer that (61) has a local center manifold

$$\mathcal{M}^c(\omega) = \{B\sin x + h^c(\omega, B\sin x)\} = \left\{ B\sin x + \sum_{n=2}^{\infty} c_n(\omega, B)\sin(nx) \right\}.$$

In this case, it is also possible to derive suitable approximation results for h^c, namely one can show that the coefficients satisfy $c_n(\omega, B) = \mathcal{O}(B^3)$ as $B \to 0$. Plugging this in (62) gives us the *nonautonomous* equation on the *center manifold*

$$\frac{\mathrm{d}}{\mathrm{d}t}(B\sin x) = B\sigma z(\theta_t \omega)\sin x - \frac{3}{4}aB^3\sin x e^{2\sigma z(\theta_t \omega)} + \mathcal{O}(B^5),$$

consequently

$$\dot{B} = \sigma z(\theta_t \omega)B - \frac{3}{4}aB^3 e^{2\sigma z(\theta_t \omega)} + \mathcal{O}(B^5).$$

Since $-u$ is also a solution for (62) we have that $c_n(\omega, B) = 0$ for n even. Therefore one has the following approximation of h as

$$h^c(\omega, B\sin x) = c_3(\omega, B)\sin 3x + \mathcal{O}(B^5).$$

Consequently, all the expressions arising in the Lyapunov-Perron method can be explicitly computed in this situation. ◆

The next step is to establish the existence of invariant sets for SPDEs *without* reducing them into PDEs with random coefficients. Using the rough path techniques presented in Section 2.6, one can immediately infer that the solution operator of (26) generates a random dynamical system, see [108]. Since the stochastic convolution / integral (28) was constructed in a pathwise way and not *almost surely*, there are no exceptional sets that can occur. Therefore, one can solve the SPDE (26) for every random input ω which is α-Hölder continuous, for $\alpha \in (1/3, 1/2]$. This case includes Brownian motion and fractional Brownian motion with Hurst index $H \in (1/3, 1/2)$, as illustrated in Section 2.6. Consequently, the results obtained using the transformation of an SPDE with linear multiplicative noise into a random PDE discussed above, will be recovered in a more general setting.

Under the assumptions on the linear operator A and drift term f specified above we can now investigate random sets for the SPDE with multiplicative noise (26). Using the notations in Section 2.6 we additionally assume that $g(0) = 0$. Regarding the deliberations above and (58), we infer that the **Lyapunov-Perron map/transform** is given by

$$J(\omega, u, u(0))[t] := S^c(s)P^c u(0) + \int_0^t S^c(t-r)P^c f(u(r)) \, \mathrm{d}r$$

$$+ \int_0^t S^c(t-r)P^c g(u(r)) \, \mathrm{d}\omega(r) + \int_{-\infty}^t S^c(t-r)P^s f(u(r)) \, \mathrm{d}r$$

$$+ \int_{-\infty}^t S^s(t-r)P^s g(u(r)) \, \mathrm{d}\omega(r), \tag{63}$$

where for notational consistency we used the identification $B^H(t, \omega) = \omega(t)$. Unlike (58), the Lyapunov-Perron transform J contains stochastic integrals and it is not clear in which function space one should formulate the fixed-point problem. However, modifying the Lyapunov-Perron method one obtains the following result, see [125].

Theorem 9. *There exists a local center manifold* $\mathcal{M}^c(\omega)$ *for* (26) *such that*

$$\mathcal{M}^c(\omega) = \{u(0) + h^c(\omega, u(0)) : u(0) \in \mathcal{H}^c \cap \mathcal{B}(0, r(\omega))\},$$

where

$$h^c(\omega, u(0)) = \int_{-\infty}^0 S^s(-r)P^s f(\Gamma(r, \omega, u(0))) \, \mathrm{d}r.$$

$$+ \int\limits_{-\infty}^{0} S^{\mathrm{s}}(-r) P^{\mathrm{s}} g(\Gamma(r, \omega, u(0))) \, \mathrm{d}\omega(r)$$

and Γ *is the fixed-point of* J.

We are going to outline the ideas of the proof. Since the estimates of the stochastic integrals appearing in the definition of J contain certain Hölder norms of the random input ω, which are uniform over the unit interval, it is meaningful to:

1) discretize the Lyapunov-Perron map (63), i.e. consider its solution at *discrete* time-points and obtain a sequence of solutions over $[0, 1]$;
2) apply the Lyapunov-Perron method in a space of *sequences*, i.e. formulate the fixed-point problem for the discrete version of J in a space of sequences (instead of $BC^{\eta, z}$ as above).

The proof essentially combines rough path techniques with the Lyapunov-Perron method for *discrete-time* dynamical systems. For further details on this topic see [125].

Remark 7. To obtain stable / unstable manifolds, one imposes different spectral assumptions on the linear part and modifies the definition of J accordingly, see [63, 64, 85, 133].

We conclude this section pointing out once more that the time-evolution of a manifold in the RDS framework is described using an appropriate shift with respect to the noise. However, there are further theories that provide bounds on the probabilities that these manifolds evolve in time, see for instance [24, 25, 28].

Similar to center manifold theory that often allows for a local reduction to finite dimensional dynamics for SPDEs, there are several other approaches that provide approximations for the solutions of SPDEs by finite-dimensional SDEs as in [25, 27]. To briefly illustrate such results, we consider similar to Example 61, for small $\varepsilon > 0$ and arbitrary $\nu, \sigma > 0$ the SPDE

$$\mathrm{d}u = [(\Delta u + u) + \nu\varepsilon^2 - u^3] \, \mathrm{d}t + \sigma\varepsilon \, \mathrm{d}W(t), \tag{64}$$

with Dirichlet boundary conditions on $[0, \pi]$. The term $\nu\varepsilon^2$ is a small linear perturbation. As seen in Example 61, $\mathcal{H}^{\mathrm{c}} = \mathrm{span}\{\sin x\}$ and $\mathcal{H}^{\mathrm{s}} = \mathrm{span}\{\sin(nx) : n \geq 2\}$. Assuming that the additive noise is acting for instance only on $\sin(2x)$ and rescaling time as $T := \varepsilon^2 t$, one can show that the rescaled solution of the SPDE (64)

$$u(t) := \varepsilon v(\varepsilon^2 t),$$

can be approximated as

$$v(T) \approx B(T)\sin x + \frac{\sigma}{3}\tilde{Z}(T)\sin(2x) + \mathcal{O}(\varepsilon^{1-}).$$

Here $\mathcal{O}(\varepsilon^{1-})$ collects the higher-order terms, B is the solution of a certain **amplitude equation** (or **modulation equation**) given by

$$\partial_T B = \left(\nu - \frac{\sigma^2}{4}\right)B - \frac{3}{4}B^3$$

and $\tilde{Z}$ is the Ornstein-Uhlenbeck process on the new time-scale given by

$$\tilde{Z}(T) = \varepsilon^{-1} \int\limits_0^T e^{-\lambda_2 \varepsilon^2 (T-\tau)} \, d\tilde{W}(\tau),$$

where λ_2 is the eigenvalue corresponding to $\sin(2x)$ and $\tilde{W}(T) := \varepsilon W(\varepsilon^{-2}T)$, is a rescaled Brownian motion. Naturally, if the additive noise acts on the other modes of $\mathcal{H}^s$, i.e. $W(t) = \sum_{k=2}^{N} \sigma_k w_k(t) e_k$ for a fixed $N > 2$ or $W(t) = \sum_{k=2}^{\infty} \sigma_k w_k(t) e_k$ together with a decay condition on the coefficients of this series, one can derive analogous amplitude equations and approximation results. For the proof of such statements and further applications, see [24, 25, 27] and [167, Section 6.6].

3.4. *Random Attractors*

As already emphasized, RDS are generalizations of nonautonomous dynamical systems. The next concept is a generalization of a *deterministic attractor* for nonautonomous systems, as considered in [5, 114, 115, 159, 161]. Intuitively, an attractor is a compact set of the phase space towards which the dynamical system evolves after a certain amount of time. Before dealing with attractors for SPDEs, we give a concise example which illustrates, why is it meaningful for nonautonomous systems to take a *pullback limit* with respect to *time* instead of a *forward* one.

Example 9. Let $a, b > 0$ and consider as in [46] the simple nonautonomous ODE

$$\begin{cases} u' = -au + bt \\ u(0) = u_0. \end{cases}$$

Its solution $u(t) = e^{-at}u_0 + \frac{bt}{a} - \frac{b}{a^2}$ obviously blows up for $t \to \infty$. However, if one considers the solution operator at time t with $u(s) = u_0$ for $t \geq s$, i.e. $\overline{\varphi}(t, s, u_0)$, and takes the pullback limit $s \to -\infty$ for a fixed $t > 0$, a straightforward computation entails that

$$|\overline{\varphi}(t, s, u_0) - \overline{\mathcal{A}}(t)| \to 0, \quad \text{as } s \to -\infty, \tag{65}$$

where $\overline{\mathcal{A}}(t) := \frac{bt}{a} - \frac{b}{a^2}$. ♦

For an SPDE, an attractor will be a *random, time-dependent* set. We present the mathematical formalism of this concept and explain what *pullback* in this framework means. Again, we assume that the solution operator of an SPDE generates a random dynamical system φ on $\mathcal{H}$ and present concrete examples later on. We recall that throughout this section $(\Omega, \mathcal{F}, (\theta_t)_{t \in \mathbb{R}}, \mathbb{P})$ denotes the metric dynamical system constructed in Example 2.

Definition 16. A random bounded set $\{\mathcal{S}(\omega)\}_{\omega \in \Omega}$ of $\mathcal{H}$ is called **tempered** with respect to $(\theta_t)_{t \in \mathbb{R}}$ if for all $\omega \in \Omega$ it holds

$$\lim_{t \to \infty} e^{-\overline{\beta}t} \sup_{x \in \mathcal{S}(\theta_{-t}\omega)} \|x\| = 0, \quad \text{for all } \overline{\beta} > 0.$$

In the sequel, $\mathcal{D}$ denotes the collection of tempered random sets in $\mathcal{H}$.

Definition 17. (Random absorbing set) A set $\{\mathcal{S}(\omega)\}_{\omega \in \Omega} \in \mathcal{D}$ is called **random absorbing set** for φ if for every $\mathcal{R} = \{\mathcal{R}(\omega)\}_{\omega \in \Omega} \in \mathcal{D}$ and $\omega \in \Omega$, there exists a random time $t_{\mathcal{R}}(\omega) > 0$ such that

$$\varphi(t, \theta_{-t}\omega, \mathcal{R}(\theta_{-t}\omega)) \subseteq \mathcal{S}(\omega), \quad \text{for all } t \geq t_{\mathcal{R}}(\omega).$$

Definition 18. (Random pullback attractor) A random set $\{\mathcal{A}(\omega)\}_{\omega \in \Omega} \in \mathcal{D}$ is called a $\mathcal{D}$-**random (pullback) attractor** for φ if the following properties are satisfied:

a) $\mathcal{A}(\omega)$ is compact for every $\omega \in \Omega$;

b) $\{\mathcal{A}(\omega)\}_{\omega \in \Omega}$ is positive invariant, i.e.

$$\varphi(t, \omega, \mathcal{A}(\omega)) = \mathcal{A}(\theta_t\omega) \text{ for all } t \geq 0;$$

c) $\{\mathcal{A}(\omega)\}_{\omega \in \Omega}$ pullback attracts every set in $\mathcal{D}$, more precisely, for every $\mathcal{R} = \{\mathcal{R}(\omega)\}_{\omega \in \Omega} \in \mathcal{D}$,

$$\lim_{t \to \infty} \text{dist}_{\mathcal{H}}(\varphi(t, \theta_{-t}\omega, \mathcal{R}(\theta_{-t}\omega)), \mathcal{A}(\omega)) = 0, \tag{66}$$

where $\text{dist}_{\mathcal{H}}$ is the **Hausdorff semi-distance**. This is given by $\text{dist}_{\mathcal{H}}(Y, Z) = \sup_{y \in Y} \inf_{z \in Z} \|y - z\|$, for any subsets $Y \subseteq \mathcal{H}$ and $Z \subseteq \mathcal{H}$.

If the set $\mathcal{A}(\omega)$ satisfies all the properties above and consists only of a singleton, then this is called **random fixed-point**.

From this abstract definition, we see that the concept pullback means in this setting, that one shifts into the past the fiber of the noise, compare (65) and (66). For a better comprehension we present the following example.

Example 10. Let $\overline{\mu} > 0$ and consider the one-dimensional SDE

$$du = -\overline{\mu}u \, dt + dW(t), \tag{67}$$

Obviously, for $t \geq t_0$, its solution is given by

$$u(t) = u(t_0)e^{-\overline{\mu}(t-t_0)} + e^{-\overline{\mu}t} \int_{t_0}^{t} e^{\overline{\mu}s} \, dW(s).$$

Note that the forward limit $t \to \infty$ does not exist. However, for a fixed $t \in \mathbb{R}$ taking the pullback limit $t_0 \to -\infty$ yields the stationary solution of this SDE

$$\lim_{t_0 \to -\infty} u(t) = \int_{-\infty}^{t} e^{-\overline{\mu}(t-s)} \, dW(s) = \int_{-\infty}^{0} e^{\overline{\mu}s} \, d\theta_t W(s),$$

which is the one-dimensional Ornstein-Uhlenbeck process $(t, \omega) \mapsto z(\theta_t \omega)$

$$z(\theta_t \omega) = \int_{-\infty}^{0} e^{\overline{\mu}s} \, d\theta_t \omega(s).$$

One can show that z is the random fixed-point of (67), i.e. in this case the attractor $\mathcal{A}(\omega) := \{z(\omega)\}$ is a singleton. The same statement holds true in the infinite-dimensional setting, recall (43) and (44). ♦

The existence of random attractors can be shown by the following criterion, see Theorem 2.1 in [157] and [42, 45].

Theorem 10. (Criterion for existence of a pullback attractor) Let φ be a continuous random dynamical system on $\mathcal{H}$ over $(\Omega, \mathcal{F}, \mathbb{P}, (\theta_t)_{t \in \mathbb{R}})$. Suppose that $\{\mathcal{K}(\omega)\}_{\omega \in \Omega}$ is a compact random absorbing set for φ in $\mathcal{D}$. Then φ has a unique $\mathcal{D}$-random attractor $\{\mathcal{A}(\omega)\}_{\omega \in \Omega}$ which is given by

$$\mathcal{A}(\omega) = \bigcap_{\tau \geq 0} \overline{\bigcup_{t \geq \tau} \varphi(t, \theta_{-t}\omega, \mathcal{K}(\theta_{-t}\omega))}.$$

This theorem indicates that one has to verify the following two aspects in order to prove the existence of a random pullback attractor.

1) The existence of an absorbing set. This usually follows by suitable a-priori estimates on the solutions. More precisely, the following condition is convenient to show the existence of an absorbing set. If for every $x \in \mathcal{R}(\theta_{-t}\omega)$, $\mathcal{R} \in \mathcal{D}$ and $\omega \in \Omega$ it holds

$$\limsup_{t \to \infty} \|\varphi(t, \theta_{-t}\omega, x)\|_{\mathcal{H}} \leq \rho(\omega), \qquad (68)$$

where $\rho(\omega) > 0$ is a *tempered random variable*, then the ball centered in 0 with radius $\rho(\omega) + \delta$, i.e. $\mathcal{S}(\omega) := \mathcal{B}(0, \rho(\omega) + \delta)$, for some constant $\delta > 0$, is a random absorbing set.

2) The compactness of the absorbing set. This follows in general by deriving estimates of the solutions in function spaces that are compactly embedded in $\mathcal{H}$.

For further details and applications see [45, 157]. We now discuss how these abstract results can be applied to concrete SPDEs, more precisely to SPDEs which are equivalent to (45). In order to obtain the necessary compactness, we additionally assume that the semigroup $(S(t))_{t \geq 0}$ is analytic and *compact*. Furthermore, we assume that there exist constants $\widetilde{M} \geq 1$ and $\bar{\mu} > 0$ such that $\|S(t)\|_{\mathcal{H}} \leq \widetilde{M}e^{-\bar{\mu}t}$ for $t > 0$. Therefore, the stationary Ornstein-Uhlenbeck process

$$Z(\theta_t\omega) = \int\limits_{-\infty}^{t} S(t - s) \, \mathrm{d}\omega(s)$$

is well-defined. An example in this sense is given by the equation

$$\mathrm{d}u = [\Delta u - \bar{\mu}u] \, \mathrm{d}t + f(u + Z(\theta_t\omega)) \, \mathrm{d}t$$

on $\mathcal{H} := L^2(D)$, where $D \subset \mathbb{R}^d$ is an open bounded domain. The most simple situation is to assume the nonlinear term $f : \mathcal{H} \to \mathcal{H}$ is globally Lipschitz continuous with Lipschitz constant $L_f > 0$ and of bounded growth with constant $0 < l_f < L_f$. In this case, letting $\varphi(t, \omega, u_0)$ denote the corresponding solution operator of (45), the Gronwall inequality immediately entails the a-priori bound

$$\|\varphi(t, \omega, u_0)\|_{\mathcal{H}} \leq \widetilde{M}e^{(l_f \widetilde{M} - \bar{\mu})t}\|u_0\|_{\mathcal{H}} + M \int\limits_{0}^{t} C(\theta_s\omega)e^{(l_f \widetilde{M} - \bar{\mu})(t-s)} \, \mathrm{d}s,$$

where $C(\omega) := l_f\|Z(\omega)\|_{\mathcal{H}} + c$ and $c > 0$ is a constant. This term occurs due to the structure and growth boundedness of f. Replacing ω by $\theta_{-t}\omega$ in the previous inequality and further assuming that $\bar{\mu} - l_f\widetilde{M} > 0$, leads to

$$\|\varphi(t, \theta_{-t}\omega, u_0)\|_{\mathcal{H}} \leq \widetilde{M}e^{(l\widetilde{M} - \bar{\mu})t}\|u_0\|_{\mathcal{H}} + \int\limits_{-t}^{0} C(\theta_\tau\omega)e^{(\bar{\mu} - l_f\widetilde{M})\tau} \, \mathrm{d}\tau$$

$$\leq \widetilde{M}\mathrm{e}^{(l\widetilde{M}-\overline{\mu})t}\|u_0\|_{\mathcal{H}} + \int\limits_{-\infty}^{0} C(\theta_\tau\omega)\mathrm{e}^{(\overline{\mu}-l_f\widetilde{M})\tau}\,\mathrm{d}\tau.$$

This bound immediately entails the existence of an absorbing set if $\mathcal{S}(\omega) := \mathcal{B}(0, \rho(\omega) + \delta)$, where $\delta > 0$ and the tempered random variable ρ is given by

$$\rho(\omega) := \int\limits_{-\infty}^{0} C(\theta_\tau\omega)\mathrm{e}^{(\overline{\mu}-l_f\widetilde{M})\tau}\,\mathrm{d}\tau.$$

Keeping this in mind, a natural candidate for a *compact* absorbing set for φ would be for instance $\mathcal{K}(\omega) := \overline{\varphi(t^*, \theta_{-t^*}\omega, \mathcal{S}(\theta_{-t^*}\omega))}$, where $t^* > 0$ is fixed and the closure is taken with respect to the topology in $\mathcal{H}$. This can be achieved showing that

$$\|\varphi(t^*, \theta_{-t^*}\omega, \mathcal{S}(\theta_{-t^*}\omega))\|_{\mathcal{H}_\gamma} < \infty,$$

where $\gamma \in (0,1)$ and $\mathcal{H}_\gamma = \mathrm{Dom}((-A)^\gamma)$ are the domains of the fractional powers of the operator, as discussed in Section 2.6. Due to the fact that $(S(t))_{t\geq 0}$ is a compact C_0-semigroup, the embedding $\mathcal{H}_\gamma \hookrightarrow \mathcal{H}$ is compact which proves the compactness of the random set $\mathcal{K}(\omega)$. Due to Theorem 10 we infer that (45) has a random pullback attractor.

Similar arguments can be employed if the nonlinear term satisfies suitable dissipativity conditions. More precisely, we consider the following equation

$$\mathrm{d}u = [\Delta u - \overline{\mu}u]\,\mathrm{d}t + f(x, u + Z(\theta_t\omega))\,\mathrm{d}t, \quad t > 0 \text{ and } x \in D. \quad (69)$$

As commonly met in the theory of reaction-diffusion equations, see [159, Section 5.1], we impose the following restrictions on the reaction terms. For $x \in D$, $s \in \mathbb{R}$ and an integer $q \geq 2$ we assume that there exist positive constants $\alpha_1, \alpha_2, \alpha_3$ and c_1, c_2, c_3 such that

$$f(x, s)s \leq -\alpha_1|s|^q + c_1$$

$$|f(x, s)| \leq \alpha_2|s|^{q-1} + c_2$$

$$\left|\frac{\partial f}{\partial s}(x, s)\right| \leq \alpha_3|s|^{q-2} + c_3.$$

This means that f can be a polynomial of odd degree with a negative leading order coefficient. In this case one can derive a-priori estimates of the solution of (69) in $L^2(D)$ to obtain the existence of an absorbing set, and in $W^{1,2}(D)$ for the compactness argument.

Major technical difficulties occur when compact embeddings are no longer available. This is for instance the case, when one considers (69) on unbounded domains, i.e. on $\mathbb{R}^d$, see [16]. In this case, one replaces the constants c_1, c_2, c_3 in the previous assumption with $L^2(\mathbb{R}^d)$-integrable functions and modifies (69) as

$$du = [\Delta u - \overline{\mu}u] \, dt + [f(x, u + Z(\theta_t\omega)) + h(x)] \, dt, \qquad (70)$$

for $t > 0$ and $x \in \mathbb{R}^d$, where $h \in L^2(\mathbb{R}^d)$. Here, one needs again estimates in $L^2(\mathbb{R}^d)$ and $W^{1,2}(\mathbb{R}^d)$. The most technical argument is to show by means of a cut-off technique that the tails of the solution of (70) will become uniformly small for large enough time. Afterwards one can use the compact embedding $W^{1,2}(D) \hookrightarrow L^2(D)$ for bounded domains $D \subset \mathbb{R}^d$ and the uniform decay for sufficiently large time outside of D. The function h is required in order to show that the attractor for (70) is set-valued. Without h, the attractor reduces to a singleton.

Another situation where compact embeddings cannot be directly employed occurs in the context of **partly dissipative systems**, i.e. coupled SPDEs with SDEs. An example in this sense is given by the following system. We assume for simplicity that $x \in [0, 1]$ and consider

$$du_1 = \Delta u_1 \, dt + [f_1(x, u_1 + Z_1(\theta_t\omega, u_2 + Z_2(\theta_t\omega))) + f_2(x, u_1 + Z_1(\theta_t\omega))] \, dt$$
$$du_2 = -\nu u_2 \, dt + f_3(x, u_1 + Z_1(\theta_t\omega)) \, dt,$$

where $\nu > 0$, Z_1, Z_2 are two Ornstein-Uhlenbeck processes and f_1, f_2, f_3 satisfy suitable dissipativity assumptions. The technical difficulty consists in the missing regularizing effect of the Laplacian in the second component. To overcome this, one splits the second component u_2 into a regular part (i.e. which belongs to $W^{1,2}(0, 1)$) and a remaining part which tends asymptotically to zero. This is sufficient to obtain the necessary compactness results. Further details on this topic can be looked up in [126].

Before pointing out some concluding remarks we would like to emphasize that the structure of random attractors can be totally different from the deterministic case. For instance, consider as in [70], the porous-media equation

$$du = [\Delta |u|^{m-1}u + u] \, dt + dW^Q(t), \qquad (71)$$

with zero Dirichlet boundary conditions on a bounded domain $D \subseteq \mathbb{R}^d$, where $d \leq 4$ and $m > 1$. The noise W^Q is a trace-class Brownian motion on $W^{-1,2}(D) = (W_0^{1,2}(D))^*$. It is known that the attractor of the *deterministic* porous-media equation has infinite fractal dimension. However, the *random* attractor of (71) reduces to a fixed-point, see [70, Section 3]. This is an example of *synchronization by noise* as investigated in [70].

Remark 8.

1) Results on the existence of random attractors for SPDEs without directly reducing the SPDE in a random PDE are available in [22]. Here one deals with semilinear *delay equations* under certain smoothness assumptions on the coefficients. These allow an *integration by parts formula* which is employed in order to get rid of the stochastic integrals and do a pathwise analysis of the equation. Apart from this, all results regarding random pullback attractors for SPDEs with *additive* or *linear multiplicative* noise employ transformations in a PDE with random coefficients as described above [16, 33, 45, 46, 70, 88–90, 157, 166]. To our best knowledge, for nonlinear multiplicative noise, and SPDEs such as (26), there are no general existence results concerning random attractors. For multiplicative fractional noise, in the more regular case, i.e. $H \in (1/2, 1)$, results on this topic are contained in [80].

2) For assertions regarding the dimension of random attractors, see [55, 129] and more recently [39].

3) There are several further concepts of random attractors, for instance *weak* [70, 156], *exponential* [39] or *mean-square* random attractors [113].

3.5. *Further Dynamical Aspects*

In this section we just point out several dynamical phenomena for SPDEs we have not discussed above due to practical space limitations within this edited volume. We also provide some references for the reader to get started in the background literature on these topics. We emphasize that some of the following aspects have been investigated especially for additive and multiplicative white-in-time noise while the influence of other type of time-correlated noise is highly challenging and remains to be explored in far more detail. Such dynamical phenomena include among others:

- traveling waves and their stability [106, 122, 144];
- pattern formation [81, 93, 118];
- stochastic bifurcations [4, 26, 127];
- early-warning signs for bifurcations in SPDEs [93, 120, 127];
- sample paths estimates for fast-slow SPDEs [18, 92];
- averaging results for fast-slow SPDEs [65, 168];
- stochastic homogenization [21, 43, 76].

• finite-time blow-up [134, 145].

Naturally, there exist further probabilistic methods that provide dynamical insights, which are based on Kolmogorov/Fokker-Planck equations associated to SPDEs [14, 29, 51]. For higher-order-in-time SPDEs, such as wave equations / dispersive equations there are available results regarding invariant manifolds [133] and random attractors [39, 166].

In summary, a lot of progress towards solution theory of SPDEs has been made over the last several decades. The development of dynamical system results for SPDEs is currently actively growing and promises to be an active area of research for many decades to come. For example, our understanding of the structure, formation, stability, and interaction of patterns for SPDEs is still far away from the level of results available in the PDE setting. A similar remark applies to bifurcation problems and the role played by singular SPDEs in this context.

Acknowledgements. CK and AN have been supported by a DFG grant in the D-A-CH framework (KU 3333/2-1). CK acknowledges support by a Lichtenberg Professorship. CK and AN would like to thank Prof. Wilfried Grecksch and Prof. Hannelore Lisei for the kind invitation to contribute with this chapter.

References

[1] H. Amann. *Linear and Quasilinear Parabolic Problems: Volume I: Abstract Linear Theory.* Springer, 1995.

[2] D.C. Antonopoulou, P.W. Bates, D. Blömker, G.D. Karali. Motion of a droplet for the stochastic mass-conserving Allen-Cahn equation. *SIAM J. Math. Anal.*, 48(1):670–708, 2016.

[3] L. Arnold. Stochastic systems: qualitative theory and Lyapunov exponents. Fluctuations and Sensitivity in Nonequilibrium Systems. Springer Proc. Phys., 1, pp. 11–18, Springer, Berlin, 1984.

[4] L. Arnold. *Random Dynamical Systems.* Springer, Berlin Heidelberg, Germany, 2003.

[5] A.V. Babin and M.I. Vishik. *Attractors of Evolution Equations.* North-Holland, Amsterdam, 1992.

[6] I. Bailleul. *Flows driven by Banach space-valued rough paths.* Séminaire de Probabilités XLVI, pp. 195–205, 2014.

[7] I. Bailleul, A. Debussche and M. Hofmanová. Quasilinear generalized parabolic Anderson model equation. *Stochastics and Partial Differential Equations: Analysis and Computation*, 7(1):40–63, 2019.

[8] I. Bailleul and M. Hoshino. Paracontrolled calculus and regularity structures. *arXiv:1812.07919*, pages 1–27, 2018.

[9] I. Bailleul, S. Riedel and M. Scheutzow. Random dynamical system, rough paths and rough flows. *J. Diffential Equat.*, 262(12):5792–5823, 2017.

[10] R.M. Balan, M. Jolis and L. Quer-Sardanyons. Intermittency for the Hyperbolic Anderson Model with rough noise in space. *Stoch. Proc. Appl.*, 127(7):2316–2338, 2017.

[11] R.M. Balan and J. Song. Second order Lyapunov exponents for parabolic and hyperbolic Anderson models. *To appear in Bernoulli.*

[12] V. Barbu, G. Da Prato and M. Röckner. Existence of strong solutions for stochastic porous media equations under general monotonicity conditions. *Ann. Probab.*, 37(2):428–452, 2009.

[13] V. Barbu and M. Röckner. An operatorial approach to stochastic partial differential equations driven by linear multiplicative noise. *J. Eur. Math. Soc.*, 17(7):1789-1915, 2015.

[14] V. Barbu and M. Röckner. Nonlinear Fokker–Planck equations driven by Gaussian linear multiplicative noise. *arXiv:1708.08768*, pages 1–42, 2017.

[15] F. Barret. Sharp asymptotics of metastable transition times for one-dimensional SPDEs. *Ann. Inst. Henri Poincaré Probab. Stat.*, 51(1):129–166, 2015.

[16] P. Bates, K. Lu and B. Wang. Random attractors for stochastic reaction-diffusion equations on unbounded domains. *J. Differential Equat.*, 246(2):845–869, 2009.

[17] N. Berglund. An introduction to singular stochastic PDEs: Allen-Cahn equations, metastability and regularity structures. *arXiv:1901.07420*, pages 1–95, 2019.

[18] N. Berglund and B. Gentz. Sharp estimates for metastable lifetimes in parabolic SPDEs: Kramers' law and beyond. *Electron. J. Probab.*, 24(18):1–58, 2013.

[19] N. Berglund and C. Kuehn. Regularity structures and renormalisation of FitzHugh–Nagumo SPDEs in three space dimensions. *Electron. J. Probab.*, 21(18):1–48, 2016.

[20] N. Berglund and C. Kuehn. Model spaces of regularity structures of space-fractional SPDEs. *J. Stat. Phys.*, 168(2):331–368, 2017.

[21] H. Bessaih, Y. Efendiev and R.F. Maris. Stochastic homogenization for a diffusion-reaction model. *Discrete Contin. Dyn. Syst.*, 39(9):5403–5429, 2019.

[22] H. Bessaih, M. Garrido-Atienza, B. Schmalfuß. Pathwise solutions and attractors for retarded SPDEs with time smooth diffusion coefficients. *Discrete Contin. Dyn. Syst.*, 34(10): 3945–3968, 2014.

[23] W. Beyn, B. Gess, P. Lescot and M. Roeckner. The global random attractor for a class of stochastic porous media equations. *Comm. Partial Differential Equations*, 36(3):446-469, 2011.

[24] D. Blömker. *Amplitude Equations for Stochastic Partial Differential Equations.* World Scientific, 2007.

[25] D. Blömker and M. Hairer. Amplitude equations for SPDEs: Approximate centre manifolds and invariant measures. *Probability and partial differential equations in modern applied mathematics.* Springer, 2005.

[26] D. Blömker, M. Hairer and G. Pavliotis. Modulation Equations: Stochastic Bifurcation in Large Domains. *Comm. Math. Phys*, 258(2):479–512, 2015.

[27] D. Blömker and W.W. Mohammed. Amplitude equations for SPDEs with cubic nonlinearities. *Stochastics: An International Journal of Probability and Stochastic Processes.* 85(2):181–215, 2013.

[28] D. Blömker and W. Wang. Qualitative properties of local random invariant manifolds for SPDEs with quadratic nonlinearity. *J. Dyn. Differ. Equations*, 22(4):677–695, 2010.

[29] V.I. Bogachev, N. V. Krylov, M. Röckner, S.V. Shaposhnikov. *Fokker-Planck-Kolmogorov Equations.* Mathematical Surverys and Monographs, vol. 207, 2015.

[30] A. Brault. Solving rough differential equations with the theory of regularity structures. *arXiv:1712.06285*, pages 1–38, 2017.

[31] Y. Bruned, A. Chandra, I. Chevyrev and M. Hairer. Renormalising SPDEs in regularity structures. *arXiv:1711.10239*, pages 1–85, 2017.

[32] Y. Bruned, M. Hairer and L. Zambotti. Algebraic renormalisation of regularity structures. *Invent. Math.*, 215(3):1039–1156, 2019.

[33] Z. Brzeźniak, B. Goldys and Q.T. Le Gia. Random Attractors for the Stochastic Navier-Stokes Equations on the 2D Unit Sphere. *J. Math. Fluid Mech.*, 20:227–253, 2018.

[34] Z. Brzeźniak and E. Motyl. The existence of martingale solutions to the stochastic Boussinesq equations. *Global Stoch. Anal.* 1(2):175–216, 2014.

[35] Z. Brzeźniak, D. Salopek and J.M.A.M. van Neerven. Stochastic evolution equations driven by Liouville fractional Brownian motion. *Czech. Math.J.*, 62(1):1–27, 2012.

[36] M. Callaway, T.S. Doan, J.S.W. Lamb and M. Rasmussen. The dichotomy spectrum for random dynamical systems and pitchfork bifurcations with additive noise. *Ann. Inst. H. Poincaré Probab. Statist.*, 53(4):1548–1574, 2017.

[37] T. Caraballo, J. Duan, K. Lu and B. Schmalfuß. Invariant manifolds for random and stochastic partial differential equations. *Adv. Nonlinear Studies*, 10(1):23–52, 2010.

[38] T. Caraballo and K. Liu. On exponential stability criteria of stochastic partial differential equations. *Stoch. Proc. Appl.*, 83(2): 289–301, 1999.

[39] T. Caraballo and S. Sonner. Random pullback exponential attractors: general existence results for random dynamical systems in Banach spaces. *Discrete Contin. Dyn. Syst.*, 37(12):6383–6403, 2017.

[40] R. Catellier and K. Chouk. Paracontrolled distributions and the 3-dimensional stochastic quantization equation. *Ann. Probab.*, 46(5):2621–2679, 2018.

[41] A. Chandra and M. Hairer. An analytic BPHZ theorem for regularity structures. *arXiv:1612.08138*, pages 1–129, 2016.

[42] H. Crauel, A. Debussche and F. Flandoli. Random attractors. *J. Dyn. Differ. Equ.*, 9(2):307–341, 1997.

[43] S. Cerrai. A Khasminskii type averaging principle for stochastic reaction-diffusion equations. *Ann. Appl. Probab.*, 19(3):899–948, 2009.

[44] S. Cerrai and M. Röckner. Large deviations for stochastic reaction-diffusion systems with multiplicative noise and non-Lipschitz reaction term. *Ann. Probab.*, 32(1B):1100–1139, 2004.

[45] H. Crauel and F. Flandoli. Attractors for random dynamical systems. *Probab. Theory Rel. Fields*, 100(3):365– 393, 1994.

[46] M.D. Chekroun, E. Simonnet and M. Ghil. Stochastic climate dynamics: Random attractors and time-dependent invariant measures. *Physica D*, 240(21):1685–1700, 2011.

[47] M.D. Chekroun, H. Liu and S. Wang. *Approximation of stochastic invariant manifolds. Stochastic manifolds for nonlinear SPDEs I.* Springer, 2015.

[48] X. Chen, A.J. Roberts and J. Duan. Centre manifolds for stochastic evolution equations. *J. Differ. Equ. Appl.*, 21(7):602–632.

[49] F. Chenal and A. Millet. Uniform large deviations for parabolic SPDEs and applications. *Stoch. Proc. Appl.*, 72(2):161–186, 1997.

[50] H. Crauel and P.E. Kloeden. Nonautonomous and random attractors. *Jahresber. Dtsch. Math.- Ver.*, 117(3):173–206, 2015.

[51] G. Da Prato and J. Zabczyk. *Stochastic Equations in Infinite Dimensions.* Cambridge University Press, 1992.

[52] R.C Dalang. Extending the martingale measure stochastic integral with applications to spatially homogeneous s.p.d.e.'s. *Electron. J. Probab.*, 4(6):1–29, 1999.

[53] R.C. Dalang and L. Quer-Sardanyons. Stochastic integrals for spdes's. A comparison. *Expositiones Mathematicae*, 29(1):67–109, 2011.

[54] K. Dareiotis and B. Gess. Nonlinear diffusion equations with nonlinear gradient noise. *arXiv:1811.08356*, pages 1–42, 2019.

[55] A. Debussche. On the finite dimensionality of random attractors. *Stoch. Anal. Appl.*, 15:473–491, 1997.

[56] A. Debussche, M. Hofmanová and J. Vovelle. Degenerate parabolic stochastic partial differential equations: quasilinear case. *Ann. Probab.*, 44(3):1916–1955, 2016.

[57] A. Deya, M. Gubinelli and S. Tindel. Non-linear rough heat equations. *Probab. Theory Related Fields.* 153:(1–2)97–147, 2012.

[58] A. Deya, M. Gubinelli, M. Hofmanova and S. Tindel. A priori estimates for rough PDEs with application to rough conservation laws. *arXiv:1604.00437*, pages 1–51, 2016.

[59] A. Deya and S. Tindel. Rate of convergence to equilibrium of fractional driven stochastic differential equations with rough multiplicative noise. *Ann. Probab.*, 47(1):464–518, 2019.

[60] G. Dhariwal, F. Huber, A. Jüngel, C. Kuehn and A. Neamţu. Global martingale solutions for quasilinear SPDEs via the boundedness-by-entropy method. *arXiv:1909.08892*, pages 1–33, 2019.

[61] H. Doss. Liens entre équations différentielles stochastiques et ordinaries. *Ann. Inst. H. Poincaré Sect B (N.S.)*, 13(2):99–125, 1977.

[62] A. Du and J. Duan. Invariant manifold reduction for stochastic dynamical systems. *Dynamic Systems and Applications* 16:681-696, 2007.

[63] J. Duan, K. Lu and B. Schmalfuß. Invariant manifolds for stochastic partial

differential equations. *Ann. Probab.*, 31(4):2109–2135, 2003.

[64] J. Duan, K. Lu and B. Schmalfuß. Smooth Stable and Unstable Manifolds for Stochastic Evolutionary Equations. *J. Dynam. Diff. Eq.*, 16(4) 949–972, 2004.

[65] J. Duan, W. Wang. *Effective Dynamics of Stochastic Partial Differential Equations*. Elsevier, 2014.

[66] T.E. Duncan, B. Pasik-Duncan and B. Maslowski. Fractional Brownian motion and stochastic equations in Hilbert spaces. *Stoch. Dyn.*, 2(2):225–250, 2002.

[67] K. Eichinger, C. Kuehn and A. Neamţu. Sample Paths Estimates for Stochastic Fast-Slow Systems driven by Fractional Brownian Motion. *arXiv:1905.06824*, pages 1–44, 2019.

[68] W.G. Faris and G. Jona-Lasinio. Large fluctuations for a nonlinear heat equation with noise. *J. Phys. A: Math. Gen.*, 15:3025–3055, 1982.

[69] B. Fehrman and B. Gess. Well-posedness of stochastic porous media equations with nonlinear, conservative noise. To appear in *Arch. Ration. Mech. Anal.*

[70] F. Flandoli, B. Gess and M. Scheutzow. Synchronization by noise for order-preserving random dynamical systems. *Ann. Probab.*, 45(2):1325–1350, 2017.

[71] F. Flandoli and K.-U. Schaumlöffel. Stochastic parabolic equations in bounded domains: random evolution operator and Lyapunov exponents. *Stochastics: An International Journal of Probability and Stochastic Processes*, 29(4):461–485, 1990.

[72] N. Fenichel. Persistence and smoothness of invariant manifolds for flows. *J. Indiana Math.*, 21(3):193–226, 1972.

[73] M. Foondun and D. Khoshnevisan. Intermittence and nonlinear stochastic partial differential equations. *Electron. J. Probab.*, 14(21):548–568, 2009.

[74] M.I. Freidlin. Random perturbations of reaction-diffusion equations: the quasideterministic approximation. *Trans. Amer. Math. Soc.*, 305(2):665–697, 1988.

[75] M.I. Freidlin and A.D. Wentzell. *Random Perturbations of Dynamical Systems*. Springer, 1998.

[76] M.I. Freidlin and A.D. Wentzell. Averaging principle for stochastic perturbations of multifrequency systems. *Stoch. Dynam.*, 3:393–408, 2003.

[77] P.K. Friz and M. Hairer. *A Course on Rough Paths*. Springer, 2014.

[78] P.K. Friz and N.B. Victoir. *Multidimensional Stochastic Processes as Rough Paths: Theory and Applications*. Cambridge Studies in Advanced Mathematics, 2010.

[79] M. Furlan and M. Gubinelli. Paracontrolled quasilinear SPDEs. *Ann. Probab.*, 47(2):1096–1135, 2019.

[80] H. Gao, M.J. Garrido-Atienza and B. Schmalfuß. Random attractors for stochastic evolution equations driven by fractional Brownian motion. *SIAM J. Math. Anal.*, 46(4):2281–2309, 2014.

[81] J. Garcia-Ojalvo and J. Sancho. *Noise in Spatially Extended Systems*. Springer, 1999.

[82] M.J. Garrido-Atienza, K. Lu and B. Schmalfuß. Random dynamical systems for stochastic evolution equations driven by multiplicative fractional Brownian noise with Hurst parametes $H \in (1/3, 1/2]$. *SIAM J. Appl. Dyn. Syst.* 15(1), 625–654, 2016.

[83] M.J. Garrido-Atienza, A. Neuenkirch and B. Schmalfuß. Asymptotical stability of differential equations driven by Hölder continuous paths. *J. Dyn. Diff. Equat.*, 30:359–377 2018.

[84] M.J. Garrido-Atienza, B. Schmalfuß. Local Stability of Differential Equations Driven by Hölder-Continuous Paths with Hölder Index in $(1/3,1/2)$. *SIAM J. Appl. Dyn. Syst.*, 17(3):2352–2380, 2018.

[85] M.J. Garrido-Atienza, K. Lu and B. Schmalfuß. Unstable invariant manifolds for stochastic PDEs driven by a fractional Brownian motion. *J. Differential Equat.*, 248(7):1637–1667, 2010.

[86] A. Gerasimovičs and M. Hairer. Hörmander's theorem for semilinear SPDEs. *arXiv:1811.06339*, pages 1–60, 2018.

[87] M. Gerencsér and M. Hairer. A solution theory for quasilinear singular SPDEs. *Comm. Pure Appl. Math.*, 2019.

[88] B. Gess. Random attractors for singular stochastic evolution equations. *J. Differential Equat.*, 255(3):524–559, 2013.

[89] B. Gess. Random attractors for stochastic porous media equations perturbed by space-time linear multiplicative noise. *Ann. Probab.*, 42(2):818–864, 2014.

[90] B. Gess, W. Liu and M. Röckner. Random attractors for a class of stochastic partial differential equations driven by general additive noise. *J. Differential Equat.*, 251(4-5):1225–1253, 2011.

[91] M. Ghani Varzaneh, S. Riedel and M. Scheutzow. A dynamical theory for singular stochastic delay differential equations with a multiplicative ergodic theorem on fields of Banach spaces. *arXiv:1903.01172*, pages 1–47, 2019.

[92] M. Gnann, C. Kuehn and A. Pein. Towards sample path estimates for fast-slow SPDEs. *Eur. J. Appl. Math.*, 30(5):1004-1024, 2019.

[93] K. Gowda and C. Kuehn. Early-warning signs for pattern formation in stochastic partial differential equations. *Commun. Nonlinear. Sci.*, 22(1):55–69, 2015.

[94] T. Grafke, T. Schäfer and E. Vanden-Eijnden. Long term effects of small random perturbations on dynamical systems: Theoretical and computational tools. In *Recent Progress and Modern Challenges in Applied Mathematics, Modeling and Computational Science*, pages 17–55. Springer, 2017.

[95] M. Gubinelli. Controlling rough paths. *J. Funct. Anal.*, 216(1):86–140, 2004.

[96] M. Gubinelli, P. Imkeller and N. Perkowski. Paracontrolled distributions and singular PDEs. *Forum of Mathematics*, Pi, Vol. 3, no. 6, 2015.

[97] M. Gubinelli and N. Perkowski. KPZ reloaded. *Comm. Math. Phys.*, 349(1):165–269, 2017.

[98] M. Gubinelli, A. Lejay and S. Tindel. Young integrals and SPDEs. *Potential Anal.*, 25(4):307–326, 2006.

[99] M. Gubinelli and S. Tindel. Rough evolution equations. *Ann. Probab.*,

38(1):1–75, 2010.

[100] M. Hairer. A theory of regularity structures. *Invent. Math.*, 198(2):269–504, 2014.

[101] M. Hairer. Solving the KPZ equation. *Ann. Math.*, 178(2):559–664, 2013.

[102] M. Hairer. Regularity structures and the dynamical Φ_3^4 model. *arXiv:1508.05261*, pages 1–46, 2015.

[103] M. Hairer and H. Shen. The dynamical Sine-Gordon model. *Commun. Math. Phys.*, 341(3):933–989, 2016.

[104] D. Henry. *Geometric Theory of Semilinear Parabolic Equations*. Springer, Berlin Heidelberg, Germany, 1981.

[105] M. Hairer. Ergodicity of stochastic differential equations driven by fractional Brownian motion. *Ann. Probab.*, 33(2):703–758, 2005.

[106] C. Hamster and H.J. Hupkes. Stability of Traveling Waves for Reaction-Diffusion Equations with Multiplicative Noise. *SIAM. J. Appl. Dyn. Syst.*, 18(1):205–278, 2017.

[107] R. Hesse and A. Neamţu. Local mild solutions for rough stochastic partial differential equations. *J. Differential Equat.*, 267(11):6480–6538, 2019.

[108] R. Hesse and A. Neamţu. Global solutions and random dynamical systems for rough evolution equations. To appear in *Discrete. Contin. Dyn. Syst.*

[109] A. Hocquet and M. Hofmanová. An energy method for rough partial differential equations. *J. Differential Equat.*, 265(4):313–345, 2018.

[110] M. Hofmanová and T. Zhang. Quasilinear parabolic stochastic partial differential equations: existence, uniqueness. *To Appear in Stoch. Proc. Appl.*

[111] W. Hu, M. Salins and K. Spiliopoulos. Large deviations and averaging for systems of slow-fast stochastic reaction-diffusion equations. To appear in *Stochastics and Partial Differential Equations: Analysis and Computations*, 2019.

[112] R.Z. Khasminskii. *Stochastic Stability of Differential Equations*. Springer, 2011.

[113] P. Kloeden and T. Lorenz. Mean-square random dynamical systems. *J. Differential Equat.*, 253(5):1422–1438, 2012.

[114] P.E. Kloeden, C. Pötzsche and Martin Rasmussen. Limitations of pullback attractors for processes. *J. Difference Equ. Appl.*, 18(4):693–701, 2012.

[115] P.E. Kloeden and M. Rasmussen. *Nonautonomous Dynamical Systems*. Mathematical Surveys and Monographs, vol. 176, 264 pp, 2011.

[116] Y. Hu, J. Huang and D. Nualart. On the intermittency front of stochastic heat equation driven by colored noises. *Electron. Commun. Probab.*, 21(21):1–13, 2016.

[117] Y. Hu, J. Huang, K. Lê, D. Nualart and S. Tindel. Stochastic heat equation with rough dependence in space.*Ann. Probab.*, 45(6B):4561–4616, 2017.

[118] A. Hutt, A. Longtin, and L. Schimansky-Geier. Additive noise-induced Turing transitions in spatial systems with application to neural fields and the Swift-Hohenberg equation. *Physica D*, 237(6):755–773, 2008.

[119] N.V. Krylov and B.L. Rozovskii. Stochastic evolution equations. *J. Sov. Math.*, 16(4):1233–1277, 1981.

[120] C. Kuehn. Warning signs for wave speed transitions of noisy Fisher-KPP

invasion fronts. *Theor. Ecol.*, 6(3):295–308, 2013.

[121] C. Kuehn. Numerical continuation and SPDE stability for the 2d cubic-quintic Allen-Cahn equation. *SIAM/ASA J. Uncertain. Quantif.*, 3(1):762–789, 2015.

[122] C. Kuehn. Travelling Waves in Monostable and Bistable Stochastic Partial Differential Equations *Jahresber. DMV, accepted / to appear.*

[123] C. Kuehn. *PDE Dynamics: An Introduction.* SIAM, 2019.

[124] C. Kuehn and A. Neamţu. Pathwise mild solutions for quasilinear stochastic partial differential equations. *arXiv:1802.10016*, pages 1–42, 2018.

[125] C. Kuehn and A. Neamţu. Rough center manifolds. *arXiv:1802.10016*, pages 1–41, 2018.

[126] C. Kuehn, A. Neamţu and A. Pein. Random attractors for stochastic partly dissipative systems. *arXiv:1906.08594*, pages 1–25, 2019.

[127] C. Kuehn and F. Romano. Scaling laws and bifurcations for SPDEs. *Eur. J. Appl. Math.*, 30(5):853-868, 2019.

[128] H. Kunita. *Stochastic Flows and Stochastic Differential Equations.* Cambridge University Press, 1990.

[129] J.A. Langa and J.C. Robinson. Fractal dimension of an invariant set. *J. Math. Pure. Appl.*, 85(2):269–294, 2006.

[130] Z. Lian and K. Lu. Lyapunov Exponents and Invariant Manifolds for Random Dynamical Systems in a Banach Space. *Mem. Amer. Math. Soc.*, 206(967):vi+106, 2010.

[131] W. Liu and M. Röckner. SPDE in Hilbert space with locally monotone coefficients. *J. Funct. Anal.*, 259(11):2902–2922, 2010.

[132] K. Lu, A. Neamţu and B. Schmalfuß. On the Oseledets-splitting for infinite-dimensional random dynamical systems. *Discrete Contin. Dyn.Syst.*, 23(3):1219–1242, 2018.

[133] K. Lu and B. Schmalfuß. Invariant manifolds for stochastic wave equations. *J. Differential Equat.*, 236(2): 460–492, 2007.

[134] G. Lv and J. Duan. Impacts of noise on a class of partial differential equations. *J. Differential Equat.*, 258(6):2196–2220, 2015.

[135] T. Lyons. Differential equations driven by rough signals. *Rev. Mat. Iberoamericana*, 14(2):215–310, 1998.

[136] T. Lyons and Z. Qian. *System control and rough paths.* Oxford Mathematical Monographs, Oxford Science Publications. Oxford University Press, Oxford, 2002.

[137] B. Mandelbrot and J. van Ness. Fractional Brownian motions, fractional noises and applications. *SIAM Review.* 10(4):1–16, 1968.

[138] X. Mao. *Exponential stability of stochastic differential equations.* Monographs and Textbooks in Pure and Applied Mathematics, 182. Marcel Dekker, Inc., New York, 1994.

[139] B. Maslowski and J. van Neerven. Equivalence of laws and null controllability for SPDEs driven by a fractional Brownian motion. *Nonlinear Differential Equations Appl.*, 20(4):1473–1498, 2013.

[140] B. Maslowski and D. Nualart. Evolution equations driven by a fractional Brownian motion. *J. Funct. Anal.*, 202(1):277–305, 2003.

[141] B. Maslowski and B. Schmalfuß. Random dynamical systems and stationary solutions of differential equations driven by the fractional Brownian motion. *Stochastic Anal. Appl.*, 22(6):1577–1607, 2004.

[142] A. Millet and M. Sanz-Solé. Approximation of rough paths of fractional Brownian motion. Seminar on Stochastic Analysis, Random Fields and Applications V, 275– 303, Progr. Probab., 59, Birkhäuser, Basel, 2008.

[143] S. Mohammed, T. Zhang, H. Zhao. *The stable manifold theorem for semilinear stochastic evolution equations and stochastic partial differential equations.* Memoirs of the AIMS, vol. 196, nr. 197, 2008.

[144] C. Mueller, L. Mytnik and J. Quastel. Effect of noise on front propagation in reaction-diffusion equations of KPP type. *Invent. Math.*, 184(2):401–453, 2011.

[145] C. Mueller and R.B. Sowers. Blowup for the heat equation with a noise term. *Prob. Theor. Rel. Fields*, 97(3):287–320, 1993.

[146] D. Nualart Stochastic calculus with respect to the fractional Brownian motion and applications. *Contemp. Math.*, 336:3–39, 2003.

[147] M. Ondreját. Uniqueness of stochastic evolution equations in Banach spaces. *Dissertationes Math.* 426, 63 pp., Warszawa, 2004.

[148] F. Otto and H. Weber. Quasilinear SPDEs via Rough Paths. *Arch. Rat. Mech. Anal.*, 232(2):873–950, 2019.

[149] A. Pazy. *Semigroups of Linear Operators and Applications to Partial Differential Equations.* Springer Applied Mathematical Series. Springer–Verlag, Berlin, 1983.

[150] C. Pötzsche. *Geometric Theory of Discrete Nonautonomous Dynamical Systems.* Lecture Notes in Mathematics, vol. 2002, Springer-Verlag, Berlin, 2010.

[151] C. Prèvôt and M. Röckner. A concise course on Stochastic Partial Differential Equations. Lecture Notes in Math., vol. 1905, Springer 2007.

[152] M. Pronk and M.C. Veraar. A new approach to stochastic evolution equations with adapted drift. *J. Differential Equat.*, 256:3634–3684, 2015.

[153] M. Röckner and Y. Wang. A Note on variational solutions to SPDE perturbed by Gaussian noise in a general class. *Infin. Dimens. Anal. Quantum Probab. Relat. Top.*, 12(2):353–358, 2009.

[154] F. Russo and P. Vallois. Forward, backward and symmetric stochastic integration. *Probab. Theory Rel.*, 97(3):403–421, 1993.

[155] M. Scheutzow. On the perfection of crude cocycles. *Random Comput. Dynam.*, 4(4):235–255, 1996.

[156] M. Scheutzow. Comparison of various concepts of a random attractor: a case study. *Arch. Math. (Basel)*, 78(3):233–240, 2002.

[157] B. Schmalfuß. *Backward Cocycles and Attractors of Stochastic Differential Equations.* Nonlinear Dynamics: Attractor Approximation and Global Behaviour, Contributions to the International. Seminar ISAM'92, 1992.

[158] H.J. Sussmann. On the gap between deterministic and stochastic ordinary differential equations. *Ann. Probab.*, 6(1):19–41, 1978.

[159] G. Sell and Y. You. *Dynamics of Evolutionary Equations.* Springer-Verlag New York, 2002.

[160] J.M.A.M. van Neerven. *Stochastic Evolution Equations.* ISEM Lecture Notes 2007/08, 2008.

[161] R. Temam. *Infinite-Dimensional Dynamical Systems in Mechanics and Physics.* Springer-Verlag, New York, 1997.

[162] J.M.A.M. van Neerven, M.C. Veraar and L. Weis. Stochastic evolution equations in UMD Banach spaces *J. Funct. Anal.* 255(4):940–993, 2008.

[163] J.M.A.M. van Neerven, M.C. Veraar and L. Weis. Stochastic maximal L^p-regularity. *Ann. Probab.*, 40(2):788–812, 2012.

[164] M.C. Veraar. Stochastic integration in Banach spaces and applications to parabolic evolution equations. PhD Thesis, Delft University, 2006.

[165] J.B. Walsh. An introduction to stochastic partial differential equations. Ecole d'Eté de Probabilités de Saint-Flour XIV - 1984, Lect. Notes Math. **1180**, 265-437, 1986.

[166] B. Wang. Random attractors for non-autonomous stochastic wave equations with multiplicative noise. Discrete Contin. Dyn. Syst., 34(1):269–300,2014.

[167] W. Wang and J. Duan. A dynamical approximation of stochastic partial differential equations. *J. Math. Phys.*, 48(10):102701-102701-14, 2007.

[168] W. Wang and A.J. Roberts. Average and deviation for slow–fast stochastic partial differential equations. *J. Differential Equat.*, 253(1):1265–1286, 2012.

[169] E. Waymire and J. Duan. *Probability and Partial Differential Equations in Modern Applied Mathematics.* Springer–Verlag, 2005.

[170] A. Yagi. *Quasilinear abstract parabolic evolution equations with applications.* Progress in Nonlinear Differential Equations and Their Applications, Vol. 50, 381–397, Birkhauser Verlag Basel/Switzerland, 2002.

[171] L.C. Young. An integration of Hölder type, connected with Stieltjes integration. *Acta Math.*, 67(1):251–282, 1936.

Chapter 2

Stochastic Itô-Volterra Backward Equations in Banach Spaces

Mahdi Azimi* and Wilfried Grecksch†

*Martin-Luther-University Halle-Wittenberg,
Faculty of Natural Sciences II, Institute of Mathematics,
D-06099 Halle (Saale), Germany*

After a short introduction to stochastic analysis in Banach spaces we present stochastic Itô-Volterra backward equations with values in a Banach space. Optimality conditions of maximum principle type are derived for an optimal control problem involving Itô-Volterra forward integral equations in Banach spaces.

1. Introduction

Stochastic partial differential equations are part of the research of infinite dimensional stochastic analysis and these equations can be interpreted as stochastic evolution equations and the solutions are defined in a generalized sense. Here we choose the so-called mild solution: The problem contains a linear operator which generates a semigroup of operators and the problem is considered as a stochastic integral equation (of Itô-Volterra type) on a given complete probability space $(\Omega, \mathcal{F}, P)$, which contains a stochastic convolution (see, for example, in the case of a Hilbert space [7] and the literature cited therein).

For example, if we consider a stochastic partial differential equation with the Laplacian in L^p (see for example [22]), then it is necessary also to consider Banach space valued stochastic integrals. For example, Banach space valued stochastic partial differential equations are discussed in the sense of mild solutions (Itô-Volterra type equations) in [6, 21, 25]. Two typical methods are common to solve optimal control problems:

- using dynamical optimization (Bellman principle);

* E-mail: azimimehd@gmail.com
† E-mail (corresponding author): wilfried.grecksch@mathematik.uni-halle.de

• using necessary optimality conditions (maximum principle).

In [19] the Bellman principle is used to solve an optimal control problem for a stochastic Banach space valued differential equation in the case of a cylindrical Wiener process as noise process. A maximum principle is proved in [10] for an optimal control of stochastic partial differential equations in Banach spaces with finite dimensional Wiener process as noise.

Optimal control problems for stochastic Itô-Volterra equations should be solved by using a maximum principle, a theory of stochastic backward equations of Itô-Volterra type must be used.

A linear backward stochastic differential equation (BSDE) was first introduced by J. Bismut [5] in the finite dimensional case as the equation for the adjoint variable in the stochastic version of the Pontryagin maximum principle. A general nonlinear backward stochastic differential equation in the finite dimensional case (see for example E. Pardoux and S. G. Peng [18]) which appears in the optimal stochastic control problem is

$$Y(t) = \xi + \int_t^1 f(s, Y(s), Z(s))ds - \int_t^1 Z(s)dW(s), \text{ for } 0 \leq t \leq 1,$$

where $(W(t))_{t \in [0,1]}$ is a Brownian motion defined on a probability space $(\Omega, \mathcal{F}, P)$ with the natural filtration $(\mathcal{F}_t)_{t \in [0,1]}$ and ξ is a given $\mathcal{F}_1$-measurable random variable such that $\mathbb{E}|\xi|^2 < \infty$. In [18] E. Pardoux and S. G. Peng considered an adapted solution as a pair of real valued adapted processes (Y, Z) which satisfy almost surely the above equation. They proved the existence and uniqueness of the adapted solution by assuming the Lipschitz continuity for the generating function f. The interest on backward stochastic differential equations has grown rapidly because of the connections of this subject with computational finance, stochastic control problem, and partial differential equations. These equations also provide a probabilistic interpretation for solutions of both elliptic and parabolic non-linear partial differential equations. Indeed, coupled with forward stochastic differential equations, such BSDEs give an extension of the Feynman-Kac formula to the nonlinear case. Numerous authors recently investigated various BSDEs and properties of their solutions, see for instance V. Anh [1], E. Essaky, K. Bahlali and Ouknine Y. [9], Y. Hu [11], [12], J. Lin [13], R. Negrea and C. Preda [16] and the references specified therein. In particular, many efforts have been made to relax the assumptions on the coefficient functions. For instance, several papers treat BSDEs with continuous or local drift. One of the first investigations treating multidimensional BSDEs with both local conditions on the drift and only square integrable terminal data is the paper of K. Bahlali [4]. This author considered BSDEs with locally Lipschitz coefficients in both variables Y and Z. Backward stochastic nonlinear integral equations have been studied by J. Lin [13] under global Lipschitz conditions on the drift term. More precisely, in this paper J. Lin proved an existence and uniqueness result for the following nonlinear BSDE of Volterra type

$$Y(t) + \int_t^T f(t,s,Y(s),Z(t,s))ds + \int_t^T [g(t,s,Y(s)) + Z(t,s)]dW(s) = \xi.$$

In 2006 J. Yong [27] started a detailed analysis of backward stochastic integral equations. The author first proved an existence and uniqueness result for general integral equation of Volterra type given by

$$Y(t) = f(t) + \int_t^T h(t,s,y(s),Z(t,s),Z(s,t))ds - \int_t^T Z(t,s)dW(s), \quad (1)$$

where $h : [0,T] \times [0,T] \times R^m \times R^{m \times d} \times R^{m \times d} \times \Omega \to R^m$ and $f : [0,T] \times \Omega \to R^m$ are given. The difficulties that arise in (1) are mainly given by the fact that the generator h depends simultaneously on t and s which implies that the equation cannot be reduced to a BSDE in general and the process f is allowed to be only $\mathcal{F}_T$ measurable (not necessarily $\mathcal{F}$-adapted). Equations of the above form often occur in various models of modern mathematical finance. In the cited paper stability statements were also formulated and a duality principle for linear forward stochastic and linear backward stochastic differential equations was obtained. As an application of the duality principle, the author presented a comparison theorem for one dimensional backward stochastic integral equations and a Pontryagin type maximum principle for optimal control of stochastic integral equations. He continued his research and the paper published in 2008 [28] generalizes the previously obtained results. This paper treats various general BSDEs and new concepts of solutions. The regularity of these solutions is studied by means of Malliavin derivative. Conditions for the Malliavin differentiability are stated and one of the most important result states that the Malliavin derivative of the process Y can be obtained by solving a BSDE.

As an extension, stochastic backward integral equations with values in a separable Hilbert space were intensively investigated by V. Anh, W. Grecksch and J. Yong [1], [2].

If we summarize, then we can say that in the above investigations on stochastic backward equations the stochastic processes are finite dimensional or Hilbert space valued.

In this paper we are interested in the study of backward stochastic differential equations for Banach space valued processes.

During the last years the stochastic analysis in Banach spaces has been developed vastly. It has only recently been realized that many results can be generalized under certain circumstances beyond the Hilbert space case. First of all one has to replace the well-known orthogonality by unconditionality. This is the essence of the UMD Banach spaces, where UMD is the abbreviation for unconditional martingale difference. In 2007, J. van Neerven, M. Veraar and L. Weis [23] gave a complete integration theory for Banach space valued processes. The UMD property is essential to obtain a two-sided estimate of the stochastic integral. In the absence of the famous Itô-isometry such estimates combined with the operator-valued version of the Burkholder-Davis-Gundy inequalities are crucial. In the cited paper, martingale representation theorems are also proved. For this paper such results are useful to solve BSDEs. Mallivian calculus is also presented in [20] for UMD Banach spaces.

However there are major differences between these statements and the well-known results from the Hilbert space case. In this setting we cannot expect to get a Banach space valued process such that a given E-valued martingale can be represented by a stochastic integral. But we always obtain an operator valued random variable with this property. Recently, M. Ondrejat and M. Veraar [17] studied weak characterizations of stochastic integrability and its connection to martingale representation theorems.

Existence, uniqueness and smoothness properties are discussed in [15] for the solution of a Banach space valued stochastic backward Itô-Volterra equation in the case of UMD Banach space valued stochastic integral with respect to a one dimensional real Wiener process. The stochastic analysis in UMD Banach spaces and the ideas from [1] and [2] are used.

In this contribution we consider stochastic processes in the Banach space $E = L^q(\mathbf{S}, \Sigma, \mu)$, where μ is a σ-finite or finite measure.

In Section 3 we introduce a forward stochastic Volterra integral equation with respect to an H-cylindrical Brownian motion in the Banach space E, where μ is a σ-finite measure and $q \geq 2$. An existence and uniqueness theorem (see Theorem 18) is formulated and a sketch of the proof is given. The Banach fixed point theorem is used. In general, the stochastic convolution has not continuous paths. In Theorem 19 conditions are given such that a solution process has continuous paths in E with probability 1, where μ is a finite measure. The case of a σ-finite measure is considered in Theorem 20. Some similarities between the assumptions in Theorem 18 and the assumptions that, for example, are given by [29] are discussed in Remark 10. The stochastic heat equation is an example.

In Section 5 we consider a general backward stochastic Volterra type integral equation given by Equation (14) in the space $E = L^q\,(\mathbf{S}, \Sigma, \mu)$, where $q \leq 2$ and μ is a σ-finite measure. The stochastic integral in E is defined with respect to an H-cylindrical Brownian motion. The solution is defined as an M-solution in the sense of Definition 24. We will see that the simple backward stochastic integral Equation (8) plays an important role to prove an existence and uniqueness theorem. After the study of Equation (8) we discuss a more general backward stochastic integral Equation (11). In a next step we introduce a stochastic Fredholm integral equation, see (16). The martingale representation theorem in Banach spaces and the Banach fixed point theorem are our main tools.

Finally, by suitable assumptions, a unique adapted M-solution is found in Theorem 22 by tracing back to this three special backward equations and using the martingale representation theorem.

We conclude this section with two duality theorems between linear forward and backward stochastic Volterra integral equations (see Theorem 23 and Theorem 21).

In Section 7 we discuss maximum principles for optimal control problems with controlled BSVIEs and a goal functional of Bolza type (see Theorem 25). We use the results of Section 6.

Finally, we consider a maximum principle for an optimal control problem involving the mild solution of a controlled heat equation.

This paper is based on the dissertation [3].

2. L^p Stochastic Integration

In this section, we summarize the required statements about stochastic integration theory in Banach spaces. More details can be found in the papers of J. M. A. M. van Neerven, M. C. Veraar, L. Weis [6, 21–23].

Definition 19. A Banach space E is said to be a $\mathbf{UMD}_p$ **space** with $1 < p < \infty$, if there exists a positive constant C_p such that for all E-valued L^p-martingale difference sequences $(d_n)_{n=1}^N$ and any choice of signs $\varepsilon_n = \pm 1$ it yields

$$\left(\mathbb{E}\left\|\sum_{n=1}^N \varepsilon_n d_n\right\|^p\right)^{\frac{1}{p}} \leq C_p \left(\mathbb{E}\left\|\sum_{n=1}^N d_n\right\|^p\right)^{\frac{1}{p}}.$$

Definition 20. (see [23, Definition, p. 1450])
Let H be a separable Hilbert space. A family $W^H = \left(W^H(t)\right)_{t\in[0,T]}$ of bounded linear operators from H to $L^2(\Omega)$ is called **H-cylindrical Brownian motion**, if

(1) $W^H h = \left(W^H(t)h\right)_{t\in[0,T]}$ is a real valued Brownian motion for each $h \in H$;

(2) $\mathbb{E}\left(W^H(s)g \cdot W^H(t)h\right) = (s \wedge t)[g,h]_H$ for all $s,t \in H$.

Definition 21. (see [23, Definition, p. 1450])
Consider an elementary process $\phi : [0,T] \times \Omega \longrightarrow \mathcal{L}(H,E)$ having the form

$$\phi(t,w) = \sum_{n=0}^N \sum_{m=1}^M \mathbf{1}_{(t_{n-1},t_n]\times A_{mn}} \sum_{k=1}^K h_k \otimes x_{kmn},$$

where $x_{kmn} \in E$, $0 \leq t_0 < \cdots < t_N \leq T$, the sets $A_{1n},\ldots,A_{Mn} \in \mathcal{F}_{n-1} = \sigma\{W^H(s) : 0 \leq s \leq t_{n-1}\}$ are disjoint for each n and $h_1,\ldots h_K$ are orthonormal elements in H. Then the **L^p-stochastic integral** is defined by

$$\int_0^T \phi(t)dW^H(t) := \sum_{n=0}^N \sum_{m=1}^M \mathbf{1}_{A_{mn}} \sum_{k=1}^K \left(W^H(t_n)h_k - W^H(t_{n-1})h_k\right) x_{kmn}.$$

Theorem 11 will show that this definition can be extended to more general processes.

Definition 22. (see [14, 25])
Let $p \in [0,2]$. A Banach space E is of **type** p, if there exists a constant $C \geq 0$ such that for all $x_1,\ldots,x_N \in E$ and any Rademacher sequence $(r_n)_{n=1}^N$, it yields

$$\left(\mathbb{E}\left\|\sum_{i=1}^N r_n x_x\right\|_E^2\right)^{\frac{1}{2}} \leq C \left(\left\|\sum_{i=1}^N x_x\right\|_E^p\right)^{\frac{1}{p}},$$

where the least possible constant C is called the type p constant of E.
Let $q \in [2,\infty]$. A Banach space E is of **cotype** q, if there exists a constant $C \geq 0$ such that for all $x_1,\ldots,x_N \in E$ and any Rademacher sequence

$(r_n)_{n=1}^N$, it yields

$$\left(\left\| \sum_{i=1}^N x_x \right\|_E^q \right)^{\frac{1}{q}} \leq C \left(\mathbb{E} \left\| \sum_{i=1}^N r_n x_x \right\|_E^2 \right)^{\frac{1}{2}},$$

where the least possible constant C is called the cotype q constant of E.

Note, that every Banach space has type 1 and cotype ∞ with constant 1. Hilbert spaces have type 2 and cotype 2 with constants 1. The L^q-spaces, $q \in [1, \infty)$, have type $\min\{p, 2\}$ and cotype $\max\{p, 2\}$.

Theorem 11. (L^p-**stochastic integrability**, see [23, Theorem 3.6, p. 1454]) *Let E be a UMD space and $1 < p < \infty$. For an H-strongly measurable and adapted process $\phi : [0, T] \times \Omega \to \mathcal{L}(H; E)$ belonging to $L^p\left(\Omega; L^2(0, T; H)\right)$, the following assertions are equivalent:*

(1) the process ϕ is L^p-stochastically integrable with respect to W^H in the sense of [23, Theorem 3.6, Condition (4), p. 1455];

(2) there exists a sequence of elementary adapted processes $\phi_n : [0, T] \times \Omega \to \mathcal{L}(H; E)$ such that

   (i) for all $h \in H$ and $x^* \in E^*$, $\lim\limits_{n \to \infty} \langle \phi_n h, x^* \rangle = \langle \phi h, x^* \rangle$ in measure on $[0, T] \times \Omega$;

   (ii) there exists a random variable $\eta \in L^p(\Omega; E)$ such that

$$\eta = \lim_{n \to \infty} \int_0^T \phi_n(t) dW^H(t) \text{ in } L^p(\Omega; E);$$

(3) there exists a strongly measurable random variable $\eta \in L^p(\Omega; E)$ such that for all $x^* \in E^*$ we have

$$\langle \eta, x^* \rangle = \int_0^T \langle \phi h, x^* \rangle \, dW^H(t) \text{ in } L^p(\Omega);$$

(4) ϕ represents an element $X \in L^p\left(\Omega; \gamma\left(L^2(0, T; H), E\right)\right)$ ($\gamma(H; E)$ stands for the γ-radonifying space of all γ-radonifying operators from H into E), see [23, p. 1443]

and we have $\eta = I^{W^h}(X) := \int_0^T \phi(t) dW^H(t)$ in $L^p(\Omega; E)$.

Definition 23. We write $A \simeq B$ if and only if there exist positive constants c, C which depend on E and p, such that $cB \leq A \leq CB$.

Theorem 12. (L^p-**stochastic integrability**, see [23, Corollary 3.11, p. 1461]) *Let E be a UMD Banach space over a σ-finite measure space $(\mathbf{S}, \Sigma, \mu)$*

and let $p \in (1, \infty)$. *Consider* $\phi : [0, T] \times \Omega \longrightarrow \mathcal{L}(H, E)$ *to be H-strongly measurable, adapted and assume that there exits a strongly measurable function* $\varphi : [0, T] \times \Omega \times \mathbf{S} \longrightarrow H$ *such that for all* $h \in H$ *and* $t \in [0, T]$ $(\phi(t)h)(\cdot) = [\varphi(t, \cdot), h]_H$ *in* E. *Then,* ϕ *is* L^p-*stochastically integrable with respect to* W^H *if and only if*

$$\mathbb{E} \left\| \left(\int_0^T \|\varphi(t, \cdot)\|_H^2 dt \right)^{\frac{1}{2}} \right\|_E^p < \infty.$$

Further we have

$$\mathbb{E} \left\| \int_0^T \phi(t) dW^H(t) \right\|_E^p \simeq \mathbb{E} \left\| \left(\int_0^T \|\varphi(t, \cdot)\|_H^2 dt \right)^{\frac{1}{2}} \right\|_E^p.$$

Theorem 13. (see [22, Lemma 2.1, p. 945]) *Let* $(\mathbf{S}, \Sigma, \mu)$ *be a σ-finite measure space and let* $1 \leq q < \infty$. *For an operator* $T \in \mathcal{L}(H, L^q(\mathbf{S}))$ *the following assertions are equivalent:*

(1) $T \in \gamma(H, L^q(\mathbf{S}))$;

(2) for some orthonormal basis $(h_n)_{n=1}^\infty$ *of* H *the function* $\left(\sum_{n \geq 1} |T h_n|^2 \right)^{\frac{1}{2}}$ *belongs to* $L^q(\mathbf{S})$;

(3) for all orthonormal basis $(h_n)_{n=1}^\infty$ *of* H *the function* $\left(\sum_{n \geq 1} |T h_n|^2 \right)^{\frac{1}{2}}$ *belongs to* $L^q(\mathbf{S})$;

(4) there exists a function $g \in L^p(\mathbf{S})$ *such that for all* $h \in H$ *we have* $|Th| \leq \|h\|_H g$ *μ-almost everywhere;*

(5) there exists a function $k \in L^q(S; H)$ *such that* $Th = [k(\cdot), h]_H$ *μ-almost everywhere*

moreover, in this situation we may take $k = \left(\sum_{n \geq 1} |T h_n|^2 \right)^{\frac{1}{2}}$ *and have*

$$\|T\|_{\gamma(H, L^q(\mathbf{S}))} \simeq \left\| \left(\sum_{n \geq 1} |T h_n|^2 \right)^{\frac{1}{2}} \right\| \leq \|g\|_{L^q(\mathbf{S})}.$$

By considering that $X \in L^p \left(\Omega; \gamma \left(L^2(0, T; H), E \right) \right)$ is the element represented by ϕ and according to the integral process

$$t \in [0, T] \longrightarrow \int_0^t \phi(s) dW^H(s),$$

the process $\xi : [0,T] \times \Omega \longrightarrow \gamma\left(L^2(0,T;H),E\right)$ can be introduced and it is associated with X; it is defined by $\xi(t,\omega)f := (X(w))(\mathbf{1}_{[0,T]}f)$, $f \in L^2((0,T);H)$. Note, that $\xi_X(T) = X$.

Theorem 14. (see [23, Proposition 4.3, p. 1462]) *Let E be a UMD space and fix $p \in (1,\infty)$. For all $X \in L^p\left(\Omega; \gamma\left(L^2(0,T;H),E\right)\right)$ the integral process $I^{W^H}(\xi_X)$ is an E-valued L^p-martingale whose p-th moment is continuous. It has a continuous adapted version which satisfies the maximal inequality*

$$\mathbb{E} \sup_{t \in [0,T]} \left\| I^{W^H}(\xi_X(t)) \right\|^p \leq q^p \left\| I^{W^H}(X) \right\|^p,$$

where $\frac{1}{p} + \frac{1}{q} = 1$.

Theorem 15. (**Burkholder-Davis-Gundy inequalities**, see [23, Theorem 4.4, p. 1463]) *Let E be a UMD space and fix $p \in (1,\infty)$. If the H-strongly measurable and adapted process $\phi : [0,T] \times \Omega \longrightarrow \mathcal{L}(H,E)$ is L^p-stochastically integrable, then*

$$\mathbb{E} \sup_{t \in [0,T]} \left\| \int_0^t \phi(s)dW^H(s) \right\|^p \simeq \mathbb{E}\|X\|_{\gamma(L^2(0,T;H),E)}^p,$$

where $X \in L^p\left(\Omega; \gamma\left(L^2(0,T;H),E\right)\right)$ is the element represented by ϕ.

Theorem 16. (**Martingale representation theorem in UMD spaces**, see [23, Theorem 5.13, p. 1476]) *Let E be a UMD space. Then every L^p-martingale $M : [0,T] \times \Omega \longrightarrow E$ adapted to the augmented filtration $\mathbb{F}^{W^H}$ has a continuous version, and there exists a unique $X \in L^p\left(\Omega; \gamma(L^2(0,T;H),E)\right)$ such that for all $t \in [0,T]$ it yields*

$$M(t) = M(0) + I^{W^H}(\xi_X(t)), \quad in \ L^p(\Omega;E).$$

For a UMD space E with cotype 2 the above representation takes the form

$$M(t) = M_0 + \int_0^t \phi(t)dW^H(t).$$

Theorem 17. (**Itô formula**, see [23, Corollary 2.6, p. 37]) *Let E_1, E_2 and F be UMD spaces and let $f : E_1 \times E_2 \longrightarrow F$ be a bilinear map. Let $(h_n)_{n \geq 1}$ be an orthonormal basis of H. For $i = 1,2$ let $\phi_i : [0,T] \times \Omega \longrightarrow \mathcal{L}(H,E_i)$ be H-strongly measurable and adapted processes which is stochastically integrable with respect to W^H. Assume that the paths of ϕ_i belong to $L^2(0,T;\gamma(H,E))$ almost surely, $\psi_i : [0,T] \times \Omega \longrightarrow E_i$ is strongly*

measurable and adapted with paths in $L^1(0, T : E)$ almost surely and $\xi_i :$
$\Omega \longrightarrow E_i$ is strongly $\mathcal{F}_0$-measurable and define

$$\zeta_i = \xi_i + \int_0^{\cdot} \psi_i(s)ds + \int_0^{\cdot} \phi_i(s)dW^H(s).$$

Then almost surely for all $t \in [0, T]$ it holds

$$f(\zeta_1(t), \zeta_2(t)) - f(\zeta_1(0), \zeta_2(0)) = \int_0^t \left(f(\zeta_1(t), \psi_2(t)) + f(\psi_1(t), \zeta_2(t))\right) ds$$

$$+ \int_0^t \left(f(\zeta_1(t), \phi_2(t)) + f(\phi_1(t), \zeta_2(t))\right) dW^H(s)$$

$$+ \int_0^t \sum_{n \geq 1} f(\phi_1(s)h_n, \phi_2(s)h_n)ds.$$

Particularly, for a UMD space E, taking $E_1 = E$, $E_2 = E^*$, $F = \mathbb{R}$ and
$f(x, x^*) = \langle x, x^* \rangle$, we have almost surely for all $t \in [0, T]$

$$\langle \zeta_1(t), \zeta_2(t) \rangle - \langle \zeta_1(0), \zeta_2(0) \rangle = \int_0^t \left(\langle \zeta_1(s), \psi_2(s) \rangle - \langle \psi_1(s), \zeta_2(s) \rangle\right) ds$$

$$+ \int_0^t \left(\langle \zeta_1(s), \phi_2(s) \rangle - \langle \phi_1(s), \zeta_2(s) \rangle\right) dW^H(s)$$

$$+ \int_0^t \sum_{n \geq 1} \langle \phi_1(s)h_n, \phi_2(s)h_n \rangle ds.$$

3. Forward Stochastic Volterra Integral Equations

In this section we are going to find a unique adapted solution for a forward
stochastic Volterra integral equation in the Banach space L^q.

Consider the following forward stochastic Volterra integral equation
(FSVIE) in the Banach space $E = L^q(\mathbf{S}, \Sigma, \mu)$, where $q \geq 2$ and μ is a
σ-finite measure

$$X(t) = \varphi(t) + \int_0^t b(t, s, X(s)) ds + \int_0^t \rho(t, s, X(s)) dW^H(s), \quad t \in [0, T].$$
$$\tag{2}$$

The stochastic integral is defined with respect to an H-cylindrical Brownian
motion W^H, see Definition 20. Our goal is to find a unique adapted E-
valued process X. Our approach will be to use the Banach fixed point
theorem in complete spaces, the definition of L^p-stochastic integrability in
the Banach space E, Definition 21 and Theorem 11; especially we use the
L^p-stochastic integrability Theorem 12 in the case $E = L^q(\mathbf{S}, \Sigma, \mu)$. We
use the concept of Nemytskii operators for b and ρ; by using L^p-stochastic

integrability it is assumed that there exists $\varrho(\cdot,\cdot,\cdot,\cdot) : \Omega \times [0,T] \times [0,T] \times$ $\mathbf{S} \longrightarrow H$ such that $(\rho(t,s,x)h)(\cdot) = [\varrho(t,s,x,\cdot),h]$, for every $h \in H$ and $x \in \mathbb{R}$. In most of this work we drop $\eta \in \mathbf{S}$ and $\omega \in \Omega$ for easiness. We are going to set some assumptions on the E-valued processes $\varphi(t)$, $b(t,s,\cdot)$ and the E-valued operator process $\rho(t,s,\cdot)$. Let $\Delta = \{(t,s) : 0 \leq s \leq t \leq T\}$ and consider the following assumptions:

(H 0) For each $t \in [0,T]$, $\varphi(t) \in L^p(\Omega;E)$ is $\mathcal{F}_t$-adapted and moreover

$$\mathbb{E}\left\| \left(\int_0^T |\varphi(t)|^2 dt \right)^{\frac{1}{2}} \right\|_E^p < \infty.$$

(H 1) $b : \Omega \times \Delta \times E \times \mathbf{S} \longrightarrow E$

$$\rho : \Omega \times \Delta \times E \times \mathbf{S} \longrightarrow \mathcal{L}(H;E)$$

are progressively measurable for each $x \in E$ and $t \in [0,T]$. We use the concept of H-strongly measurability for $\rho(\cdot,\cdot,\cdot)$, i.e. for every $h \in H$, $\rho(\cdot,\cdot,\cdot)h$ is strongly measurable.

(H 2) There exist some positive constants K_1 and K_2 such that for every $t,s \in [0,T]$ and $x,y \in \mathbb{R}$

$$|b(t,s,x) - b(t,s,y)| \leq K_1 |x - y|,$$

$$\|\varrho(t,s,x) - \varrho(t,s,y)\|_H \leq K_2 |x - y|.$$

(H 3) $\mathbb{E}\left\| \sup_{t \in [0,T]} \left(\int_0^t |b(t,s,0)|^2 ds \right)^{\frac{1}{2}} \right\|_E^p < \infty$

$$\mathbb{E}\left\| \sup_{t \in [0,T]} \left(\int_0^t \|\varrho(t,s,0)\|_H^2 ds \right)^{\frac{1}{2}} \right\|_E^p < \infty.$$

In this paper we consider the case $2 \leq q \leq p$. Define the following Banach space

$$\mathcal{M}_T = \left\{ X(\cdot) \in L_{\mathbb{F}}^p(\Omega;E) : \|X(\cdot)\|_{\mathcal{M}_T}^p = \mathbb{E}\left\| \left(\int_0^T |X(t)|^2 ds \right)^{\frac{1}{2}} \right\|_E^p < \infty \right\}.$$

First we are going to prove some lemmas and finally the existence and uniqueness theorem will be given.

Lemma 1. *If the assumptions* (H 0), (H 1), (H 2) *and* (H 3) *hold and* $x(\cdot) \in \mathcal{M}_T$, *then* $X(\cdot) \in \mathcal{M}_T$, *where* $X(\cdot)$ *is defined by the operator* $\mathcal{A}$:

$\mathcal{M}_T \longrightarrow \mathcal{M}_T$ as follows

$$X(t) = \mathcal{A}(x(\cdot))(t) = \varphi(t) + \int_0^t b(t,s,x(s))ds + \int_0^t \rho(t,s,x(s))dW^H(s)$$

for $t \in [0,T]$.

Proof. The adaptedness follows clearly by definition. Let $x(t) \in L^p_{\mathbb{F}}(\Omega; E)$. Then we have

$$\mathbb{E}\|X(t)\|^p = \mathbb{E}\left\|\varphi(t) + \int_0^t b(t,s,x(s))ds + \int_0^t \rho(t,s,x(s))dW^H(s)\right\|^p$$

$$\leq c\left\{\mathbb{E}\|\varphi(t)\|^p + \mathbb{E}\left\|\int_0^t b(t,s,x(s))ds\right\|^p \right.$$

$$\left. + \mathbb{E}\left\|\int_0^t \rho(t,s,x(s))dW^H(s)\right\|^p\right\},$$

where we used the norm property and Young's inequality and c is a positive constant.[a] Now we consider the summands from the above equation separately. The first term is bounded by assumption (H 0). We can write for the second term, by using the norm property, Hölder's inequality and the assumptions given in (H 2)

$$\mathbb{E}\left\|\int_0^t b(t,s,x(s))ds\right\|^p$$

$$\leq c\left\{\mathbb{E}\left\|\sup_{t\in[0,T]}\left(\int_0^t |b(t,s,0)|^2 ds\right)^{\frac{1}{2}}\right\|^p + \mathbb{E}\left\|\left(\int_0^T |x(s)|^2 ds\right)^{\frac{1}{2}}\right\|^p\right\} \quad (3)$$

and it results that $\mathbb{E}\left\|\int_0^t b(t,s,x(s))ds\right\|^p$ is bounded for every $t \in [0,T]$. For the third part we use the L^p-stochastic integrability from Theorem 12. We have

$$\mathbb{E}\left\|\int_0^t \rho(t,s,x(s))dW^H(s)\right\|^p \leq c\mathbb{E}\left\|\left(\int_0^t \|\varrho(t,s,x(s))\|^2_H ds\right)^{\frac{1}{2}}\right\|^p$$

$$\leq c\left\{\mathbb{E}\left\|\sup_{t\in[0,T]}\left(\int_0^t \|\varrho(t,s,0))\|^2_H ds\right)^{\frac{1}{2}}\right\|^p + \mathbb{E}\left\|\left(\int_0^T |x(s)|^2 ds\right)^{\frac{1}{2}}\right\|^p\right\},$$

which yields that $\mathbb{E}\left\|\int_0^t \rho(t,s,x(s))dW^H(s)\right\|^p$ is bounded for every $t \in [0,T]$. By using the above calculations it follows that for all

---

[a]In the whole paper we assume that c is a positive universal constant and it could change its values.

$t \in [0,T]$, $X(t) \in L^p(\Omega; E)$. Finally, it remains to show that
$$\mathbb{E}\left\| \left(\int_0^T |X(t))|^2 dt \right)^{\frac{1}{2}} \right\|_E^p < \infty.$$ For simplicity we set for all $\eta \in \mathbf{S}$

$$\|X(\cdot)\|_{L^2(0,T)} = \|X(\cdot,\eta)\|_{L^2(0,T)} = \left(\int_0^T |X(t,\eta))|^2 dt \right)^{\frac{1}{2}}.$$

Then we can write by using the norm property
$$\mathbb{E}\left\| \|X(\cdot)\|_{L^2(0,T)} \right\|_E^p$$

$$\leq c \left\{ \mathbb{E}\left\| \|\varphi(\cdot)\|_{L^2(0,T)} \right\|_E^p + \mathbb{E}\left\| \left\| \int_0^\cdot b(\cdot,s,x(s))ds \right\|_{L^2(0,T)} \right\|_E^p \right.$$

$$\left. + \mathbb{E}\left\| \left\| \int_0^\cdot \rho(\cdot,s,x(s))dW^H(s) \right\|_{L^2(0,T)} \right\|_E^p \right\}.$$

Now we consider the above summands separately. The first summand is by assumption bounded, since

$$\mathbb{E}\left\| \|\varphi(\cdot)\|_{L^2(0,T)} \right\|_E^p = \mathbb{E}\left\| \left(\int_0^T |\varphi(t)|^2 dt \right)^{\frac{1}{2}} \right\|_E^p < \infty.$$

For the second summand it yields

$$\mathbb{E}\left\| \left\| \int_0^\cdot b(\cdot,s,x(s))ds \right\|_{L^2(0,T)} \right\|_E^p$$

$$\leq c \left\{ \mathbb{E}\left\| \sup_{t\in[0,T]} \left(\left| \int_0^t b(t,s,0)ds \right|^2 \right)^{\frac{1}{2}} \right\|_E^p + \mathbb{E}\left\| \left(\int_0^T |x(s)|^2 ds \right)^{\frac{1}{2}} \right\|_E^p \right\} < \infty.$$

For the third summand, which contains a stochastic integral, we can write

$$\mathbb{E}\left\| \left\| \int_0^\cdot \rho(\cdot,s,x(s))dW^H(s) \right\|_{L^2(0,T)} \right\|_E^p$$

$$= \mathbb{E}\left(\int_\mathbf{S} \left(\int_0^T \left| \int_0^t \rho(t,s,x(s))dW^H(s) \right|^2 T\frac{dt}{T} \right)^{\frac{q}{2}} d\mu \right)^{\frac{p}{q}},$$

since $2 \leq q$ we use Jensen's inequality for the Lebesgue measure dt on the finite interval $[0,t]$, therefore it holds

$$\leq T^{(\frac{q}{2}-1)\frac{p}{q}} \mathbb{E}\left(\int_\mathbf{S} \left(\int_0^T \left| \int_0^t \rho(t,s,x(s))dW^H(s) \right|^q dt \right) d\mu \right)^{\frac{p}{q}}$$

by applying Fubini's theorem to the last term, it yields

$$= T^{(\frac{q}{2}-1)\frac{p}{q}} \mathbb{E} \left(\int_0^T \int_S \left| \int_0^t \rho(t,s,x(s))dW^H(s) \right|^q d\mu dt \right)^{\frac{p}{q}}$$

$$\leq c \int_0^T \mathbb{E} \left\| \int_0^t \rho(t,s,x(s))dW^H(s) \right\|_E^p dt.$$

Since $2 \leq q \leq p$ we used again Jensen's inequality for dt and Fubini's theorem for the above calculations and by applying the L^p-stochastic integrability Theorem 12 and assumption (H 2), the last term can be estimated by

$$\leq c \int_0^T c\mathbb{E} \left\| \left(\int_0^t \|\rho(t,s,x(s))\|_H^2 ds \right)^{\frac{1}{2}} \right\|_E^p dt$$

$$\leq \int_0^T c \left\{ \mathbb{E} \left\| \sup_{t \in [0,T]} \left(\int_0^t |\varrho(t,s,0)|^2 ds \right)^{\frac{1}{2}} \right\|^p + \mathbb{E} \left\| \left(\int_0^T |x(s)|^2 ds \right)^{\frac{1}{2}} \right\|^p \right\} dt$$

$$< \infty.$$

By combining the above calculations it results

$$\mathbb{E} \left\| \left(\int_0^T |X(t)|^2 dt \right)^{\frac{1}{2}} \right\|_E^p < \infty.$$

$\square$

Lemma 2. *For a sufficiently small value $\tau > 0$ the map $\mathcal{A}: \mathcal{M}_\tau \to \mathcal{M}_\tau$ is contractive, where $\mathcal{A}(x(\cdot))(t) = X(t)$ is given in Lemma 1 and $t \in [0,\tau]$.*

Proof. Let $x(\cdot), y(\cdot) \in \mathcal{M}_\tau$ and $X(t), Y(t)$ are the processes defined by $\mathcal{A}(\cdot)$. By using the norm property it results

$$\|X(\cdot) - Y(\cdot)\|_{\mathcal{M}_\tau}^p \leq c \left\{ \mathbb{E} \left\| \left\| \int_0^{\cdot} (b(\cdot,s,x(s)) - b(\cdot,s,y(s))) ds \right\|_{L^2(0,\tau)} \right\|_E^p \right.$$

$$\left. + \mathbb{E} \left\| \left\| \int_0^{\cdot} (\rho(\cdot,s,x(s)) - \rho(\cdot,s,y(s))) dW^H(s) \right\|_{L^2(0,\tau)} \right\|_E^p \right\}.$$

By the properties of b and ρ we get

$$\|X(\cdot) - Y(\cdot)\|_{\mathcal{M}_\tau}^p \leq \left(K_1^p \tau^p + cK_2^p \tau^{\frac{p}{2}} \right) \|x(\cdot) - y(\cdot)\|_{\mathcal{M}_\tau}^p.$$

By taking τ sufficiently small it follows that $\mathcal{A}: \mathcal{M}_\tau \to \mathcal{M}_\tau$ is contractive.

$\square$

Theorem 18. *By the assumptions* (H 0), (H 1), (H 2) *and* (H 3) *the FSVIE* (2) *has a unique adapted solution* $X(\cdot) \in \mathcal{M}_T$.

Proof. By using Lemma 1 and Lemma 2, it is clear that the map $\mathcal{A}$ defined in Lemma 1 is contractive on the interval $[0, \tau]$. Then, by using the Banach fixed point theorem in the complete space $\mathcal{M}_\tau$, there exists a unique solution defined on the interval $[0, \tau]$ and, by induction, it can be extended to the whole interval $[0, T]$. Hence there exists a unique solution $X(t)$ for $t \in [0, T]$. □

Next we want to study the path continuity of the solution derived in Theorem 18. For this reason we consider the following assumptions:

(G 1) there exist positive constants K_3, K_4, K_5 such that for every $t_1, t_2 \in [0, T]$

$$|\varphi(t_1) - \varphi(t_2)| \le K_3|t_1 - t_2|, \quad |b(t_1, s, x) - b(t_2, s, x)| \le K_4|t_1 - t_2|,$$

$$\|\varrho(t_1, s, x) - \varrho(t_2, s, x)\|_H \le K_5|t_1 - t_2|;$$

(G 2) $\qquad \mathbb{E}\left\|\left(\int_0^T |\varphi(t)|^4 ds\right)^{\frac{1}{4}}\right\|_E^p < \infty, \quad \mathbb{E}\left\|\left(\int_0^T |X(t)|^4 ds\right)^{\frac{1}{4}}\right\|_E^p < \infty,$

$$\mathbb{E}\left\|\sup_{t\in[0,T]}\left(\int_0^t |b(t, s, 0)|^4 ds\right)^{\frac{1}{4}}\right\|_E^p < \infty,$$

$$\mathbb{E}\left\|\sup_{t\in[0,T]}\left(\int_0^t \|\varrho(t, s, 0)\|_H^4 ds\right)^{\frac{1}{4}}\right\|_E^p < \infty.$$

In the following we write $\|\cdot\|$ instead of $\|\cdot\|_E$. We can easily prove:

Lemma 3. *If* $q \ge 4$ *and the assumptions* (H 0), (H 1), (H 2) *and* (H 3) *hold, then the assumptions* (G 2) *hold.*

By using Kolmogrov's continuity theorem we can also prove:

Theorem 19. *Let* μ *be a finite measure over* $(\mathbf{S}, \Sigma)$ *and consider that the assumptions* (H 0), (H 1), (H 2), (H 3), (G 1) *and* (G 2) *hold. Then the solution of the FSVIE* (2) *has continuous path in* $E = L^q(S, \mathcal{S}, \mu)$ *for some value p.*

Remark 9. It is essential for the path continuity in Theorem 19 that μ is a finite measure. But in the case of a σ-finite measure we could also have path continuity by changing our assumptions in (G 1).

(G′ 1) there exist nonnegative functions $k_3(\cdot), k_4(\cdot), k_5(\cdot) \in E$ such that for every $t_1, t_2 \in [0, T]$ and $\eta \in \mathbf{S}$,

$$|\varphi(t_1, \eta) - \varphi(t_2, \eta)| \le k_3(\eta)|t_1 - t_2|,$$

$$|b(t_1, s, x, \eta) - b(t_2, s, x, \eta)| \le k_4(\eta)|t_1 - t_2|,$$

$$\|\varrho(t_1, s, x, \eta) - \varrho(t_2, s, x, \eta)\|_H \le k_5(\eta)|t_1 - t_2|.$$

Similarly to Theorem 19 we can show:

Theorem 20. *If the assumptions* (H 0), (H 1), (H 2), (H 3), (G′ 1), (G 2) *hold and if μ is a σ-finite measure, then the solution of the FSVIE* (2) *has continuous path in E for some $p > 4$.*

Remark 10. Theorem 18 is also true in the case of $\mu(\mathbf{S}) < \infty$, if we replace the conditions (H 0), (H 1) and (H 2) as follows:

We denote $\mathcal{H}_0$ the set of all measurable functions $\kappa : \Delta \longrightarrow \mathbb{R}^+$ with

$$t \longrightarrow \int_0^t \kappa(t, s)ds \in L^\infty(0, T) \text{ and } \lim_{\varepsilon \downarrow 0} \sup \left\| \int_{\cdot}^{\cdot + \varepsilon} \kappa(\cdot + t, s)ds \right\|_{L^\infty(0,T)} = 0. \text{ We}$$

assume:

(H 0 ′) $\kappa_1, \kappa_2 \in \mathcal{H}_0$, $\operatorname{ess\,sup}_{t \in [0,T]} \displaystyle\int_0^t \kappa_i^2(t, s)ds < \infty$ and

$$\operatorname{ess\,sup}_{t \in [0,T]} \int_0^t (\kappa_1^2(t, s) + \kappa_2^2(t, s))\mathbb{E}\|\varphi(s)\|_E^2 < \infty;$$

(H 1 ′) for all $t, s \in [0, T]$ and $x, y \in \mathbb{R}$

$$|b(t, s, x) - b(t, s, y)| \le \kappa_1(t, s)|x - y|,$$

$$\|\varrho(t, s, x) - \varrho(t, s, y)\|_H \le \kappa_2(t, s)|x - y|;$$

(H 2 ′) for all $t, s \in [0, T]$ and $x \in \mathbb{R}$

$$|b(t, s, x)| \le \kappa_1(t, s)(|x| + 1),$$

$$\|\varrho(t, s, x)\|_H \le \kappa_2(t, s)(|x| + 1).$$

These conditions are similar to those given by Zhang in [29].

If $(\mathbf{S}, \Sigma, \mu)$ is a σ-finite measure space, then we need the following conditions such that Theorem 18 holds also in this case:

If $x, y \in E$, then we can replace (H 1 $'$) and (H 2 $'$) by:

(H 1 $''$) for all $t, s \in [0, T]$

$$\|b(t, s, x) - b(t, s, y)\|_E \leq \kappa_1(t, s)\|x - y\|_E,$$

$$\|\|\varrho(t, s, x) - \varrho(t, s, y)\|_H\|_E \leq \kappa_2(t, s)\|x - y\|_E$$

(H 2 $''$) for all $t, s \in [0, T]$

$$\|b(t, s, x)\|_E \leq \kappa_1(t, s)(\|x\|_E + 1), \quad \|\|\varrho(t, s, x)\|_H\|_E \leq \kappa_2(t, s)(\|x\|_E + 1).$$

Exemplary we consider the second term (3) in the proof of Lemma 1

$$\mathbb{E}\left\|\int_0^t b(t, s, x(s))ds\right\|^p \leq \mathbb{E}\left\|\int_0^t \kappa_1(t, s)(|x(s)| + 1)^2 ds\right\|^p$$

$$\leq \mathbb{E}\left\|\left(\int_0^t \kappa_1^2(t, s)ds\right)^{\frac{1}{2}}\left(\int_0^t |x(s) + 1|^2 ds\right)^{\frac{1}{2}}\right\|^p$$

$$\leq \left(\underset{t \in [0,T]}{\text{ess sup}}\int_0^t \kappa_1^2(t, s)ds\right)^{\frac{p}{2}}\left\{c\sqrt{T}\mu(\mathbf{S}) + \mathbb{E}\left\|\left(\int_0^T |x(s)|^2 ds\right)^{\frac{1}{2}}\right\|^p\right\}$$

and in the proof of Lemma 2 we have

$$\mathbb{E}\left\|\left(\int_0^\tau \left|\int_0^t (b(t, s, x(s)) - b(t, s, y(s)))ds\right|^2 dt\right)^{\frac{1}{2}}\right\|_E^p$$

$$\leq \mathbb{E}\left\|\left(\int_0^\tau \left(\int_0^t \kappa_1^2(t, s)ds \int_0^t |x(s) - y(s)|^2 ds\right) dt\right)^{\frac{1}{2}}\right\|_E^p$$

$$\leq \tau^{\frac{p}{2}}\left(\underset{t \in [0,T]}{\text{ess sup}}\int_0^t \kappa_1^2(t, s)ds\right)^{\frac{p}{2}}\mathbb{E}\left\|\left(\int_0^\tau |x(s) - y(s)|^2 ds\right)^{\frac{1}{2}}\right\|_E^p.$$

As an example, we consider some form of the stochastic heat equation with homogeneous Dirichlet boundary conditions in the Banach space $E = L^q(0, 1)$, where $q \geq 2$.

3.1. Stochastic Heat Equation with Lipschitz Nonlinearities

We introduce

$$dX(t,\xi) = \Delta X(t,\xi) + F(X(t,\xi))dt + G(X(t,\xi))dW^H(t)$$

with $t \in [0,T], \xi \in (0,1)$, $X(0,\xi) = X_0(\xi)$, $\xi \in (0,1)$, $X_0(\cdot) \in E$ and $X(t,0) = X(t,1) = 0$ for $t \in [0,T]$. We define the solution in the sense of a mild solution, that is

$$X(t) = S(t)X_0 + \int_0^t S(t-s)F(X(s))ds + \int_0^t S(t-s)G(X(s))dW^H(s), \quad (4)$$

$t \in [0,T]$, where $S(t)$ ($t \in [0,T]$) is the C_0-semigroup generated by the one dimensional Laplace operator Δ with homogenous Dirichlet boundary conditions. We assume that there exist a constant $L > 0$ and $F : \mathbb{R} \longrightarrow \mathbb{R}$ with $F(0) = 0$ and $|F(x) - F(y)| \le L|x - y|$ for all $x, y \in \mathbb{R}$; moreover, $G : \mathbb{R} \longrightarrow H$ with $G(0) = 0$, $\|G(x) - G(y)\|_H \le L|x - y|$ for all $x, y \in \mathbb{R}$. Obviously the assumptions of Theorem 18 are fulfilled, if we set $\varphi(t) = S(t)X_0$, $b(t,s,X(s)) = S(t-s)F(X(s))$ and $\rho(t,s,X(s)) = S(t-s)G(X(s))$.

4. More Smoothness of the Solution Process from Section 3.1

In general, the mild solution is not a strong solution, where a strong solution of (4) is defined by $X(t) \in D(A)$ and

$$X(t) = X_0 + \int_0^t (AX(s) + F(X(s)))ds + \int_0^t G(X(s))dW^H(s), \quad t \in [0,T]$$

for all $t \in [0,T]$ with probability one.

Let A be the generator of an analytical C_0-semigroup $(S(t))_{t\ge 0}$ in a Banach space E. Then, it is possible to define the fractional power $A^{-\alpha}$ for $\alpha > 0$, see [24, formula (6.9), p. 70] and $A^\alpha = (A^{-\alpha})^{-1}$ for $\alpha > 0$, see [24, Definition 6.7, p. 72]. If $\alpha \in (0,1)$, then an explicit formula for $A^\alpha x$ ($x \in D(A) \subset D(A^\alpha)$) is known, see [24, formula (6.16), p. 72]. The operator norm of $A^{-\alpha}$ is bounded, see [24, Lemma 6.3, p. 71]. A^α is a closed operator with the following domain

$$D(A^\alpha) = R(A^{-\alpha}), \quad \overline{D(A^\alpha)} = E, \quad (5)$$

by [24, Theorem 6.8, p. 72]. If additionally $0 \in \rho(A)$, then by [24, Theorem 6.13, p. 74] the operator $A^\alpha S(t)$ is bounded for every $t > 0$ and $\|A^\alpha S(t)\| \le M_\alpha t^{-\alpha} e^{-\theta t}$.

We consider F and G as defined in Section 3.1 and we want to choose conditions such that at least $X(t) \in D(A^\alpha)$. We introduce for $\alpha \in (0, \frac{1}{2})$ and $X_0 \in D(A^\alpha)$ the following FSVIE

$$
\begin{aligned}
Y(t) = A^\alpha S(t) X_0 &+ \int_0^t A^\alpha S(t-s) F(A^{-\alpha} Y(s)) ds \\
&+ \int_0^t A^\alpha S(t-s) G(A^{-\alpha} Y(s)) dW^H(s), t \in [0, T].
\end{aligned}
$$
(6)

Obviously the conditions of Remark 10 are fulfilled with $\kappa_1(t, s) = \kappa_2(t, s) = \frac{1}{(t-s)^\alpha}$ for $\alpha \in (0, \frac{1}{2})$. If we apply $A^{-\alpha}$ on both sides of (6), then we have the following equation

$$
\begin{aligned}
A^{-\alpha} Y(t) = S(t) X_0 &+ \int_0^t S(t-s) F(A^{-\alpha} Y(s)) ds \\
&+ \int_0^t S(t-s) G(A^{-\alpha} Y(s)) dW^H(s), t \in [0, T],
\end{aligned}
$$
(7)

since $A^{-\alpha}$ is linear and bounded. We set $X(t) := A^{-\alpha} Y(t)$, then we get with (5) that $X(t) \in D(A^\alpha)$.

5. Backward Stochastic Volterra Type Integral Equation

In this section we introduce a backward stochastic Volterra integral equation (BSVIE) in the Banach space $E = L^q(\mathbf{S}, \Sigma, \mu)$ with respect to an H-cylindrical Brownian motion. We are going to prove the existence of a unique adapted solution.

5.1. *A Special BSVIE*

First we consider the following simple BSVIE in the Banach space $E = L^q(\mathbf{S}, \Sigma, \mu)$, where $1 < q \le 2$ and μ is a σ-finite measure,

$$
Y(t) = \varphi(t) - \int_t^T Z(t, s) dW^H(s), \, t \in [0, T],
$$
(8)

while $\varphi(\cdot)$ is given and satisfies $\mathbb{E}\left(\sup_{t \in [0,T]} \|\varphi(t)\|^p \right) < \infty, 1 < p \le q \le 2$, and $\varphi(t)$ is $\mathcal{F}_t^H$-measurable, $\mathcal{F}_t^H$ denotes the σ-algebra generated by $W^H(s)$ (the H-cylindrical Brownian motion) for $s \le t$. We are interested in finding a unique adapted solution $(Y(t), Z(t, s))$, where $Z(t, s)$ is a linear bounded operator such that $Z(t, \cdot) : \Omega \times [0, T] \longrightarrow \mathcal{L}(H; E)$ and $Z(t, s)$ is $\mathcal{F}_s^H$-measurable for $s \leq t$. Since $\mathbb{E}\left(\sup_{t \in [0,T]} \|\varphi(t)\|^p\right) < \infty$, we have especially $\mathbb{E}\|\varphi(t)\|^p < \infty$ for all $t \in [0, T]$ and $\mathbb{E}\left(\int_0^T \|\varphi(t\|^p dt\right) < \infty$. By using Hölder's inequality it yields that $\mathbb{E}\|\varphi(t)\| < \infty$ for all $t \in [0, T]$. Therefore, the conditional expectation could be defined as

$$\psi_t(r) := \mathbb{E}\left(\varphi(t)|\mathcal{F}_r^H\right) \quad \text{for all } t \in [0, T], 0 \leq r \leq T, \tag{9}$$

$\psi_t(r)$ is only $\mathcal{F}_r$-measurable and is a L^p-martingale with respect to r (for simplicity we set $\mathcal{F}_r := \mathcal{F}_r^H$), because

$$\mathbb{E}\|\psi_t(r)\|^p = \mathbb{E}\|\mathbb{E}\left(\varphi(t)|\mathcal{F}_r\right)\|^p \leq \mathbb{E}(\mathbb{E}\{\|\varphi(t)\|^p|\mathcal{F}_r\}) = \mathbb{E}\|\varphi(t)\|^p < \infty$$

and for $s < r$

$$\mathbb{E}(\psi_t(r)|\mathcal{F}_s) = \mathbb{E}(\mathbb{E}(\varphi(t)|\mathcal{F}_r)|\mathcal{F}_s) = \mathbb{E}(\varphi(t)|\mathcal{F}_s) = \psi_t(s).$$

Now we can use Theorem 16, the martingale representation theorem in Banach spaces. Then we can find a unique $X_t \in L_{\mathbb{F}}^p\left(\Omega; \gamma\left(L^2(0, T; H), E\right)\right)$ such that

$$\psi_t(r) = \mathbb{E}(\psi_t(r)) + I^{W_H}(\xi_{X_t}(r)),$$

where for $f \in H$, $\xi_{X_t}(r, \omega)f := (X(t, \omega))(1_{[0,r]}f)$ and I^{W_H} is the integral process for ξ_{X_t}, $I^{W_H} : r \longrightarrow I^{W_H}(\xi_{X_t}(r))$. In the special case, when E has cotype 2 (see Definition 22), then this integral process is represented by a L^p-stochastically integrable process $Z_t(\cdot)$ that is unique and we have

$$I^{W_H}(\xi_{X_t}(r)) = \int_0^r Z_t(s)dW^H(s).$$

For simplicity we denote $Z_t(s) := Z(t, s)$. In our case $E = L^q(\mathbf{S}, \Sigma, \mu)$, $1 < q \leq 2$, this space has cotype 2, then we can find a unique process $Z(t, s) \in L^p(\Omega; L^2(0, T; \gamma(H, E)))$. Moreover, we can use Theorem 13 and Theorem 12 and find $\mathfrak{Z} : \Omega \times [0, T] \times \mathbf{S} \longrightarrow H$ such that for each $h \in H$, $(Z_t(s)h)(\cdot) = [\mathfrak{Z}_t(s, \cdot), h]_H$ is an element in E, where $[\cdot, \cdot]_H$ denotes the scalar product in the Hilbert space H) and $\mathbb{E}\left\|\left(\int_0^T \|\mathfrak{Z}_t(s, \cdot)\|_H^2 ds\right)^{\frac{1}{2}}\right\|_E^p < \infty$. Theorem 12 indicates that Z_t is L^p-stochastically integrable with respect to W^H if and only if

$$\mathbb{E}\left\|\left(\int_0^T \|\mathfrak{Z}_t(s, \cdot)\|_H^2 ds\right)^{\frac{1}{2}}\right\|_E^p < \infty$$

and we have

$$\mathbb{E}\left\|\int_0^T Z_t(s)dW^H(s)\right\|_E^p \simeq \mathbb{E}\left\|\left(\int_0^T \|3_t(s,\cdot)\|_H^2 ds\right)^{\frac{1}{2}}\right\|_E^p.$$

Again for simplicity we write $3_t(s) := 3(t,s)$ similar to $Z_t(s) := Z(t,s)$. Therefore we have

$$\psi_t(r) = \mathbb{E}(\varphi(t)) + \int_0^r Z(t,s)dW^H(s);$$

by letting $r = T$ it yields

$$\mathbb{E}(\varphi(t)|\mathcal{F}_T) = \mathbb{E}(\varphi(t)) + \int_0^T Z(t,s)dW^H(s)$$

and since $\varphi(t)$ is $\mathcal{F}_T$ measurable we obtain

$$\varphi(t) = \mathbb{E}(\varphi(t)) + \int_0^T Z(t,s)dW^H(s).$$

We can write

$$\mathbb{E}(\varphi(t)) + \int_0^t Z(t,s)dW^H(s) = \varphi(t) - \int_t^T Z(t,s)dW^H(s).$$

By defining $Y(t) = \mathbb{E}(\varphi(t)) + \int_0^t Z(t,s)dW^H(s)$ for $t \in [0,T]$ it results that $Y(t)$ is an $\mathcal{F}_t$-adapted process and $(Y(\cdot), Z(\cdot,\cdot))$ is an adapted solution for the BSVIE. This pair has the following properties:

$$\int_0^T Z(t,s)dW^H(s) = \varphi(t) - \mathbb{E}(\varphi(t)),$$

$$\mathbb{E}\left\|\int_0^T Z(t,s)dW^H(s)\right\|^p = \mathbb{E}\|\varphi(t) - \mathbb{E}(\varphi(t))\|^p \leq c(\mathbb{E}\|\varphi(t)\|^p + \mathbb{E}\|\mathbb{E}(\varphi(t))\|^p).$$

By Jensen's inequality it yields

$$\mathbb{E}\left\|\int_0^T Z(t,s)dW^H(s)\right\|^p \leq c(\mathbb{E}\|\varphi(t)\|^p + \mathbb{E}\mathbb{E}(\|\varphi(t)\|^p) \leq c\mathbb{E}\|\varphi(t)\|^p$$

and we have

$$\mathbb{E}\left\|\left(\int_0^T \|3(t,s)\|_H^2 ds\right)^{\frac{1}{2}}\right\|^p \leq c\mathbb{E}\|\varphi(t)\|^p.$$

By construction $Y(t)$ is $\mathcal{F}_t$-adapted, hence we can write

$$\mathbb{E}(Y(t)|\mathcal{F}_t) = \mathbb{E}\left(\left(\varphi(t) - \int_t^T Z(t,s)dW^H(s)\right)\bigg|\mathcal{F}_t\right),$$

$$Y(t) = \mathbb{E}(\varphi(t)|\mathcal{F}_t) - \mathbb{E}\left(\int_t^T Z(t,s)dW^H(s)\right) = \mathbb{E}(\varphi(t)|\mathcal{F}_t),$$

and

$$\mathbb{E}\|Y(t)\|^p = \mathbb{E}\|\mathbb{E}(\varphi(t)|\mathcal{F}_t)\|^p \leq \mathbb{E}\left(\mathbb{E}\{\|\varphi(t)\|^p|\mathcal{F}_t\}\right) = \mathbb{E}\|\varphi(t)\|^p < \infty.$$

By using the assumptions on $\varphi(\cdot)$ it yields

$$\mathbb{E}\left(\int_0^T \|Y(t)\|^p dt\right) \leq \mathbb{E}\left(\sup_{t\in[0,T]} \|\varphi(t)\|^p\right) < \infty.$$

Then $Y(\cdot) \in L_{\mathbb{F}}^p(\Omega \times [0,T]; E)$, $Z(\cdot,\cdot) \in L_{\mathbb{F}}^p\left(\Omega \times [0,T]; \gamma\left(L^2(0,T;H), E\right)\right)$ and $3(\cdot,\cdot) \in L_{\mathbb{F}}^p\left(\Omega \times [0,T]; L^q\left(\mathbf{S}, L^2(0,T;H)\right)\right)$. We also get

$$\sup_{t\in[0,T]} \|Y(t)\|^p \leq \mathbb{E}\left(\sup_{t\in[0,T]} \|\varphi(t)\|^p|\mathcal{F}_t\right)$$

and

$$\mathbb{E}\left(\sup_{t\in[0,T]} \|Y(t)\|^p\right) \leq \mathbb{E}\left(\sup_{t\in[0,T]} \|\varphi(t)\|^p\right) < \infty.$$

By putting these results together, it yields

$$\mathbb{E}\left(\sup_{t\in[0,T]} \|Y(t)\|^p\right) + \mathbb{E}\left(\int_0^T \left\|\int_0^T Z(t,s)dW^H(s)\right\|^p dt\right) \tag{10}$$

$$\leq c\mathbb{E}\left(\sup_{t\in[0,T]} \|\varphi(t)\|^p\right).$$

The following inequalities also result

$$\mathbb{E}\left(\int_0^T \|Y(t)\|^p dt\right) + \mathbb{E}\left(\int_0^T \left\|\int_0^T Z(t,s)dW^H(s)\right\|^p dt\right)$$

$$\leq c\mathbb{E}\left(\int_0^T \|\varphi(t)\|^p dt\right) \leq c\mathbb{E}\left(\sup_{t\in[0,T]} \|\varphi(t)\|^p\right).$$

In what follows we show the uniqueness of the solution. Let $(Y(\cdot), Z(\cdot,\cdot))$ and $(\bar{Y}(\cdot), \bar{Z}(\cdot,\cdot))$ be solutions of the BSVIE (8). Then we have

$$Y(t) - \bar{Y}(t) = -\int_t^T \left(Z(t,s) - \bar{Z}(t,s)\right) dW^H(s)$$

by letting $\delta^Y(\cdot) := Y(\cdot) - \bar{Y}(\cdot)$ and $\delta^Z(\cdot,\cdot) := Z(\cdot,\cdot) - \bar{Z}(\cdot,\cdot)$ and $\delta^3(\cdot,\cdot) := 3(\cdot,\cdot) - \bar{3}(\cdot,\cdot)$, we get the following BSVIE

$$\delta^Y(t) = -\int_t^T \delta^Z(t,s)dW^H(s)$$

and we set for this problem the process $\varphi(t)$ to be 0. Then from Equation (10) we obtain

$$\mathbb{E}\left(\sup_{t\in[0,T]} \|\delta^Y(t)\|^p\right) + \mathbb{E}\left(\int_0^T \left\|\int_0^T \delta^Z(t,s)dW^H(s)\right\|^p dt\right)$$

$$\leq \mathbb{E}\left(\sup_{t\in[0,T]} \|\delta^Y(t)\|^p + c\int_0^T \left\|\left(\int_0^T \|\delta^3(t,s)ds\|_H^2\right)^{\frac{1}{2}}\right\|^p dt\right)$$

$$\leq 0.$$

Subsequently, $\delta^Y(\cdot) = 0$ a.s., $\delta^Z(\cdot,\cdot) = 0$ a.s. and $3(\cdot,\cdot) \equiv \bar{3}(\cdot,\cdot)$ a.s. So, we have shown the following result:

Theorem 21. *The special BSVIE (8) has a unique adapted solution* $(Y(\cdot), Z(\cdot,\cdot)) \in \left(L_{\mathbb{F}}^p(\Omega \times [0,T]; E) \times L_{\mathbb{F}}^p\left(\Omega \times [0,T]; \gamma\left(L^2(0,T;H), E\right)\right)\right).$

5.2. *A More General BSDE*

We study a more general case: we consider the following backward stochastic differential equation (BSDE)

$$Y(t) = X + \int_t^T f(s, Y(s), Z(s))ds - \int_t^T Z(s)dW^H(s), \; t \in [0,T],$$

where X is $\mathcal{F}_T$-measurable and is given such that $\mathbb{E}\|X\|_E^p < \infty$,

$$Z : \Omega \times [0,T] \longrightarrow \mathcal{L}(H,E)$$

$$Y : \Omega \times [0,T] \longrightarrow E$$

and the generator function f is $f : \Omega \times [0,T] \times E \times E \times \mathbf{S} \longrightarrow E$; by using the stochastic integrability Theorems 13,

$$3 : \Omega \times [0,T] \times \mathbf{S} \longrightarrow H$$

to be a strongly measurable function such that for all $h \in H$ and $t \in [0,T]$ $(Z(t)h)(\cdot) = [3(t,\cdot), h]_H$, and by this definition f could be defined as

$$f : \Omega \times [0,T] \times E \times H \times \mathbf{S} \longrightarrow E$$

and the BSDE can be written as

$$Y(t) = X + \int_t^T f(s, Y(s), 3(s))ds - \int_t^T Z(s)dW^H(s), \; t \in [0,T]. \quad (11)$$

We consider the following assumptions:

(H1) f is $\mathcal{F}_t$-adapted and $\mathbb{E}\left\|\int_0^T |f(t,0,0,\cdot)|dt\right\|^p < \infty$;

(H2) for all $t \in [0,T], y, \bar{y} \in \mathbb{R}, \mathfrak{z}, \bar{\mathfrak{z}} \in H$,

$$|f(t,y,\mathfrak{z},.) - f(t,\bar{y},\bar{\mathfrak{z}},.)| \leq L_y(t)|y - \bar{y}| + L_\mathfrak{z}(t)\|\mathfrak{z} - \bar{\mathfrak{z}}\|_H,$$

where $L_y(\cdot) \in L^\infty(0,T)$ and $L_\mathfrak{z}(\cdot)$ are positive deterministic functions such that $\int_0^T L_\mathfrak{z}(t)^{2+\varepsilon}dt < \infty$ for $\varepsilon > 0$.

We define the space

$$H^p[R,S] := L^p_\mathbb{F}([R,S] \times \Omega; E) \times L^p_\mathbb{F}\left(\Omega; L^q\left(\mathbf{S}; L^2([R,S];H)\right)\right),$$

which is a Banach space equipped with the norm

$$\|(Y(\cdot),\mathfrak{z}(\cdot))\|_{H^p[R,S]} := \left\{\mathbb{E}\left(\sup_{t \in [R,S]} \|Y(t)\|_E^p\right) + \mathbb{E}\left\|\left(\int_R^S \|\mathfrak{z}(s,\cdot)\|_H^2 ds\right)^{\frac{1}{2}}\right\|_E^p\right\}^{\frac{1}{p}}.$$

Proposition 1. *Assume that* $X \in L^p(\Omega; E)$ *and that* (H1) *and* (H2) *hold. Then, the BSDE* (11) *admits a unique solution* $(Y(\cdot),\mathfrak{z}(\cdot)) \in H^p[0,T]$, *correspondingly* $(Y(\cdot), Z(\cdot))$, *where for all* $h \in H$, $(Z(t)h)(\cdot) = [\mathfrak{z}(t,\cdot),h]_H$ *in* E. *The following estimate holds*

$$\mathbb{E}\left(\sup_{t \in [0,T]} \|Y(t)\|^p\right) + \mathbb{E}\left\|\left(\int_0^T \|\mathfrak{z}(s,\cdot)\|_H^2 ds\right)^{\frac{1}{2}}\right\|^p$$

$$\leq c\mathbb{E}\left\{\|X\|^p + \left\|\int_0^T |f(s,Y(s),\mathfrak{z}(s),\cdot)|ds\right\|^p\right\}$$

$$\leq c\mathbb{E}\left\{\|X\|^p + \left\|\int_0^T |f(s,0,0,\cdot)|ds\right\|^p\right\}.$$

Proof. Let $(y(\cdot),\mathfrak{z}(\cdot)) \in H^p[0,T]$. We want to find the unique solution $(Y(\cdot), Z(\cdot,\cdot))$ for the following special backward equation

$$Y(t) = X + \int_t^T f(s,y(s),\mathfrak{z}(s))ds - \int_t^T Z(s)dW^H(s), \quad t \in [0,T]. \quad (12)$$

One can show that $\mathbb{E}\left\|\int_0^T f(s,y(s),\mathfrak{z}(s),\cdot)ds\right\|^P < \infty.$

Then, we get $\mathbb{E} \left\| X + \int_0^T f(s, y(s), \mathfrak{z}(s), \cdot) ds \right\|^p < \infty$. Now we define the E-valued L^p-martingale similar to Equation (9) by

$$\psi(r) = \mathbb{E}\left(\left(X + \int_0^T f(s, y(s), \mathfrak{z}(s)) ds \right) \Big| \mathcal{F}_r \right), \quad r \in [0, T].$$

By the martingale representation Theorem 16, there exists a unique process $Z(\cdot) \in L_{\mathbb{F}}^p \left(\Omega, \gamma(L^2((0, T), H); E) \right)$, where

$$\psi(r) = \mathbb{E}\left(X + \int_0^T f(s, y(s), \mathfrak{z}(s)) ds \right) + \int_0^r Z(s) dW^H(s).$$

Let $r = T$ and by the $\mathcal{F}_T$-measurability of $X + \int_0^T f(s, y(s), \mathfrak{z}(s)) ds$, we have

$$X + \int_0^T f(s, y(s), \mathfrak{z}(s)) ds$$
$$= \mathbb{E}\left(X + \int_0^T f(s, y(s), \mathfrak{z}(s)) ds \right) + \int_0^T Z(s) dW^H(s),$$

i.e.

$$X + \int_0^t f(s, y(s), \mathfrak{z}(s)) ds + \int_t^T f(s, y(s), \mathfrak{z}(s)) ds$$
$$= \mathbb{E}\left(X + \int_0^T f(s, y(s), \mathfrak{z}(s)) ds \right) + \int_0^t Z(s) dW^H(s) + \int_t^T Z(s) dW^H(s).$$

By taking

$$Y(t) = \mathbb{E}\left(X + \int_0^T f(s, y(s), \mathfrak{z}(s)) ds \right)$$
$$- \int_0^t f(s, y(s), \mathfrak{z}(s)) ds + \int_0^t Z(s) dW^H(s)$$

we can find analogously to the simple BSVIE (8), the unique $(Y(\cdot), Z(\cdot))$ such that Equation (12) holds. By the L^p-martingale property and adaptedness of the generator function, we can write

$$Y(t) = \mathbb{E}\left(Y(t) \mid \mathcal{F}_t \right) = \mathbb{E}\left(\left(X + \int_t^T f(s, y(s), \mathfrak{z}(s)) ds \right) \Big| \mathcal{F}_t \right).$$

This yields

$$|Y(t)| \le \mathbb{E}\left(\left(|X| + \int_0^T |f(s,y(s),\mathfrak{z}(s))|\, ds \right) \Big| \mathcal{F}_t \right)$$

and consequently

$$\mathbb{E}\left(\sup_{t \in [0,T]} \|Y(t)\|^p \right) < \infty.$$

Further we get

$$\mathbb{E}\left\| \int_0^T Z(s)dW^H(s) \right\|^p$$

$$= \mathbb{E}\left\| X + \int_0^T f(s,y(s),\mathfrak{z}(s))ds - \mathbb{E}\left(X + \int_0^T f(s,y(s),\mathfrak{z}(s))ds \right) \right\|^p$$

$$\le 2^p \mathbb{E}\left\| X + \int_0^T f(s,y(s),\mathfrak{z}(s))ds \right\|^p < \infty.$$

By using the theorem for stochastic integrability of stochastic processes in $L^q(\mathbf{S},\Sigma,\mu)$, Theorem 12, where $1 < q \le 2$, it results

$$\mathbb{E}\left\| \left(\int_0^T \|\mathfrak{z}(s)\|_H^2 ds \right)^{\frac{1}{2}} \right\|^p \le c\mathbb{E}\left\| X + \int_0^T f(s,y(s),\mathfrak{z}(s))ds \right\|^p$$

and, finally, we have

$$\mathbb{E}\left(\sup_{t \in [0,T]} \|Y(t)\|^p \right) + \mathbb{E}\left\| \left(\int_0^T \|\mathfrak{z}(s,\cdot)\|_H^2 ds \right)^{\frac{1}{2}} \right\|^p$$

$$\le c\mathbb{E}\left\{ \|X\|^p + \left\| \int_0^T |f(s,y(s),\mathfrak{z}(s),\cdot)|ds \right\|^p \right\}.$$

Here $\mathfrak{z}(s)$ corresponds to $Z(s)$ and it is the solution. Therefore, for each $S \in [0,T]$ we define the map $\Phi : H^p[S,T] \longrightarrow H^p[S,T]$ by $\Phi(y(\cdot),\mathfrak{z}(\cdot)) = (Y(\cdot),\mathfrak{z}(\cdot))$, where $(Z(t)h)(\cdot) = [\mathfrak{z}(t,\cdot),h]_H$. For finding the unique solution in $[S,T]$ we use the Banach fixed point theorem and we show that Φ is contractive on $[S,T]$ for some S. Let $(\bar{y}(\cdot),\bar{\mathfrak{z}}(\cdot)) \in H^p[S,T]$; according to the above calculations let $(\bar{Y}(\cdot),\bar{\mathfrak{z}}(\cdot)) \in H^p[S,T]$ be the unique solution of the related BSDE, then we have for $t \in [S,T]$

$$Y(t) - \bar{Y}(t) = \int_t^T (f(s,y(s),\mathfrak{z}(s)) - f(s,\bar{y}(s),\bar{\mathfrak{z}}(s)))\, ds$$

$$- \int_t^T \left(Z(s) - \bar{Z}(s) \right) dW^H(s).$$

This is a BSDE and by defining $\delta_Y(t) = Y(t) - \bar{Y}(t)$, $\delta_Z(t) = Z(t) - \bar{Z}(t)$, $\delta_3(t) = 3(t) - \bar{3}(t)$ and by using again the previous calculations we have

$$\mathbb{E} \left\{ \sup_{t \in [0,T]} \|\delta_Y(t)\|^p + \left\| \left(\int_0^T \|\delta_3(s)\|_H^2 ds \right)^{\frac{1}{2}} \right\|^p \right\}$$

$$= \left\| (Y(\cdot), 3(\cdot)) - (\bar{Y}(\cdot), \bar{3}(\cdot)) \right\|_{H^p[S,T]}^p$$

$$\leq c \left\{ \mathbb{E} \left\| \int_S^T L_y(s) |y(s) - \bar{y}(s)| ds \right\|^p + \left\| \int_S^T L_3(s) \|3(s) - \bar{3}(s)\|_H ds \right\|^p \right\}.$$

Then,

$$\left\| (Y(\cdot), 3(\cdot)) - (\bar{Y}(\cdot), \bar{3}(\cdot)) \right\|_{H^p[S,T]}^p$$

$$\leq cF(T - S) \left\| (y(\cdot), 3(\cdot)) - (\bar{y}(\cdot), \bar{3}(\cdot)) \right\|_{H^p[S,T]}^p,$$

where $F(S - T) = K^p (T - S)^p + K'^{\frac{p}{p+\varepsilon}} (T - S)^{\frac{p\varepsilon}{2(2+\varepsilon)}}$ and $K' = \left(\int_S^T L_3(s)^{2+\varepsilon} ds \right)^{\frac{p}{2+\varepsilon}}$. $F(T - S)$ is a polynomial function of $T - S$ and $F(0) = 0$, then we can choose S such that $F(T - S) < \frac{1}{c}$, it means that the map Φ is contractive and we can find the unique fixed point $(Y(\cdot), 3(\cdot))$ for $t \in [S, T]$, correspondingly $(Y(\cdot), Z(\cdot))$, and by induction we can find the unique solution on $[0, T]$. $\square$

5.3. The General BSVIE

In this section we consider the following generalized BSVIE

$$Y(t) = \varphi(t) + \int_t^T f(t, s, Y(s), Z(t, s), Z(s, t)) ds - \int_t^T Z(t, s) dW^H(s),$$

$$t \in [0, T]. \tag{13}$$

By similar arguments that led to Equation (11), it can be rewritten as

$$Y(t) = \varphi(t) + \int_t^T f(t, s, Y(s), 3(t, s), 3(s, t)) ds - \int_t^T Z(t, s) dW^H(s),$$

$$t \in [0, T], \tag{14}$$

where $f : \Omega \times \Delta^c \times E \times H \times H \times \mathbf{S} \longrightarrow E$, $\varphi : \Omega \times [0, T] \longrightarrow E$ are given functions and $\Delta = \{(t, s) : 0 < s < t < T\}$. Before generalizing the solution to (14), it must be mentioned that without considering some additional assumptions on f, φ and especially on the solutions for this equation, we could not expect to obtain a unique solution (see for example (15)). The most important restriction on the solutions is to require that they should be adapted M-solutions.

Definition 24. $(Y(\cdot), Z(\cdot, \cdot))$ is called **adapted M-solution** of Equation (13) with respect to the H-cylindrical Brownian motion W^H, if $(Y(\cdot), Z(\cdot, \cdot))$ is an adapted solution of the BSVIE (13) and the following equation holds for $0 \leq S' \leq T$

$$Y(t) = \mathbb{E}(Y(t)|\mathcal{F}_{S'})) + \int_{S'}^{t} Z(t, s) dW^H(s), \quad t \in [S', T]. \tag{15}$$

Remark 11. For the previous BSDE (11) studied in Proposition 1, since $Y(\cdot)$ is adapted process, we can find a unique adapted M-solution by using the martingale representation theorem as follows

$$Y(t) = \mathbb{E}(Y(t)) + \int_{0}^{t} \zeta(t, s) dW^H(s)$$

$$Z(t, s) = \begin{cases} \zeta(t, s) & (t, s) \in \Delta \\ Z(s) & (t, s) \in \Delta^c \end{cases}.$$

Remark 12. We will see, that we can not easily find the solution by induction similar to Proposition 1.

Now consider for all $R, S \in [0, T]$

$$\lambda(t, r) = \varphi(t) + \int_{r}^{T} f(t, s, \varrho(t, s)) ds - \int_{r}^{T} \rho(t, s) dW^H(s), \tag{16}$$

$$\text{for all } r \in [R, T], \ t \in [S, T],$$

where $(\rho(t, s)h)(\cdot) = [\varrho(t, s, \cdot), h]_H$ in E and f is given. This equation is a stochastic Fredholm integral equation (SFIE) on $[S, T]$, parametrized by $r \in [R, T]$. We want to find the unique process $(\lambda(\cdot, \cdot), \rho(\cdot, \cdot))$ for which $(\lambda(t, \cdot), \rho(t, \cdot))$ is adapted for each $t \in [S, T]$. Consider the following assumptions:

(F1) $R, S \in [0, T]$ and $f : \Omega \times [S, T] \times [R, T] \times H \times \mathbf{S} \longrightarrow E$ be $\mathcal{B}([S, T] \times [R, T]) \otimes \mathcal{B}(H) \otimes \mathcal{B}(E) \times \mathcal{F}_T$-measurable such that $s \longrightarrow f(t, s, \mathfrak{z}, \cdot)$ is progressively measurable for all $(t, \mathfrak{z}) \in [S, T] \times H$ and

$$\int_{S}^{T} \mathbb{E} \left\| \int_{R}^{T} |f(t, s, 0, \cdot)| ds \right\|_{E}^{p} dt < \infty;$$

(F2) for all $(t, s) \in [S, T] \times [R, T]$ and $\mathfrak{z}, \bar{\mathfrak{z}} \in H$

$$|f(t, s, \mathfrak{z}, \cdot) - f(t, s, \bar{\mathfrak{z}}, \cdot)| \leq L(t, s) \|\mathfrak{z}(\cdot) - \bar{\mathfrak{z}}(\cdot)\|_H,$$

where $L : [S, T] \times [R, T] \longrightarrow \mathbb{R}^+$ is a deterministic function such that

$$\sup_{t \in [S,T]} \int_R^T L(t, s)^{2+\varepsilon} ds < \infty \text{ for some } \varepsilon > 0.$$

Proposition 2. *If* (F1), (F2) *hold and* $\varphi(\cdot) \in L^p_{\mathcal{F}_T}(\Omega \times [S, T]; E)$ *such that* $\mathbb{E} \left(\sup_{t \in [S,T]} \|\varphi(t)\|^p \right) < \infty$ *then Equation* (16) *has for every* $t \in [S, T]$ *a unique adapted solution* $(\lambda(t, \cdot), \varrho(t, \cdot)) \in H^p[R, T]$ *and the following estimates hold*

$$\|(\lambda(t, \cdot), \varrho(t, \cdot))\|^p_{H^p[R,T]}$$

$$= \mathbb{E} \left\{ \sup_{r \in [R,T]} \|\lambda(t, r))\|^p + \left\| \left(\int_R^T \|\varrho(t, s, \cdot)\|^2_H ds \right)^{\frac{1}{2}} \right\|^p \right\}$$

$$\leq c \mathbb{E} \left\{ \|\varphi(t)\|^p + \left\| \int_R^T |f(t, s, \varrho(t, s), \cdot| ds \right\|^p \right\}$$

$$\leq c \mathbb{E} \left\{ \|\varphi(t)\|^p + \left\| \int_R^T |f(t, s, 0, \cdot| ds \right\|^p \right\}.$$

Proof. This is derived from Proposition 1, if we consider Equation (16) for every fixed $t \in [0, T]$. $\qquad\square$

At first let us consider for our purpose some special cases of Equation (16). First let $r = S$ and we define $\psi^S(t) = \lambda(t, S)$, $Z(t, s) = \rho(t, s)$, $\mathfrak{z}(t, s) = \varrho(t, s)$ for all $t \in [R, S]$, $s \in [S, T]$. Then, Equation (16) yields

$$\psi^S(t) = \varphi(t) + \int_S^T f(t, s, \mathfrak{z}(t, s)) ds - \int_S^T Z(t, s) dW^H(s), \tag{17}$$

$$t \in [R, S], \; R, \; S \in [0, T].$$

The following result is a consequence of Proposition 2:

Proposition 3. *If the assumptions* (F1) *and* (F2) *hold and* $\varphi(\cdot) \in L^p_{\mathcal{F}_S}(\Omega \times [R, S]; E)$ *such that* $\mathbb{E} \left(\sup_{t \in [R,S]} \|\varphi(t)\|^p \right) < \infty$, *then Equation* (17) *has a* *unique adapted solution $(\psi^S(t), \mathfrak{Z}(t, \cdot)) \in H^p[R, T]$ for all $t \in [R, S]$ and the following estimates hold*

$$
\left\| (\psi^S(t), \mathfrak{Z}(t, \cdot)) \right\|_{H^p[R,T]}^p = \mathbb{E} \left\{ \|\psi^S(t))\|^p + \left\| \left(\int_R^T \|\mathfrak{Z}(t, s, \cdot)\|_H^2 ds \right)^{\frac{1}{2}} \right\|^p \right\}
$$

$$
\leq c\mathbb{E} \left\{ \|\varphi(t)\|^p + \left\| \int_R^T |f(t, s, \mathfrak{Z}(t, s), \cdot)| ds \right\|^p \right\}
$$

$$
\leq c\mathbb{E} \left\{ \|\varphi(t)\|^p + \left\| \int_R^T |f(t, s, 0, \cdot)| ds \right\|^p \right\},
$$

$$
\forall t \in [R, S].
$$

Here it must be mentioned that $\psi^S(t)$ is $\mathcal{F}_S$-measurable for each t. Another representation of Equation (17) is the following: let $S = R$ and

$$
\begin{cases}
Y(t) = \lambda(t, t) & t \in [S, T] \\
Z(t, s) = \rho(t, s) & (t, s) \in \Delta^c[S, T] = \{(t, s) : S < t < s < T\}) \\
\mathfrak{Z}(t, s) = \varrho(t, s) & (t, s) \in \Delta^c[S, T].
\end{cases}
$$

Then Equation (16) yields

$$
Y(t) = \varphi(t) + \int_t^T f(t, s, \mathfrak{Z}(t, s)) ds - \int_t^T Z(t, s) dW^H(s), \ t \in [S, T]. \quad (18)
$$

We confine ourselves to M-solutions as

$$
Y(t) = \mathbb{E}(Y(t)|\mathcal{F}_S) + \int_S^t Z(t, s) dW^H(s),
$$

where $(t, s) \in \Delta[S, T]$, $(S < s \leq t < T)$. Notice that in the case $(t, s) \in \Delta[S, T]$, $Z(t, s)$ and $\rho(t, s)$ could be different. Now we can state the following proposition:

Proposition 4. *Suppose that the assumptions* (F1), (F2) *hold and let $\varphi(t)$ be $\mathcal{F}_T$ measurable and $\mathbb{E}\left(\sup_{t \in [S, T]} \|\varphi(t)\|^p \right) < \infty$. Then, Equation* (18) *has a unique adapted M-solution*

$$
(Y(\cdot), \mathfrak{Z}(\cdot, \cdot)) \in L_{\mathbb{F}}^p(\Omega \times [S, T]; E) \times L_{\mathbb{F}}^p(\Omega \times [S, T]; L^q(\mathbf{S}; L^2(S, T; H)))
$$

and following inequalities hold

$$
\mathbb{E} \left\{ \|Y(t)\|^p + \left\| \left(\int_t^T \|\mathfrak{Z}(t, s, \cdot)\|_H^2 ds \right)^{\frac{1}{2}} \right\|^p \right\}
$$

$$\leq c\mathbb{E}\left\{\|\varphi(t)\|^p + \left\|\int_t^T |f(t,s,\mathfrak{z}(t,s),\cdot)|ds\right\|^p\right\}$$

$$\leq c\mathbb{E}\left\{\|\varphi(t)\|^p + \left\|\int_t^T |f(t,s,0,\cdot)|ds\right\|^p\right\}, \quad t \in [S,T]$$

and

$$\mathbb{E}\left\{\int_S^T \|Y(t)\|^p dt + \int_S^T \left\|\left(\int_S^T \|\mathfrak{z}(t,s,\cdot)\|_H^2 ds\right)^{\frac{1}{2}}\right\|^p dt\right\}$$

$$\leq c\mathbb{E}\left\{\int_S^T \|\varphi(t)\|^p dt + \int_S^T \left\|\int_S^T |f(t,s,0,\cdot)|ds\right\|^p dt\right\}.$$

Proof. First, let $(y(\cdot),\mathfrak{z}(\cdot,\cdot)) \in H^p[S,T]$ be such that $y(t) = \mathbb{E}(y(t)|\mathcal{F}_S) + \int_S^t z(t,s)dW^H(s)$. Then, similar to Proposition 1, and by using the Banach fixed point theorem, we can find the unique adapted M-solution. □

Remark 13. Without the assumption $\mathbb{E}\left(\sup_{t\in[S,T]} \|\varphi(t)\|^p\right) < \infty$, the Propositions 2, 3, 4 are also satisfied; the assumption $\varphi(\cdot) \in L^p_{\mathcal{F}_T}(\Omega \times [S,T]; E)$ is sufficient.

Now we can deal with the generalized BSVIE (14). First we formulate some assumptions:

(G1) Let $f : \Omega \times \Delta^c \times E \times H \times H \times \mathbf{S} \longrightarrow E$ be $\mathcal{B}(\Delta^c) \otimes \mathcal{B}(E \times H \times H) \otimes \Sigma \otimes \mathcal{F}_T$-measurable such that $s \longrightarrow f(t,s,y,\mathfrak{z},\varrho,\zeta)$ is $\mathbb{F}$-progressively measurable for $(t,y,\mathfrak{z},\varrho,\zeta) \in [0,T] \times E \times H \times H \times \mathbf{S}$ and

$$\mathbb{E}\int_0^T \left\|\int_t^T |f(t,s,0,0,0,\cdot)|ds\right\|^p dt < \infty;$$

for simplicity we define $f_0(t,s,\cdot) := f(t,s,0,0,0,\cdot)$;

(G2) The following Lipschitz condition for every $(t,s) \in \Delta^c$, $y,\bar{y} \in \mathbb{R}$, $\mathfrak{z},\bar{\mathfrak{z}},\varrho,\bar{\varrho} \in H$ holds

$$|f(t,s,y,\mathfrak{z},\varrho) - f(t,s,\bar{y},\bar{\mathfrak{z}},\bar{\varrho})| \leq L_1(t,s)|y - \bar{y}|$$
$$+ L_2(t,s)\|\mathfrak{z} - \bar{\mathfrak{z}}\|_H + L_3(t,s)\|\varrho - \bar{\varrho}\|_H,$$

where $L_i(t,s), i = 1,2,3$ are positive deterministic functions such that

$$\sup_{t\in[0,T]} L_1(t,s) \in L^\infty[0,T] \text{ and } \sup_{t\in[0,T]} \int_t^T L_i(t,s)^{2+\varepsilon}ds < \infty, i = 2,3$$

for some $\varepsilon > 0$.

We define the Banach space

$$\mathcal{H}^p[S,T] = \left\{(y(\cdot),\mathfrak{z}(\cdot,\cdot)) : \|(y(\cdot),\mathfrak{z}(\cdot,\cdot))\|_{\mathcal{H}^p[S,T]} < \infty\right\}$$

equipped with the norm

$$\|(y(\cdot),\mathfrak{z}(\cdot,\cdot))\|^p_{\mathcal{H}^p[S,T]} = \mathbb{E}\left\{\int_S^T \|y(t)\|^p dt + \int_S^T \left\|\left(\int_S^T \|\mathfrak{z}(t,s)\|_H^2 ds\right)^{\frac{1}{2}}\right\|^p dt\right\}.$$

Theorem 22. *If* (G1), (G2) *hold and* $\varphi(\cdot) \in L^p_{\mathcal{F}_T}(\Omega \times [0,T]; E)$, *then there exists a unique adapted M-solution* $(Y(\cdot),\mathfrak{Z}(\cdot,\cdot)) \in \mathcal{H}[0,T]$ *for the BSVIE* (14), *such that the following estimate holds*

$$\|(Y(\cdot),\mathfrak{Z}(\cdot,\cdot))\|_{\mathcal{H}^p[S,T]} \le c\mathbb{E}\left\{\int_0^T \|\varphi(t)\|^p dt + \int_0^T \left\|\int_0^T |f_0(t,s,\cdot)| ds\right\|^p dt\right\}.$$

Proof. We prove this theorem in several steps:

Step(1) First we find the solution in $[S,T]$ for some $S \in [0,T]$, we define the space $\mathcal{M}^p[S,T]$ of all $(y(\cdot),\mathfrak{z}(\cdot,\cdot)) \in \mathcal{H}^p[0,T]$ such that

$$Y(t) = \mathbb{E}(Y(t)|\mathcal{F}_S)) + \int_S^t Z(t,s) dW^H(s), \quad t \in [S,T],$$

and we consider the following equation

$$Y(t) = \varphi(t) + \int_t^T f(t,s,y(s),\mathfrak{Z}(t,s),\mathfrak{z}(s,t)) ds - \int_t^T Z(t,s) dW^H(s),$$
$$t \in [0,T],$$

where $(y(\cdot),\mathfrak{z}(\cdot,\cdot)) \in \mathcal{M}^p[S,T]$ is given. This equation is similar to Equation (18) and we can use Proposition 4 to find the unique adapted M-solution $(Y(\cdot),\mathfrak{Z}(\cdot,\cdot))$. We define the map

$$\Upsilon : \mathcal{M}^p[S,T] \longrightarrow \mathcal{M}^p[S,T]$$

by

$$\Upsilon(y(\cdot),\mathfrak{z}(\cdot,\cdot)) = (Y(\cdot),\mathfrak{Z}(\cdot,\cdot)), \quad \forall (y(\cdot),\mathfrak{z}(\cdot,\cdot)) \in \mathcal{M}^p[S,T].$$

We show that this map is contractive and we use the Banach fixed point theorem to obtain the existence of a unique solution in $[S,T]$. Let $(\bar{y}(\cdot),\bar{\mathfrak{z}}(\cdot,\cdot)) \in \mathcal{M}^p[S,T]$ be such that $\Upsilon(\bar{y}(\cdot),\bar{\mathfrak{z}}(\cdot,\cdot)) = (\bar{Y}(\cdot),\bar{\mathfrak{Z}}(\cdot,\cdot))$. Then by using Proposition 4 and by elementary calculations we obtain

$$\|(Y(\cdot),\mathfrak{Z}(\cdot,\cdot)) - (\bar{Y}(\cdot),\bar{\mathfrak{Z}}(\cdot,\cdot))\|^p_{\mathcal{H}^p[S,T]}$$

$$\leq c\left((T-S)^p k_1 + (T-S)^{\frac{p}{2}\cdot\frac{\varepsilon}{2+\varepsilon}} k_3\right) \|(y(\cdot),\mathfrak{z}(\cdot,\cdot)) - (\bar{y}(\cdot),\bar{\mathfrak{z}}(\cdot,\cdot))\|^p_{\mathcal{H}^p[S,T]},$$

where k_1, and k_3 are positive constants. By taking $T - S > 0$ small enough the contraction property holds. The unique adapted M-solution $(Y(t), Z(t,s))$, for $(t,s) \in [S,T] \times [S,T]$ is found by the Banach fixed point theorem.

Step(2) Now we want to find the values $Z(t,s)$ for $(t,s) \in [S,T] \times [R,S]$. As in the previous step we obtain $Y(\cdot) \in L_{\mathbb{F}}(\Omega \times [S,T]; E)$, and it results that

$$\forall t \in [S,T] \quad \psi_t(r) = \mathbb{E}(Y(t)|\mathcal{F}_r), \quad 0 < r < T$$

is an L^p-martingale. By the martingale representation theorem in UMD spaces ($1 < q \leq 2$, E has cotype 2), we can find a unique adapted process $Z(t,s)$ such that

$$\mathbb{E}(Y(t)|\mathcal{F}_S) = \mathbb{E}(Y(t)) + \int_0^S Z(t,s)dW^H(s)$$

or by conditional expectation on $\mathcal{F}_R$, for $R < S$ we have

$$\mathbb{E}(Y(t)|\mathcal{F}_S) = \mathbb{E}(Y(t)|\mathcal{F}_R) + \int_R^S Z(t,s)dW^H(s), \quad t \in [S,T].$$

In other words we could find the values $Z(t,s)$ for $(t,s) \in [S,T] \times [R,S]$, $R \in [0,S]$. By combining these values with Step(1) we have $Z(t,s)$ for $(t,s) \in [S,T] \times [R,T]$.

Step(3) It follows from Step(2) that we have the values $Y(s)$, $Z(s,t)$ for $t \in [R,S]$ and $s \in [S,T]$. Let $\mathfrak{z} \in H$ and define $f^S(t,s,\mathfrak{z}) = f(t,s,Y(s),\mathfrak{z},\mathfrak{z}(s,t))$, for $(t,s,\mathfrak{z}) \in [R,S] \times [S,T] \times H$. Then we can write

$$\int_S^T f(t,s,Y(s),\mathfrak{z},\mathfrak{z}(s,t))ds = \int_S^T f^S(t,s,\mathfrak{z})ds$$

and we consider the following SFIE

$$\varphi^S(t) = \varphi(t) + \int_S^T f^S(t,s,\mathfrak{z})ds - \int_S^T Z(t,s)dW^H(s).$$

By Proposition 4 we can find a unique solution of the above SFIE,

$$\begin{cases} \varphi(\cdot)^S \in L^p_{\mathcal{F}_S}(\Omega \times [R,S]; E) \\ \mathcal{Z}(\cdot,\cdot) \in L^p_{\mathbb{F}}\left(R,S; L^q(\mathbf{S}; L^2(S,T;H))\right). \end{cases}$$

It means that we found $Z(t,s)$ for $(t,s) \in [R,S] \times [S,T]$ and by combining the previous results we found the unique adapted M-solution $(Y(\cdot), Z(\cdot,\cdot))$, for $Y(t)$, $t \in [S,T]$ and $Z(t,s)$ for $(t,s) \in [S,T] \times [R,T] \bigcup [R,S] \times [S,T]$.

Step(4) By using the SFIE from Step(3), we can write our BSVIE as follows

$$Y(t) = \varphi^S(t) + \int_t^S f(t, s, Y(s), \mathfrak{Z}(t,s), \mathfrak{Z}(s,t))ds - \int_t^S Z(t,s)dW^H(s),$$

$$t \in [0, S].$$

From Step(3) we know that $\varphi^S(t)$ is $\mathcal{F}_S$-measurable for each $t \in [0, S]$ and this equation is a BSVIE and it could be easily solved as in the Steps (1)-(2)-(3) on the interval $[S', S]$. Finally, by induction we can find the unique adapted M-solution on the whole interval. $\qquad\square$

6. Duality Principles

In this section, duality principles between a linear forward stochastic Volterra integral equation and a linear backward stochastic Volterra integral equation are derived. The Itô formula in UMD Banach spaces was derived by Brzezniak, Neerven, Veraar and Weis in [6]. It is a crucial tool for the proof of the duality principle.

6.1. *A First Duality Principle*

We consider the following FSVIE in the Banach space $E = L^q(\mathbf{S}, \Sigma, \mu)$, where $q \geq 2$ and μ is a σ-finite measure

$$X(t) = \varphi(t) + \int_0^t A_0(t,s)X(s)ds + \int_0^t A_1(t,s)X(s)dW^H(s), \quad t \in [0,T],$$

$$(19)$$

where $W^H(\cdot)$ is an H-cylindrical Brownian motion, $\varphi(\cdot) \in L_{\mathbb{F}}^p([0,T] \times \Omega; E)$ and for each $(t, s) \in [0, T] \times [0, T]$ let $A_i(t, s)$, $i = 0, 1$ be linear bounded operators defined as follows

$$A_0(t,s) : \Omega \times E \longrightarrow E, \quad A_1(t,s) : \Omega \times E \longrightarrow \mathcal{L}(H; E)$$

$$\begin{cases} A_0(\cdot, \cdot) \in L^\infty\left(0, T; L_{\mathbb{F}}^\infty(0, T; \mathcal{L}(E; E))\right) \\ A_1(\cdot, \cdot) \in L^\infty\left(0, T; L_{\mathbb{F}}^\infty(0, T; \mathcal{L}(E; \mathcal{L}(H; E)))\right). \end{cases}$$

For each $t \in [0, T]$ $A_i(t, \cdot)$ is $\mathcal{F}_s$-adapted, $s \geq 0$, $\sup\limits_{t\in[0,T]} \sup\limits_{s\in[0,T]} \|A_i(t,s)\| < \infty$, $i = 0, 1$. According to these assumptions and by using Theorem 18 the FSVIE (19) admits a unique adapted solution $X(\cdot) \in L_{\mathbb{F}}^p([0,T] \times \Omega; E)$, where $q \leq p$.

Now we consider the following BSVIE in the dual space of E, i.e. $E^* = L^{q'}(\mathbf{S}, \Sigma, \mu)$, $1 < q' \leq 2$ and $\frac{1}{q} + \frac{1}{q'} = 1$; let $A_0^*(t, s)$ be the adjoint operator of $A_0(t,s)$ and $A_1^*(t,s)h := (A_1(t,s)h)^*$ be the adjoint operator of $A_1(t,s)h$ for every $(t,s) \in [0,T] \times [0,T]$ and $h \in H$; consider

$$
\begin{aligned}
Y(t) = & \psi(t) + \int_t^T \left(A_0^*(s,t)Y(s) + \sum_{n \geq 1} A_1^*(s,t)h_n Z(s,t)h_n \right) ds \\
& - \int_t^T Z(t,s)dW^H(s), \quad t \in [0,T].
\end{aligned} \tag{20}
$$

The H-cylindrical Wiener process $W^H(\cdot)$ is the same process for the two equations and is defined on the Hilbert space H. $\{h_n\}_{n \geq 1}$ is the orthonormal basis in H and for each $\omega \in \Omega$ and $(t,s) \in [0,T] \times [0,T]$

$$
Z(t,s) \in \mathcal{L}(H;E^*), \ A_0^* \in \mathcal{L}(E^*;E^*), \ A_1^*h \in \mathcal{L}(E^*;E^*),
$$

where $Z(t,s)$, A_0^* and A_1^*h are bounded linear operators and for each $h \in H$, $Z(t,s)h \in E^*$ and $A_1^*(t,s)h \in \mathcal{L}(E^*;E^*)$, then $A_1^*(t,s)hZ(t,s)h := A_1^*(t,s)h(Z(t,s)h) \in E^*$. We can find $\mathfrak{Z} : [0,T] \times [0,T] \times \Omega \times \mathbf{S} \longrightarrow H$ such that $(Z(t,s)h)(\cdot) = [\mathfrak{Z}(t,s,\cdot),h]_H$ is a function on $\mathbf{S}$ which belongs to E^* for all $h \in H$. We also assume that $\sum_{n \geq 1} \|A_1^*(s,t)h_n\|$ is bounded for every $(t,s) \in [0,T] \times [0,T]$. If $\psi(\cdot) \in L_{\mathcal{F}_T}^p([0,T] \times \Omega; E^*)$ by considering

$$
f(t,s,Y(s),\mathfrak{Z}(t,s),\mathfrak{Z}(s,t)) = A_0^*(s,t)Y(s) + \sum_{n \geq 1} A_1^*(s,t)h_n Z(s,t)h_n
$$

we can find a unique adapted M-solution $(Y(\cdot),Z(\cdot,\cdot))$ of (20).

For every $x^* \in E^*$ let the duality pairing be given by $x^*(x) = \langle x, x^* \rangle$.

Theorem 23. *If the FSVIE (19) and the BSVIE (20) are fulfilled, then the following duality principle holds*

$$
\mathbb{E} \left\{ \int_0^T \langle X(t), \psi(t) \rangle \, dt \right\} = \mathbb{E} \left\{ \int_0^T \langle \varphi(t), Y(t) \rangle \, dt \right\}.
$$

Proof. Since $Y(\cdot)$ is $\mathcal{F}_t$-adapted and $1 < q' \leq 2$, we can use the martingale representation theorem in Banach spaces, Theorem 16. Then, there exists a unique adapted process $Z(\cdot,\cdot)$ such that $Y(t) = \mathbb{E}(Y(t)) + \int_0^t Z(t,s)dW^H(s)$ and we obtain by using the definition for adjoint operators and the properties of the stochastic integral that

$$
\mathbb{E} \int_0^T \langle \varphi(t), Y(t) \rangle
$$

$$= \mathbb{E} \int_0^T \langle X(t), Y(t)\rangle dt - \mathbb{E} \int_0^T \int_t^T \langle X(t), A_0^*(s,t) Y(s)\rangle ds dt$$

$$- \mathbb{E} \int_0^T \left\langle \int_0^t A_1(t,s) X(s) dW^H(s), \int_0^t Z(t,s) dW^H(s) \right\rangle dt.$$

Now for the last term we use the Itô formula in Banach spaces by Brzezniak, Neerven, Veraar and Weis, Theorem 17.

Let for each (fixed) $t \in [0,T]$, $Q(r) = \int_0^r A_1(t,s) X(s) dW^H(s)$ in E and $V(r) = \int_0^r Z(t,s) dW^H(s)$ in E^*, where $r \in [0,t]$. Then, we observe by Theorem 17 that

$$\langle Q(r), V(r)\rangle - \langle Q(0), V(0)\rangle = \int_0^r \left(\langle Q(s), 0\rangle + \langle 0, V(s)\rangle \right) ds$$

$$+ \int_0^r \left(\langle Q(s), Z(t,s)\rangle + \langle A_1(t,s) X(s), V(s)\rangle \right) dW^H(s)$$

$$+ \int_0^r \sum_{n \geq 1} \langle (A_1(t,s) X(s))(h_n), (Z(t,s))(h_n)\rangle ds.$$

By taking expectation of the above equations for $r = t$ and by knowing that the expectation of the stochastic integral is zero, we have

$$\mathbb{E} \left\langle \int_0^t A_1(t,s) X(s) dW^H(s), \int_0^t Z(t,s) dW^H(s) \right\rangle$$

$$= \mathbb{E} \int_0^t \sum_{n \geq 1} \langle (A_1(t,s) X(s)) h_n, (Z(t,s)) h_n\rangle ds.$$

We write $(A_1(t,s) X(s))(h_n) = (A_1(t,s) h_n)(X(s)) = A_1(t,s) h_n X(s)$, since $X(\cdot) \in E$ and $A_1(t,s) \in \mathcal{L}(E; \mathcal{L}(H;E))$. By integrating the above result over $[0,T]$ with respect to the Lebesgue measure dt and by using Fubini's theorem and the property of the adjoint operator, it yields

$$\mathbb{E} \int_0^T \left\langle \int_0^t A_1(t,s) X(s) dW^H(s), \int_0^t Z(t,s) dW^H(s) \right\rangle dt$$

$$= \mathbb{E} \int_0^T \int_0^t \sum_{n \geq 1} \langle X(s), A_1^*(t,s) h_n Z(t,s) h_n\rangle ds dt.$$

Therefore, by substitution, by the adaptedness of $X(\cdot)$, by knowing that the expectation of the stochastic integral is zero and by Fubini's theorem, by Equation (20) and by using elementary calculus, it yields

$$\mathbb{E} \int_0^T \langle \varphi(t), Y(t)\rangle =$$

$$= \int_0^T \mathbb{E} \left\langle X(t), \psi(t) - \int_t^T Z(t,s)dW^H(s) \right\rangle dt$$

$$= \int_0^T \mathbb{E} \left\langle X(t), \psi(t) \right\rangle dt - \int_0^T \mathbb{E} \left\langle X(t), \mathbb{E} \left(\int_t^T Z(t,s)dW^H(s)|\mathcal{F}_t \right) \right\rangle dt$$

$$= \int_0^T \mathbb{E} \left\langle X(t), \psi(t) \right\rangle dt .$$

$\square$

6.2. A More General Duality Principle

Now we want to show another duality principle between the FSVIE (19) and another general BSVIE. Therefore, we consider the following BSVIE in the dual space E^*

$$Y(t) = \psi(t) + A_0^*(T,t)\eta + \sum_{n \geq 1} A_1^*(T,t)h_n \rho(t)h_n$$

$$+ \int_t^T \left(A_0^*(s,t)Y(s) + \sum_{n \geq 1} A_1^*(s,t)h_n Z(s,t)h_n \right) ds \qquad (21)$$

$$- \int_t^T Z(t,s)dW^H(s), \quad t \in [0,T] ,$$

where $\eta \in L_{\mathcal{F}_T}^{p'}(\Omega; E^*)$, $\frac{1}{p} + \frac{1}{p'} = 1$, $q \leq p$ and ρ is a unique adapted process that is defined by the martingale representation theorem as follows: since η is $\mathcal{F}_T$-measurable and $1 < q' \leq 2$, E^* has cotype 2, according to Theorem 16, we have the following representation result

$$\eta = \mathbb{E}(\eta) + \int_0^T \rho(t)dW^H(t) .$$

It means that we can find a unique $\rho(\cdot) \in L_{\mathbb{F}}^{p'} \left(\Omega; \gamma \left(L^2(0,T;H); E^* \right) \right)$ and correspondingly $\varrho(\cdot) \in L_{\mathbb{F}}^{p'} \left(\Omega; L^{q'} \left(\mathbf{S}; L^2(0,T;H) \right) \right)$ such that $(\rho(t)h)(\cdot) = [\varrho(t,\cdot),h]_H$ in E^*. Therefore, ρ is well defined and also Equation (21); by the assumptions it has a unique adapted M-solution

$$(Y(\cdot), Z(\cdot)) \in L_{\mathbb{F}}^{p'} \left([0,T] \times \Omega; E^* \right) \times L_{\mathbb{F}}^{p'} \left([0,T]; \gamma \left(L^2(0,T;H); E^* \right) \right)$$

or correspondingly

$$(Y(\cdot), \mathfrak{Z}(\cdot)) \in L_{\mathbb{F}}^{p'} \left([0,T] \times \Omega; E^* \right) \times L_{\mathbb{F}}^{p'} \left([0,T]; L^{q'} \left(\mathbf{S}; L^2(0,T;H) \right) \right) ,$$

where $(Z(t,s)h)(\cdot) = [\mathfrak{Z}(t,s,\cdot),h]_H$ in E^* .

Theorem 24. *If the FSVIE* (19) *and the BSVIE* (21) *are fulfilled, then the following duality holds for every* $\eta \in L^{p'}_{\mathcal{F}_T}(\Omega; E^*)$

$$\mathbb{E}\left\{ \langle X(T), \eta \rangle + \int_0^T \langle X(t), \psi(t) \rangle \, dt \right\} = \mathbb{E}\left\{ \langle \varphi(T), \eta \rangle + \int_0^T \langle \varphi(t), Y(t) \rangle \, dt \right\}.$$

Proof. We show only the idea of proof: We can write the BSVIE (21) as

$$Y(t) = \hat{\psi}(t) + \int_t^T \left(A_0^*(s,t)Y(s) + \sum_{n \geq 1} A_1^*(s,t)h_n Z(s,t)h_n \right) ds$$

$$- \int_t^T Z(t,s) dW^H(s),$$

where $\hat{\psi}(t) = \psi(t) + A_0^*(T,t)\eta + \sum_{n \geq 1} A_1^*(T,t)h_n \rho(t)h_n$. By Theorem 23 it results

$$\mathbb{E} \int_0^T \langle \varphi(t), Y(t) \rangle \, dt = \mathbb{E} \int_0^T \left\langle X(t), \hat{\psi}(t) \right\rangle dt$$

$$= \mathbb{E} \int_0^T \langle X(t), \psi(t) \rangle \, dt + \mathbb{E} \int_0^T \langle A_0(T,t)X(t), \eta \rangle \, dt$$

$$+ \mathbb{E} \int_0^T \sum_{n \geq 1} \langle A_0(T,t)h_n X(t), \rho(t)h_n \rangle \, dt.$$

$$(22)$$

The following equation, where the Itô formula in UMD Banach spaces, Theorem 17, is used, holds

$$\mathbb{E} \langle X(T) - \varphi(T), \eta \rangle =$$

$$\mathbb{E} \int_0^T \left\{ \langle A_0(T,t)X(t), \eta \rangle + \sum_{n \geq 1} \langle A_1(T,t)h_n X(t), \rho(t)h_n \rangle \right\} dt.$$

Now by replacing it into Equation (22) it yields

$$\mathbb{E} \int_0^T \langle \varphi(t), Y(t) \rangle \, dt = \mathbb{E} \int_0^T \left\langle X(t), \hat{\psi}(t) \right\rangle dt$$

$$= \mathbb{E} \int_0^T \langle X(t), \psi(t) \rangle \, dt + \mathbb{E} \langle X(T) - \varphi(T), \eta \rangle$$

and the statement of this theorem holds obviously. $\qquad \square$

Remark 14. It must be mentioned that the martingale representation theorem can be used since $L^{q'}$ spaces, $q' \leq 2$, have cotype 2. Our proof holds for every FSVIE (19) in every UMD space E, in the case when we can find unique solutions for the FSVIE (19). Second, to consider the BSVIE (21) in the dual space E^* we need a martingale representation theorem. This is known only in the case when E^* has cotype 2. If E has these properties, then we get also duality principles.

7. Optimality Conditions of Maximum Principle Type

We consider in this section an optimal control problem for a FSVIE in the Banach space E, where $E = L^q(\mathbf{S}, \Sigma, \mu)$, μ is a finite measure and the stochastic integral is defined with respect to an H-cylindrical Brownian motion. We prove an optimality condition of maximum principle type.

7.1. *Stochastic Optimal Control Problem*

Consider the following FSVIE in the Banach space $E = L^q(\mathbf{S}, \Sigma, \mu)$ $(q \geq 2)$

$$X(t) = \varphi(t) + \int_0^t b\left(t, s, X(s), u(s)\right) ds + \int_0^t \rho\left(t, s, X(s), u(s)\right) dW^H(s),$$

$$t \in [0, T]$$

where $X(\cdot)$, $u(\cdot)$ are the state and the control processes respectively, and $W^H(\cdot)$ is an H-cylindrical Brownian motion.

We set the following assumptions: $\varphi(\cdot) \in L^p_{\mathbb{F}}(\Omega \times [0, T]; E)$, where $q \leq p$, and b, ρ are measurable (measurability is defined similarly to (3)) as

$$b : \Omega \times [0, T] \times [0, T] \times E \times U \times \mathbf{S} \longrightarrow E$$
$$\rho : \Omega \times [0, T] \times [0, T] \times E \times U \times \mathbf{S} \longrightarrow \mathcal{L}(H; E)$$

$\varphi(\cdot)$ and $X(\cdot)$ are E-valued random variables and $u(\cdot)$ is a real valued process

$$u : [0, T] \times \Omega \times \mathbf{S} \longrightarrow U,$$

where U is a bounded closed interval of $\mathbb{R}$. We define

$$\mathcal{U} = \{u : [0, T] \times \Omega \times \mathbf{S} \longrightarrow U \mid u(\cdot) \text{ is } \mathbb{F}\text{-progressively measurable}\}.$$

Clearly we can interpret $\mathcal{U} \subset E$.

We define the following cost function in Bolza form

$$J(u(\cdot)) = \mathbb{E} \int_0^T \int_{\mathbf{S}} h(t, X(t), u(t)) d\mu dt + \mathbb{E} \int_{\mathbf{S}} g(X(T)) d\mu, \qquad (23)$$

where

$$h : \Omega \times [0, T] \times E \times U \times \mathbf{S} \longrightarrow E$$
$$g : \Omega \times E \times \mathbf{S} \longrightarrow E.$$

We use the concept of Nemytskii operator for b, h, g and we assume that their first derivatives with respect to x and u are continuous and bounded, in other words we require

$$\left| \frac{\partial b(t, s, x, u)}{\partial x} \right| := |b_x(t, s, x, u)| \leq K_x, \quad \left| \frac{\partial b(t, s, x, u)}{\partial u} \right| := |b_u(t, s, x, u)| \leq K_u$$

for all $t, s \in [0, T]$, $\eta \in \mathbf{S}$, for all $x \in \mathbb{R}, u \in U$ almost surely, where K_x, K_u are positive constants. Since $\rho \in \mathcal{L}(H; E)$ is a linear bounded operator for all $t, s \in [0, T]$, $\eta \in \mathbf{S}$, $\omega \in \Omega$, $x \in E$, $u \in \mathcal{U}$, we assume that it has continuous and bounded first Fréchet derivative with respect to x. By defintion $\rho_x(t, s, x, u) : E \longrightarrow \mathcal{L}(H; E)$ is a linear continuous operator from E to $\mathcal{L}(H; E)$. We also assume that it has first continuous Fréchet derivative with respect to u. Consequently $\rho_u(t, s, x, u) : \mathcal{U} \longrightarrow \mathcal{L}(H; E)$ is linear and continuous.

Since $\mathcal{U}$ is convex, for each $\varepsilon \in (0, 1)$, $u_\varepsilon(\cdot) \in \mathcal{U}$ where $u_\varepsilon(\cdot) := \overline{u}(\cdot) + \varepsilon(u(\cdot) - \overline{u}(\cdot))$ and let $(\overline{X}(\cdot), \overline{u}(\cdot))$ be an optimal state and control processes. Let $\overline{X}_\varepsilon(\cdot)$ be the solution of the following FSVIE, when $\overline{u}_\varepsilon(\cdot)$ is chosen as control process

$$\overline{X}_\varepsilon(t) = \varphi(t) + \int_0^t b\left(t, s, \overline{X}_\varepsilon(s), \overline{u}_\varepsilon(s)\right) ds + \int_0^t \rho\left(t, s, \overline{X}_\varepsilon(s), \overline{u}_\varepsilon(s)\right) dW^H(s),$$

$t \in [0, T]$ and we define $\xi_\varepsilon(t) := \frac{X_\varepsilon(t) - \overline{X}(t)}{\varepsilon}$ for each $t \in [0, T]$.

Lemma 4. *The process $\xi_\varepsilon(t)$ is uniformly bounded in $L^p(\Omega; E)$ with respect to t and ε and correspondingly $\xi_\varepsilon(\cdot)$ is uniformly bounded in $L^p(\Omega \times [0, T]; E)$ with respect to ε for $2 \leq q \leq p$.*

Proof. By using the definition of $\xi_\varepsilon(\cdot)$, substituting the corresponding FSVIE for $X(\cdot)$ and $\overline{X}(\cdot)$, applying the triangle inequality and Young's inequality it yields

$$\mathbb{E} \|\xi_\varepsilon(t)\|^p \leq \frac{c}{\varepsilon^p} \left\{ \mathbb{E} \left\| \int_0^t \left\{ b\left(t, s, X_\varepsilon(s), u_\varepsilon(s)\right) - b\left(t, s, \overline{X}(s), \overline{u}(s)\right) \right\} ds \right\|^p \right.$$

$$\left. + \mathbb{E} \left\| \int_0^t \left\{ \rho\left(t, s, X_\varepsilon(s), u_\varepsilon(s)\right) - \rho\left(t, s, \overline{X}(s), \overline{u}(s)\right) \right\} dW^H(s) \right\|^p \right\}.$$

Now we consider each summand from the above equation separately, namely A and B (the stochastic integral). For the first summand, A, since there exists the first derivative of $b(t, s, x, u)$ with respect to x, u for every $\eta \in \mathbf{S}$ and it is bounded μ-a.e., there exist $r_1 := r_1(t, s, x, u) \in (0, 1)$ and $r_2 := r_2(t, s, x, u) \in (0, 1)$ such that the following equations hold

$$b\left(t, s, X_\varepsilon(s), u_\varepsilon(s)\right) - b\left(t, s, \overline{X}(s), u_\varepsilon(s)\right)$$
$$= b_x\left(t, s, \overline{X}(s) + r_1(X_\varepsilon(s) - \overline{X}(s)), u_\varepsilon(s)\right)\left(X_\varepsilon(s) - \overline{X}(s)\right)$$

and

$$b\left(t, s, \overline{X}(s), u_\varepsilon(s)\right) - b\left(t, s, \overline{X}(s), \overline{u}(s)\right)$$
$$= b_x\left(t, s, \overline{X}(s), \overline{u}(s) + r_2(u_\varepsilon(s) - \overline{u}(s))\right)\left(u_\varepsilon(s) - \overline{u}(s)\right).$$

For simplicity we set $\widetilde{X}_\varepsilon(s) := \overline{X}(s) + r_1(X_\varepsilon(s) - \overline{X}(s))$ and $\widetilde{u}_\varepsilon(s) := \overline{u}(s) + r_2(u_\varepsilon(s) - \overline{u}(s))$. Then, by the boundedness of the derivatives and by the assumptions, K_x, K_u are positive upper bounds for the derivatives, which do not depend on η, t, s, w. By using Jensen's inequality for the measure ds and by the boundedness of $\mathcal{U}$ and by $p > 1$, we have

$$A \leq \frac{c}{\varepsilon^p}\left\{\mathbb{E}\left\|\int_0^t K_x\left|X_\varepsilon(s) - \overline{X}(s)\right| ds + \int_0^t K_u\left|u_\varepsilon(s) - \overline{u}(s)\right| ds\right\|^p\right\}$$
$$\leq ct^{p-1}\left\{\mathbb{E}\int_0^t\left\|\frac{X_\varepsilon(s) - \overline{X}(s)}{\varepsilon}\right\|^p ds + \mathbb{E}\int_0^t\left\|u(s) - \overline{u}(s)\right\|^p ds\right\}.$$

For the stochastic integral part, B, by using the L^p-stochastic integrability in L^q spaces, Theorem 12, it results

$$B \leq \frac{c}{\varepsilon^p}\mathbb{E}\left\|\left(\int_0^t\left\|\varrho\left(t, s, X_\varepsilon(s), u_\varepsilon(s)\right) - \varrho(t, s, \overline{X}(s), \overline{u}(s))\right\|_H^2 ds\right)^{\frac{1}{2}}\right\|^p,$$

where $(\rho(t, s, x, u)h)(\cdot) = [\varrho(t, s, x, u, \cdot), h]_H$ by Theorem 13 and Theorem 12. Since we assumed $2 \leq q \leq p$, we can apply Jensen's inequality, which yields

$$B \leq \frac{c}{\varepsilon^p}\mathbb{E}\left(\int_{\mathbf{S}}\left(\int_0^t\left\|\varrho\left(t, s, X_\varepsilon(s), u_\varepsilon(s)\right) - \varrho(t, s, \overline{X}(s), \overline{u}(s))\right\|_H^2 ds\right)^{\frac{q}{2}} d\mu\right)^{\frac{p}{q}}$$
$$\leq \frac{c}{\varepsilon^p}t^{\frac{p}{2}-1}\mathbb{E}\int_0^t\left\|\left\|\varrho(t, s, X_\varepsilon(s), u_\varepsilon(s)) - \varrho(t, s, \overline{X}(s), \overline{u}(s))\right\|_H\right\|^p ds.$$

By the definition of $\widetilde{X}_\varepsilon(s)$ and $\widetilde{u}_\varepsilon(s)$ there exist $r_1, r_2 \in (0, 1)$ such that we can write

$$\varrho\left(t, s, X_\varepsilon(s), u_\varepsilon(s)\right) - \varrho\left(t, s, \overline{X}(s), \overline{u}(s)\right)$$

$$= \left(\varrho\left(t, s, X_\varepsilon(s), u_\varepsilon(s)\right) - \varrho\left(t, s, \overline{X}(s), u_\varepsilon(s)\right)\right)$$
$$\quad + \varrho\left(t, s, \overline{X}(s), u_\varepsilon(s)\right) - \varrho\left(t, s, \overline{X}(s), \overline{u}(s)\right)$$
$$= \varrho_x(t, s, \widetilde{X}_\varepsilon(s), u_\varepsilon(s))(X_\varepsilon(s) - \overline{X}(s)) + \varrho_u\left(t, s, \overline{X}(s), \widetilde{u}_\varepsilon(s)\right)(u_\varepsilon(s) - \overline{u}(s)).$$

By substituting this equation into B, it results

$$B \le c\left\{ \mathbb{E} \int_0^t \left\|\left\| \varrho_x(t, s, \widetilde{X}_\varepsilon(s), u_\varepsilon(s)) \xi_\varepsilon(s) \right\|_H \right\|^p ds \right.$$
$$\left. + \mathbb{E} \int_0^t \left\|\left\| \varrho_u(t, s, \overline{X}(s), \widetilde{u}_\varepsilon(s))(u(s) - \overline{u}(s)) \right\|_H \right\|^p ds \right\}.$$

Now we consider $\varrho_x(t, s, \widetilde{X}_\varepsilon(s), u_\varepsilon(s))$. We have for every $\eta \in \mathbf{S}$ and $h \in \{h_n\}_{n\ge 1}$

$$\left(\rho_x(t, s, \widetilde{X}_\varepsilon(s), u_\varepsilon(s)) \xi_\varepsilon(s) h \right)(\eta) = \left[\left(\varrho_x(t, s, \widetilde{X}_\varepsilon(s), u_\varepsilon(s)) \xi_\varepsilon(s) \right)(\eta), h \right]_H.$$

Consequently, ϱ_x defines also a linear operator from E to H and it is a function with respect to η, that belongs to E too. We can consider this operator as a Nemytskii operator for every $\eta \in \mathbf{S}$, this means that $\varrho_x(t, s, \widetilde{X}_\varepsilon(s, \eta), u_\varepsilon(s, \eta), \eta) \xi_\varepsilon(s, \eta)$ for every fixed t, s and η is an element of $\mathbb{R}$, therefore we have the following norm inequality

$$\left\| \varrho_x(t, s, \widetilde{X}_\varepsilon(s, \eta), u_\varepsilon(s, \eta), \eta) \xi_\varepsilon(s, \eta) \right\|_H \le$$
$$\left\| \varrho_x(t, s, \widetilde{X}_\varepsilon(s, \eta), u_\varepsilon(s, \eta), \eta) \right\|_{\mathcal{L}(\mathbb{R};H)} |\xi_\varepsilon(s, \eta)|.$$

By a similar explanation we have the following norm inequality for $\varrho_u(t, s, \overline{X}(s), \widetilde{u}_\varepsilon(s))$

$$\left\| \varrho_u(t, s, \overline{X}(s, \eta), \widetilde{u}_\varepsilon(s, \eta), \eta)(u(s, \eta) - \overline{u}(s, \eta)) \right\|_H$$
$$\le \left\| \varrho_u(t, s, \overline{X}(s, \eta), \widetilde{u}_\varepsilon(s, \eta), \eta) \right\|_{\mathcal{L}(\mathbb{R};H)} |u(s, \eta) - \overline{u}(s, \eta)|.$$

It must be mentioned that $\xi_\varepsilon(s)$, $(u(s) - \overline{u}(s))$ are elements of E. By replacing the above calculations in B and by the assumptions, the expressions $\left\| \varrho_x(t, s, \widetilde{X}_\varepsilon(s, \eta), u_\varepsilon(s, \eta), \eta) \right\|_{\mathcal{L}(\mathbb{R};H)}$ and $\left\| \varrho_u(t, s, \overline{X}(s, \eta), \widetilde{u}_\varepsilon(s, \eta), \eta) \right\|_{\mathcal{L}(\mathbb{R};H)}$ are uniformly a.s. bounded. Finally, for B, it yields

$$B \le c\left\{ \mathbb{E} \int_0^t \|\xi_\varepsilon(s)\|^p ds + \mathbb{E} \int_0^t \|(u(s) - \overline{u}(s))\|^p ds \right\}.$$

By combining the inequalities for A and B, we obtain

$$\mathbb{E} \|\xi_\varepsilon(t)\|^p \le c\left\{ \mathbb{E} \int_0^t \|\xi_\varepsilon(s)\|^p ds + \mathbb{E} \int_0^t \|u(s) - \overline{u}(s)\|^p ds \right\}, \quad \text{for all } t \in [0, T]$$

where C is a universal constant. By denoting $v(t) = \mathbb{E} \|\xi_\varepsilon(t)\|^p$, $\alpha(t) = C\mathbb{E} \int_0^t \|u(s) - \overline{u}(s)\|^p \, ds$, $F = C$, we have

$$v(t) \le \alpha(t) + F \int_0^t v(s) ds, \quad 0 \le t \le T.$$

Then, by Gronwall's inequality, it follows $v(t) \le \alpha(t) \exp(Ft)$, $0 \le t \le T$; since $u(\cdot), \overline{u}(\cdot) \in \mathcal{U}$ and $\mathcal{U}$ is bounded, $\alpha(t)$ is bounded by some positive constant C too. It results

$$\mathbb{E} \left\| \frac{X_\varepsilon(t) - \overline{X}(t)}{\varepsilon} \right\|^p = \mathbb{E} \|\xi_\varepsilon(t)\|^p \le C \exp(ct) \le K, \quad 0 \le t \le T,$$

where K is a positive constant. Moreover, it yields

$$\mathbb{E} \int_0^T \left\| \frac{X_\varepsilon(t) - \overline{X}(t)}{\varepsilon} \right\|^p dt = \mathbb{E} \int_0^T \|\xi_\varepsilon(t)\|^p \, dt \le KT \text{ for all } \varepsilon \ge 0.$$

It results that $\mathbb{E} \int_0^T \|\xi_\varepsilon(t)\|^p \, dt$ is uniformly bounded with respect to ε. $\quad\square$

Lemma 5. *If ε tends to zero, then $\xi_\varepsilon(\cdot)$ tends to $\xi(\cdot)$ in $L^p([0,T] \times \Omega; E)$ and $\xi_\varepsilon(t)$ tends to $\xi(t)$ in $L^p(\Omega; E)$ for all $t \in [0,T]$, where $\xi(t)$ satisfies the following FSVIE*

$$\xi(t) = \int_0^t \left\{ b_x \left(t, s, \overline{X}(s), \overline{u}(s) \right) \xi(s) + b_u \left(t, s, \overline{X}(s), \overline{u}(s) \right) (u(s) - \overline{u}(s)) \right\} ds$$

$$+ \int_0^t \left\{ \rho_x \left(t, s, \overline{X}(s), \overline{u}(s) \right) \xi(s) + \rho_u \left(t, s, \overline{X}(s), \overline{u}(s) \right) (u(s) - \overline{u}(s)) \right\} dW^H(s),$$

$0 \le t \le T.$

Proof. Sketch of the proof: similar to the previous lemma and using the dominated convergence theorem, it results

$$\mathbb{E} \|\xi_\varepsilon(t) - \xi(t)\|^p \le cK\mathbb{E} \int_0^t \|\xi_\varepsilon(s) - \xi(s)\|^p \, ds + cF_\varepsilon(t),$$

where

$$F_\varepsilon(t) = \mathbb{E} \left\| \int_0^t \left(b_u \left(t, s, \overline{X}(s), \widetilde{u}_\varepsilon(s) \right) - b_u \left(t, s, \overline{X}(s), \overline{u}(s) \right) \right) (u(s) - \overline{u}(s)) \, ds \right\|^p$$

$$+ \mathbb{E} \int_0^t \left\| \left\| \varrho_u \left(t, s, \overline{X}(s), \widetilde{u}_\varepsilon(s) \right) - \varrho_u \left(t, s, \overline{X}(s), \overline{u}(s) \right) \right\|_{\mathcal{L}(\mathbb{R};H)} |u(s) - \overline{u}(s)| \right\|_E^p ds$$

$$+ \mathbb{E} \int_0^t \left\| \left\{ b_x \left(t, s, \widetilde{X}_\varepsilon(s), u_\varepsilon(s) \right) - b_x \left(t, s, \overline{X}(s), u_\varepsilon(s) \right) \right\} \xi(s) \right\|^p ds$$

$$+ \mathbb{E} \int_0^t \left\| \left\{ b_x \left(t, s, \overline{X}(s), u_\varepsilon(s) \right) - b_x \left(t, s, \overline{X}(s), \overline{u}(s) \right) \right\} \xi(s) \right\|^p ds$$

$$+ \mathbb{E} \int_0^t \left\| \left\| \varrho_x \left(t, s, \tilde{X}_\varepsilon(s), u_\varepsilon(s) \right) - \varrho_x \left(t, s, \overline{X}(s), u_\varepsilon(s) \right) \right\|_{\mathcal{L}(\mathbb{R};H)} |\xi(s)| \right\|_E^p ds$$

$$+ \mathbb{E} \int_0^t \left\| \left\| \varrho_x \left(t, s, \overline{X}(s), u_\varepsilon(s) \right) - \varrho_x \left(t, s, \overline{X}(s), \overline{u}(s) \right) \right\|_{\mathcal{L}(\mathbb{R};H)} |\xi(s)| \right\|_E^p ds .$$

We denote $v(t) = \mathbb{E} \left\| \xi_\varepsilon(t) - \xi(t) \right\|^p$, $\alpha(t) = c F_\varepsilon(t)$, $A = cK$ and have

$$v(t) \leq \alpha(t) + A \int_0^t v(s) ds, \quad 0 \leq t \leq T .$$

Then, by using Gronwall's inequality, we get

$$v(t) \leq F_\varepsilon(t) \exp \left\{ \int_0^t cK ds \right\}, \quad 0 \leq t \leq T ,$$

i.e.

$$\mathbb{E} \left\| \xi_\varepsilon(t) - \xi(t) \right\|^p \leq F_\varepsilon(t) e^{cKt}$$

for each $t \in [0, T]$. If ε tends to zero, then $F_\varepsilon(t)$ converges to zero, and it results that $\xi_\varepsilon(\cdot)$ tends to $\xi(\cdot)$ in $L^p ([0, T] \times \Omega; E)$. $\qquad \Box$

Now we consider our control problem. Since $E = L^q(\mathbf{S}, \Sigma, \mu)$ is reflexive, it is a Radon-Nikodyn space (see [8]) and the dual space of $L^p ([0, T] \times \Omega; L^q (\mathbf{S}, \Sigma, \mu))$ is $L^{p'} \left([0, T] \times \Omega; L^{q'} (\mathbf{S}, \Sigma, \mu) \right)$, where $\frac{1}{p} + \frac{1}{p'} = 1$, $\frac{1}{q} + \frac{1}{q'} = 1$. For every $f \in L^p ([0, T] \times \Omega; E)$ and $g \in L^{p'} ([0, T] \times \Omega; E^*)$, where E^* is dual space of E, the duality paring is given by

$$\langle f(\cdot), g(\cdot) \rangle_{L^{p'} ([0,T] \times \Omega; E^*)} = \mathbb{E} \int_0^T \langle f(t), g(t) \rangle_{E^*} dt = \mathbb{E} \int_0^T \int_{\mathbf{S}} f(t, \cdot) g(t, \cdot) d\mu dt .$$

Based on Lemmas 4, 5 and since $g_x(T) \in L^{p'} (\Omega; E^*)$, where $\frac{1}{p} + \frac{1}{p'} = 1$ and $E^* = L^{q'} (\mathbf{S}, \Sigma, d\mu)$, $\frac{1}{q} + \frac{1}{q'} = 1$, we obtain the following result:

Lemma 6. *If ε tends to zero, then*

$$\lim_{\varepsilon \to 0} \left| \frac{J (u_\varepsilon(\cdot)) - J (\overline{u}(\cdot))}{\varepsilon} \right.$$

$$- \mathbb{E} \int_0^T \int_{\mathbf{S}} \left\{ h_x \left(t, \overline{X}(t), \overline{u}(t) \right) \xi(t) + h_u \left(t, \overline{X}(t), \overline{u}(t) \right) (u(t) - \overline{u}(t)) \right\} d\mu dt$$

$$\left. - \mathbb{E} \int_{\mathbf{S}} \left(g_x \left(\overline{X}(T) \right) \right) (\xi(T)) d\mu \right| = 0 .$$

We know that $\xi(t)$ solves the FSVIE in the space $E = L^q\left(\mathbf{S}, \Sigma, \mu\right)$ and it has the property $\xi(\cdot) \in L^p\left([0,T] \times \Omega; E\right)$. Since $h_x\left(t, \overline{X}(\cdot), \overline{u}(\cdot)\right)$ is bounded, then for every $\overline{u}(\cdot) \in \mathcal{U}$, $h_x\left(\cdot, \cdot, \overline{u}(\cdot)\right) \in L^{p'}\left([0,T] \times \Omega; E^*\right)$ and by similar arguments it results that $g_x(\cdot), h_u(\cdot, \cdot, \overline{u}(\cdot)) \in L^{p'}\left([0,T] \times \Omega; E^*\right)$. Since $(\overline{u}(\cdot), \overline{X}(\cdot))$ is an optimal pair, by Lemma 6, we can write

$$0 \leq \frac{J\left(u_\varepsilon(\cdot)\right) - J\left(\overline{u}(\cdot)\right)}{\varepsilon} \xrightarrow{\varepsilon \to 0} \mathbb{E}\left\langle \xi(T), g_x\left(\overline{X}(T)\right)\right\rangle$$

$$+ \mathbb{E}\left\{\int_0^T \left(\left\langle \xi(t), h_x\left(t, \overline{X}(t), \overline{u}(t)\right)\right\rangle + \left\langle u(t) - \overline{u}(t), h_u\left(t, \overline{X}(t), \overline{u}(t)\right)\right\rangle\right) dt\right\}.$$

It follows from Lemma 5 that

$$\xi(t) = \overline{\varphi}(t) + \int_0^t b_x\left(t, s, \overline{X}(s), \overline{u}(s)\right)\xi(s)ds$$

$$+ \int_0^t \rho_x\left(t, s, \overline{X}(s), \overline{u}(s)\right)\xi(s)dW^H(s), \quad t \in [0,T],$$

where

$$\overline{\varphi}(t) = \int_0^t b_u\left(t, s, \overline{X}(s), \overline{u}(s)\right)\left(u(s) - \overline{u}(s)\right)ds$$

$$+ \int_0^t \rho_u\left(t, s, \overline{X}(s), \overline{u}(s)\right)\left(u(s) - \overline{u}(s)\right)dW^H(s).$$

(24)

By the boundedness assumptions we can define the following bounded linear operators

$$B_u(t,s)u(s) = b_u\left(t, s, \overline{X}(s), \overline{u}(s)\right)u(s), \quad B_x(t,s)\xi(s) = b_x\left(t, s, \overline{X}(s), \overline{u}(s)\right)\xi(s),$$

$$A_u(t,s)u(s) = \rho_u\left(t, s, \overline{X}(s), \overline{u}(s)\right)u(s), \quad A_x(t,s)\xi(s) = \rho_x\left(t, s, \overline{X}(s), \overline{u}(s)\right)\xi(s),$$

where for every $(t,s) \in [0,T] \times [0,T]$

$$B_u(t,s) \in \mathcal{L}\left(E; E\right), \quad B_x(t,s) \in \mathcal{L}\left(E; E\right),$$

$$A_u(t,s) \in \mathcal{L}\left(E; \mathcal{L}\left(H; E\right)\right), \quad A_x(t,s) \in \mathcal{L}\left(E; \mathcal{L}\left(H; E\right)\right).$$

We can define the adjoint operators and similar to Section 6 we let $A_x^*\left(\cdot, \cdot\right)h_n := \left(A_x\left(\cdot, \cdot\right)h_n\right)^*$ and $A_x^*\left(\cdot, \cdot\right)h_n := \left(A_u\left(\cdot, \cdot\right)h_n\right)^*$.

Theorem 25. *If* $(\overline{X}(\cdot), \overline{u}(\cdot))$ *is an optimal solution and*

$$\sum_{n \geq 1}\left\|A_x\left(t, s\right)h_n\right\|_{\mathcal{L}(E;E)}, \quad \sum_{n \geq 1}\left\|A_u\left(t, s\right)h_n\right\|_{\mathcal{L}(E;E)}$$

are bounded a.s. for every $t, s \in [0, T] \times [0, T]$, *then there exists a unique adapted M-solution* $(Y(\cdot), Y_0(\cdot), v(\cdot); Z(\cdot, \cdot), Z_0(\cdot, \cdot), \zeta(\cdot))$ *of the following BSVIEs:*

$$Y(t) = h_x\left(t, \overline{X}(t), \overline{u}(t)\right) + B_x^*\left(T, t\right) g_x\left(\overline{X}(T)\right) + \sum_{n \geq 1} A_x^*\left(T, t\right) h_n \zeta(t) h_n$$

$$+ \int_t^T \left\{ B_x^*\left(s, t\right) Y(s) + \sum_{n \geq 1} A_x^*\left(s, t\right) h_n Z(s, t) h_n \right\} ds$$

$$- \int_t^T Z(t, s) dW^H(s),$$

$$v(t) = g_x\left(\overline{X}(T)\right) - \int_t^T \zeta(s) dW^H(s),$$

$$Y_0(t) = B_u^*\left(T, t\right) g_x\left(\overline{X}(T)\right) + \sum_{n \geq 1} A_u^*\left(T, t\right) h_n \zeta(t) h_n$$

$$+ \int_t^T \left\{ B_u^*\left(s, t\right) Y(s) + \sum_{n \geq 1} A_u^*\left(s, t\right) h_n Z(s, t) h_n \right\} ds$$

$$- \int_t^T Z_0(t, s) dW^H(s), \qquad t \in [0, T],$$

where $Y(\cdot), Y_0(\cdot), v(\cdot) \in L_{\mathbb{F}}^p(\Omega \times [0, T]; E)$ *and*
$Z(\cdot, \cdot), Z_0(\cdot, \cdot), \zeta(\cdot) \in L_{\mathbb{F}}^p\left(\Omega \times [0, T]; \gamma\left(L^2(0, T; H), E\right)\right)$ *such that*

$$\left\langle u(t) - \overline{u}(t), Y_0(t) + h_u\left(t, \overline{X}(t), \overline{u}(t)\right) \right\rangle \geq 0, \text{ for all } u(\cdot) \in \mathcal{U}, t \in [0, T] \text{ a.s.}$$

Proof. By the boundedness of $\mathcal{U}$, it results that $\mathbb{E} \left\| \overline{\varphi}(t) \right\|^p$ is bounded with respect to $t \in [0, T]$, where $\overline{\varphi}(t)$ is defined by Equation (24). Since $g_x(\cdot)$ is bounded, we have $\mathbb{E} \left\| g_x\left(\overline{X}(T)\right) \right\|_{E^*} < \infty$, and by the assumptions we know that $g_x\left(\overline{X}(T)\right)$ is $\mathcal{F}_T$-measurable, therefore we can use the martingale representation theorem in Banach spaces, Theorem 16 ($E = L^q\left(\mathbf{S}, \Sigma, \mu\right)$, $2 \leq q \leq p$, hence $E^* = L^{q'}\left(\mathbf{S}, \Sigma, \mu\right)$, $1 < p' \leq q' \leq 2$ and E^* has cotype 2). We can find a unique adapted process $\zeta(\cdot)$ such that for every $r \in [0, T]$

$$\gamma(r) = \mathbb{E}\left(g_x\left(\overline{X}(T)\right) | \mathcal{F}_r\right) = \mathbb{E}\left(g_x\left(\overline{X}(T)\right)\right) + \int_0^r \zeta(s) dW^H(s)$$

and from the above equation it results

$$g_x\left(\overline{X}(T)\right) - \int_t^T \zeta(s) dW^H(s) = \mathbb{E}\left(g_x\left(\overline{X}(T)\right)\right) + \int_0^t \zeta(s) dW^H(s).$$

Let $v(t) := \mathbb{E}\left(g_x\left(\overline{X}(T)\right)\right) + \int_0^t \zeta(s)dW^H(s)$. So we can define the BSVIE in $E^* = L^{q'}(\mathbf{S}, \Sigma, \mu)$ as follows

$$Y(t) = h_x\left(t, \overline{X}(t), \overline{u}(t)\right) + B_x^*\left(T, t\right)g_x\left(\overline{X}(T)\right) + \sum_{n \geq 1} A_x^*\left(T, t\right)h_n\zeta(t)h_n$$

$$+ \int_t^T \left\{ B_x^*\left(s, t\right)Y(s) + \sum_{n \geq 1} A_x^*\left(s, t\right)h_n Z(s, t)h_n \right\} ds$$

$$- \int_t^T Z(t, s)dW^H(s), \qquad t \in [0, T].$$

Let

$$\hat{\psi}(t) = h_x\left(t, \overline{X}(t), \overline{u}(t)\right) + B_x^*\left(T, t\right)g_x\left(\overline{X}(T)\right) + \sum_{n \geq 1} A_x^*\left(T, t\right)h_n\zeta(t)h_n,$$

$t \in [0, T]$. By the assumptions it holds $\mathbb{E}\left\|\hat{\psi}(t)\right\|_{E^*}^{p'} < \infty$ for each t in $[0, T]$. Then, the BSVIE has a unique M-adapted solution

$$(Y(\cdot), Z(\cdot, \cdot)) \in L_{\mathbb{F}}^{p'}\left([0, T] \times \Omega; E^*\right) \times L_{\mathbb{F}}^{p'}\left([0, T]; L^{q'}\left(\mathbf{S}; L^2(0, T; H)\right)\right).$$

Now we can apply the duality principle from Theorem 14 and we have the following relation

$$\mathbb{E}\left\{ \left\langle \xi(T), g_x\left(\overline{X}(T)\right) \right\rangle + \int_0^T \left\langle \xi(t), h_x\left(t, \overline{X}(t), \overline{u}(t)\right) \right\rangle dt \right\}$$

$$= \mathbb{E}\left\{ \left\langle \overline{\varphi}(T), g_x\left(\overline{X}(T)\right) \right\rangle + \int_0^T \left\langle \overline{\varphi}(t), Y(t) \right\rangle dt \right\}.$$

We replace $\overline{\varphi}(T)$ and $\overline{\varphi}(t)$ from the FSVIE, and we denote the summands in the last term by F_1 and F_2. By using adjoint operators and Itô's formula in Banach spaces (see Theorem 17) it yields for F_1

$$F_1 = \mathbb{E}\left\langle \xi(T), g_x\left(\overline{X}(T)\right) \right\rangle$$

$$= \mathbb{E}\int_0^T \left\langle u(t) - \overline{u}(t), B_u^*(T, t)g_x\left(\overline{X}(T)\right) + \sum_{n \geq 1} A_u^*(T, t)h_n\zeta(t)h_n \right\rangle dt.$$

Now we consider the term F_2. For the following calculations, we use the properties of the adjoint operators, the martingale representation theorem for $Y(t)$ (see Theorem 16), Itô's formula in Banach spaces (see Theorem 17) and the properties of the stochastic integral to obtain

$$F_2 = \mathbb{E}\int_0^T \left\langle \overline{\varphi}(t), Y(t) \right\rangle dt$$

$$= \mathbb{E} \int_0^T \left\langle \int_0^t b_u\left(t, s, \overline{X}(s), \overline{u}(s)\right)(u(s) - \overline{u}(s))\, ds, Y(t) \right\rangle dt$$

$$+ \mathbb{E} \int_0^T \left\langle \int_0^t \rho_u\left(t, s, \overline{X}(s), \overline{u}(s)\right)(u(s) - \overline{u}(s))\, dW^H(s), \right.$$

$$\left. \mathbb{E}(Y(t) + \int_0^t Z(t,s) dW^H(s) \right\rangle dt.$$

Moreover, it results that F_2 is equal to

$$\mathbb{E} \int_0^T \left\langle u(t) - \overline{u}(t), \int_t^T \left\{ B_u^*(s,t)Y(s) + \sum_{n \geq 1} A_u^*(s,t) h_n Z(s,t) h_n \right\} ds \right\rangle dt.$$

By combining the above two results for F_1 and F_2, we have

$$\mathbb{E} \left\{ \langle \overline{\varphi}(T), g_x\left(\overline{X}(T)\right) \rangle + \int_0^T \langle \overline{\varphi}(t), Y(t) \rangle\, dt \right\}$$

$$= \mathbb{E} \int_0^T \left\langle u(t) - \overline{u}(t), B_u^*(T,t) g_x\left(\overline{X}(T)\right) + \sum_{n \geq 1} A_u^*(T,t) h_n \zeta(t) h_n \right.$$

$$\left. + \int_t^T \left\{ B_u^*(s,t)Y(s) + \sum_{n \geq 1} A_u^*(s,t) h_n Z(s,t) h_n \right\} ds \right\rangle dt.$$

We define another simple BSVIE in E^* as follows

$$Y_0(t) = B_u^*(T,t)\, g_x\left(\overline{X}(T)\right) + \sum_{n \geq 1} A_u^*(T,t)\, h_n \zeta(t) h_n$$

$$+ \int_t^T \left\{ B_u^*(s,t)\, Y(s) + \sum_{n \geq 1} A_u^*(s,t)\, h_n Z(s,t) h_n \right\} ds$$

$$- \int_t^T Z_0(t,s) dW^H(s), \qquad t \in [0,T].$$

It is well defined and there exists a unique M-solution

$$(Y_0(\cdot), Z_0(\cdot, \cdot)) \in L_{\mathbb{F}}^{p'}([0,T] \times \Omega; E^*) \times L_{\mathbb{F}}^{p'}\left([0,T]; L^{q'}\left(\mathbf{S}; L^2(0,T;H)\right)\right).$$

Finally, we consider again the following relation by the minimality of the cost function with respect to the control process $\overline{u}(\cdot)$

$$0 \leq \frac{J\left(u_\varepsilon(\cdot)\right) - J\left(\overline{u}(\cdot)\right)}{\varepsilon} \xrightarrow{\varepsilon \to 0} \mathbb{E} \left\{ \langle \xi(T), g_x\left(\overline{X}(T)\right) \rangle \right.$$

$$+ \int_0^T \left(\left\langle \xi(t), h_x\left(t, \overline{X}(t), \overline{u}(t)\right) \right\rangle + \left\langle u(t) - \overline{u}(t), h_u\left(t, \overline{X}(t), \overline{u}(t)\right) \right\rangle \right) dt \right\}$$

$$= \mathbb{E} \left\{ \int_0^T \left\langle u(t) - \overline{u}(t), Y_0(t) + \int_t^T Z_0(t,s) dW^H(s) + h_u\left(t, \overline{X}(t), \overline{u}(t)\right) \right\rangle dt \right\}$$

$$= \mathbb{E} \left\{ \int_0^T \left\langle u(t) - \overline{u}(t), Y_0(t) + h_u\left(t, \overline{X}(t), \overline{u}(t)\right) \right\rangle dt \right\}, \ \forall u(\cdot) \in \mathcal{U}.$$

Therefore it results

$$\left\langle u(t) - \overline{u}(t), Y_0(t) + h_u\left(t, \overline{X}(t), \overline{u}(t)\right) \right\rangle \geq 0, \ \text{for all } u(\cdot) \in \mathcal{U}, t \in [0, T] \text{ a.s.}$$

$\square$

If we consider instead of the cost function (23) the cost function

$$J(u(\cdot)) = \mathbb{E} \int_0^T \int_S h(t, X(t), u(t)) d\mu dt, \tag{25}$$

then we can prove the following optimality condition with similar considerations by using of the duality principle from Theorem 13:

Theorem 26. *If* $(\overline{X}(\cdot), \overline{u}(\cdot))$ *is an optimal solution with the cost function* (25) *and* $\sum_{n \geq 1} \|A_x(t, s) h_n\|_{\mathcal{L}(E;E)}$, $\sum_{n \geq 1} \|A_u(t, s) h_n\|_{\mathcal{L}(E;E)}$ *are bounded a.s. for every* $t, s \in [0, T] \times [0, T]$, *then there exists a unique adapted M-solution* $(Y(\cdot), Z(\cdot, \cdot))$ *of the following BSVIE*

$$Y(t) = - h_x\left(t, \overline{X}(t), \overline{u}(t)\right)$$

$$+ \int_t^T \left\{ B_x^*(s, t) Y(s) + \sum_{n \geq 1} A_x^*(s, t) h_n Z(s, t) h_n \right\} ds$$

$$- \int_t^T Z(t, s) dW^H(s),$$

where
$Y(\cdot) \in L_{\mathbb{F}}^p(\Omega \times [0, T]; E)$ *and* $Z(\cdot, \cdot) \in L_{\mathbb{F}}^p\left(\Omega \times [0, T]; \gamma\left(L^2(0, T; H), E\right)\right)$
such that

$$\left\langle u(t) - \overline{u}(t), \right.$$

$$\mathbb{E}\left(\int_t^T \left\{ B_u^*(s, t) Y(s) + \sum_{n \geq 1} A_u^*(s, t) h_n Z(s, t) h_n \right\} ds \middle| \mathcal{F}_t \right) - h_u(t, \overline{X}(t), \overline{u}(t)) \right\rangle$$

$$\leq 0, \ \text{for all } u(\cdot) \in \mathcal{U}, \ t \in [0, T] \text{ a.s.}$$

7.2. *An Optimal Control Problem for a Heat Equation with Multiplicative Noise*

In this section we consider an example for a stochastic optimal control problem by using a maximum principle. We use the form of stochastic heat equation with homogeneous Dirichlet boundary conditions in the Banach space $E = L^q(0,1)$, $q \geq 2$. Let the following stochastic heat equation be controlled by $u(\cdot, \cdot)$

$$\begin{cases} dX(t,\xi) = \Delta X(t,\xi)dt + u(t,\xi)dt + \psi X(t,\xi)dW^H(t), \ t \in [0,T], \xi \in (0,1) \\ X(0,\xi) = X_0(\xi), \ \xi \in (0,1), X_0(\cdot) \in L^p(E) \\ X(t,0) = X(t,1) = 0, \ t \in [0,T], \end{cases}$$

where Δ is the Laplacian, $H = L^2(0,1)$ and let $u(t,\xi)$ be a real valued, with respect to t adapted, control process which takes values in a closed bounded interval of $\mathbb{R}$ and $\psi \in \mathcal{L}(E; \gamma(H;E))$. By these assumptions, the heat equation can be reformulated as

$$\begin{cases} dX(t) = AX(t)dt + F(t,X(t))dt + B(t,X(t))dW^H(t), \ t \in [0,T], \\ X(0) = X_0 \end{cases}$$

where A is the Dirichlet Laplacian on E with homogenous boundary condition and it is generated by the analytic C_0-semigroup $S(t)$, and $F(t,X(t)) := u(t)$, $B(t,X(t)) := \psi X(t)$ are defined as

$$\begin{cases} F : [0,T] \times E \longrightarrow E \\ B : [0,T] \times E \longrightarrow \gamma(H;E) . \end{cases}$$

The assumptions in [22, Theorem 1.1] are satisfied and there exists a unique mild solution which satisfies the following equation

$$X(t) = S(t)X_0 + \int_0^t S(t-s)F(s,X(s))ds + \int_0^t S(t-s)B(s,X(s))dW^H(s)$$

for $t \in [0,T]$. The above equation can be written as the following FSVIE

$$X(t) = S(t)X_0 + \int_0^t b(t,s,X(s),u(s))ds + \int_0^t \rho(t,s,X(s),u(s))dW^H(s)$$

for $t \in [0,T]$, where $b(t,s,X(s),u(s)) := S(t-s)u(s)$ and $\rho(t,s,X(s),u(s)) := S(t-s)\psi X(s)$. The cost function is given by (25)

We have $\left| \frac{\partial b(t,s,x,u)}{\partial u} \right| = \left| \frac{dS(t-s)u}{du} \right| = \|S(t-s)\| \leq c$ for every $t,s \in [0,T]$, $\left| \frac{\partial b(t,s,x,u)}{\partial x} \right| = 0$ and the Fréchet derivative of $\rho(t,s,X(s),u(s))$ with respect to u is zero and the Fréchet derivative with respect to x is $\rho_x(t,s,X(s),u(s)) = S(t-s)\psi$. It means that in Theorem 25 $A_x(t,s) =$

$S(t-s)\psi$. By applying Theorem 25, it results that, if $(\overline{X}(\cdot), \overline{u}(\cdot))$ is an optimal stochastic solution and $\sum_{n\geq 1} \|S(t-s)\psi h_n\|_{\mathcal{L}(E;E)}$ is bounded a.s., then there exists a unique adapted M-solution $Y(\cdot), Z(\cdot, \cdot)$ of the following BSVIE in E^*

$$Y(t) = -h_x(t, \overline{X}(t), \overline{u}(t)) + \int_t^T \sum_{n\geq 1}(S(t-s)\psi h_n)^* Z(s,t) h_n ds$$

$$- \int_0^T Z(t,s) dW^H(s), \; t \in [0,T],$$

such that

$$\left\langle u(t) - \overline{u}(t), \int_t^T S(t-s)^* Y(s) ds - h_u(t, \overline{X}(t), \overline{u}(t)) \right\rangle \leq 0,$$

for all $u(\cdot) \in \mathcal{U}, t \in [0,T]$ a.s.

or, correspondingly,

$$\int_0^1 \left\{ \int_t^T S(t-s)^* Y(s) ds - h_u(t, \overline{X}(t), \overline{u}(t)) \right\} (u(t) - \overline{u}(t)) \, d\mu \leq 0,$$

for all $u(\cdot) \in \mathcal{U}, t \in [0,T]$ a.s.

and Theorem 26 leads to

$$\left\{ \mathbb{E}\left(\int_0^1 \int_t^T S(t-s)^* Y(s) ds \Big| \mathcal{F}_t \right) - h_u(t, \overline{X}(t), \overline{u}(t)) \right\} (u(t) - \overline{u}(t)) \, d\mu \leq 0,$$

for all $u(\cdot) \in \mathcal{U}, t \in [0,T]$ a.s.

If $h(t, \overline{X}(t,\xi), \overline{u}(t,\xi)) = h_1(t, \overline{X}(t,\xi)) + \overline{u}(t,\xi)$, then it follows from the last inequality that $\overline{u}(t,\xi) = \text{sign}\left(\mathbb{E}\left(\int_t^T S(t-s)^* Y(s,\xi) ds \Big| \mathcal{F}_t \right) - 1 \right)$ for all $t \in [0,T]$ and $\xi \in [0,1]$.

References

[1] V. Anh and J. Yong. Backward stochastic Volterra integral equations in Hilbert spaces. *Differential and Difference Equations and Applications*, 57–66. Hindawi, New York, 2006.

[2] V. Anh, W. Grecksch and J. Yong. Regularity of backward stochastic Volterra integral Equations in Hilbert Spaces. *Stoch. Anal. Appl.*, Vol. 29, Issue 1, 146–168, 2011.

[3] M. Azimi. *Banach space valued stochastic integral equations and their op-
     timal control.* Ph.D. Thesis, Martin-Luther-University of Halle-Wittenberg,
     Institute of Mathematics, 2018.

[4] K. Bahlali. Backward stochastic differential equations with locally Lipschitz
     coefficient. *C.R.A.S. Paris, Série I*, 33, 481–486, 2001.

[5] J. M. Bismut. Conjugate convex functions in optimal stochastic control, *J.
     Math. Anal. Appl.* 44, 384–404, 1973.

[6] Z. Brzezniak, J.M.A.M. van Neerven, M.C. Veraar and L. Weis. Itô's formula
     in UMD Banach spaces and regularity of solution of the Zakai equation. *J.
     Differ. Equations* 245, No. 1, 30–58, 2008.

[7] G. Da Prato and J. Zabczyk. *Stochastic equations in infinite dimensions.*
     Cambridge University Press, 1992.

[8] N. Dunford and B. J. Pettis. Linear operators on summable functions. *Trans.
     Amer. Math. Soc.* 47, 323–392, 1940.

[9] E. Essaky, K. Bahlali and Y. Ouknine. Reflected backward stochastic differ-
     ential equation with jumps and locally Lipschitz coefficient. *Random Oper.
     Stoch. Equ.*, 10, 335–350, 2002.

[10] M. Fuhrmann, Y. Hu and G. Tessitore. Stochastic maximum principle for
     optimal controls of SPDE's. *arxiv.org/pdf/1302.0286vI*, 2013.

[11] Y. Hu, D. Nualart and X. Song. Malliavian calculus for backward stochastic
     differential equations and application to numerical solutions. *Ann. Appl.
     Probab.*, Vol. 21, No. 6, 2379–2423, 2011.

[12] Y. Hu and S. Peng. Adapted solution of backward stochastic evolution equa-
     tion. *Stoch. Anal. Appl.*, 9, 445–459, 1991.

[13] J. Lin. Adapted solution of backward stochastic nonlinear Volterra integral
     equation. *Stoch. Anal. Appl.*, 20 No. 1, 165–183, 2002.

[14] B. Maurey. *Type, cotype and K-convexity.* In Handbook of the Geometry of
     Banach Spaces, Vol. 2, North-Holland, Amsterdam, 2003.

[15] A. Neamtu. *Backward stochastic Itô-Volterra differential equations in UMD
     Banach spaces.* Master Thesis. Martin-Luther-University Halle-Wittenberg,
     Institute of Mathematics, 2013.

[16] R. Negrea and C. Preda. Fixed point technique for a class of backward
     stochastic differential equations. *J. Nonlinear Sci. Appl.*, 6, 41–50, 2013.

[17] M. Ondrejat and M.C. Veraar. Weak characterisations of stochastic integra-
     bility and Dudley's theorem in infinite dimensions. *J. Theoret. Probab.*, 27,
     No. 4, 1350–1374, 2014.

[18] E. Pardoux and S. G. Peng. Adapted solution of a backward stochastic
     differential equation. *Systems Control Lett.*, 14, 55–61, 1990.

[19] R. A. S. Perdomo. *Optimal control of stochastic partial differential equations
     in Banach Spaces.* PhD Thesis, University of York, Heslington, 2010.

[20] M. Pronk and M.C. Veraar. Tools for Malliavin calculus in UMD Banach
     spaces. *Potential Anal.*, 40, No. 4, 307–344, 2014.

[21] J.M.A.M van Neerven. *Stochastic evolution equations.* ISEM Lecture Notes,
     2007/08.

[22] J.M.A.M. van Neerven, M.C. Veraar and L. Weis, Stochastic evolution equa-
     tions in UMD Banach spaces. *J. Funct. Anal.*, 255, 940–993, 2008.

[23] J.M.A.M. van Neerven, M.C. Veraar, L. Weis. Stochastic integration in UMD Banach spaces. *Ann. Probab.*, 35, 1438–1478, 2007.

[24] A. Pazy. *Semigroups of linear operators and applications to partial differential equations.* Applied Mathematical Sciences, Vol. 44, Springer-Verlag, New York, 1983.

[25] M.C. Veraar. *Stochastic integrals in Banach spaces and applications to parabolic evolution equations.* Ph.D. Thesis, Delft University of Technology, 2006.

[26] M.C. Veraar. Continuous local martingales and stochastic integration in UMD Banach spaces. *Stochastics*, 79, No. 6, 601–618, 2007.

[27] J. Yong. Backward stochastic Volterra integral equations and some related problems. *Stochastic Process. Appl.*, 116, 779–795, 2006.

[28] J. Yong. Well-posedness and regularity of backward stochastic Volterra integral equations. *Probab. Theory Related Fields*, 142, 21–77, 2008.

[29] X. Zhang. Stochastic Volterra equations in Banach spaces and stochastic partial differential equation. *J. Funct. Anal.*, 258, 1361–1425, 2010.

Chapter 3

Stochastic Schrödinger Equations

Wilfried Grecksch*

*Martin-Luther-University Halle-Wittenberg,
Faculty of Natural Sciences II, Institute of Mathematics,
D-06099 Halle (Saale), Germany*

Hannelore Lisei†

*Babeş-Bolyai University, Faculty of Mathematics and Computer Science
1, Kogălniceanu Street, 400084 Cluj-Napoca, Romania*

In the first part of this contribution we discuss an optimal control problem for a stochastic linear homogeneous Schrödinger equation in a bounded domain driven by a linear multiplicative cylindrical fractional Brownian motion with Hurst index $1/2 < h < 1$. The state equation has such a structure that the method of diagonalisable operators can be used such that a sequence of scalar problems arises. These problems can be split in a pathwise equation and a process constructed by the noise term. In the second part of this contribution an optimal control problem for a nonlinear stochastic Schrödinger equation is considered, where the noise process is a linear multiplicative fractional Brownian motion with Hurst index $1/2 < h < 1$. The noise structure is chosen so that the transformation into a pathwise and a stochastic process can be used. Optimality conditions of maximum principle type are proven for both cases. The case of an additive cylindrical fractional noise term is also discussed.

Contents

---

*E-mail: wilfried.grecksch@mathematik.uni-halle.de
†E-mail (corresponding author): hanne@math.ubbcluj.ro

1. Introduction

In physics, specifically in quantum mechanics, the Schrödinger equation is an equation that describes how the quantum state of a physical system changes in time. For a single particle the time-dependent linear Schrödinger equation with potential takes the form

$$i\hbar\frac{\partial}{\partial t}X(t,\,x) = -\frac{\hbar^2}{2m}\nabla^2 X(t,\,x) + \Psi(x)X(t,\,x) \quad x\in\mathbb{R}, t>0 \quad (1)$$

with $X(0,\,\cdot) = X_0$, where i is the imaginary unit, $-\dfrac{\hbar^2}{2m}\nabla^2$ is the kinetic energy operator, m is the mass of the particle, $\hbar$ is Planck's constant, Ψ is a time-independent potential energy, the complex valued function X is the wave function at position x at time t, X_0 is the initial condition.

The Schrödinger equation takes several different forms, depending on the physical situation. Nonlinear Schrödinger equations appear, for example, in laser beams [40], in the theory of solids [5] and crystals [16], in electromechanical systems [12], in saturating laws and laser produced plasmas [11] or Bose-Einstein condensates [34, 55].

Stochastic partial differential equations belong to the modern research of infinite dimensional stochastic analysis. Such equations can be interpreted as stochastic evolution equations and the solutions are defined in a generalized sense. In the literature, there are mainly three approaches: The *mild solution*: The problem contains a linear operator which generates a semigroup of operators or a group of operators in a Hilbert space and the problem is considered as a stochastic integral equation (Itô-Volterra type), which contains a stochastic convolution (see for example [23] and the literature cited therein). The *generalized weak solution* (weak solution, analytically weak solution): The solution of the partial differential equation satisfies a scalar product equation (see [23, 56]). The *variational solution* (generalized solution, (V, H)-solution): The problem is defined by a stochastic evolution equation over a triplet of rigged Hilbert spaces (V, H, V^*) (see for example [56, 59]).

In stochastic analysis the solution can be defined on a given probability space, then the solution is called (probabilistic) strong solution, or by constructing the probability space, then the solution is called (probabilistic) weak solution. Here we consider problems on a given complete probability space and the solutions of the studied equations are closely related to the the mild solution.

Using stochastic processes in Schrödinger equations one can model spontaneous emissions or thermic fluctuations or general random disturbances. Many authors investigated stochastic equations of Schrödinger type. We give only some examples and refer to the literature in the corresponding papers. The case of an additive noise is considered in [22, 27, 50], while the case of multiplicative noise is discussed in [6, 18, 19, 57], where the noise terms are defined by stochastic integrals with respect to a Wiener process. The solution processes are defined in the mild sense. Variational solutions are considered for stochastic Schrödinger equations for example in [30, 44].

For nonlinear stochastic Schrödinger equations we distinguish, as in the deterministic case, between Lipschitz nonlinearities (see for example, [30]) and Kerr nonlinearities with their generalizations (see for example, [18, 44, 51]).

A method for solving stochastic partial differential equations consists in their transformation into pathwise problems. So-called splitting methods involve time discretizations, such that certain pathwise problems and specific stochastic problems can be solved alternately. A splitting method is used for stochastic equations of Schrödinger type in [32]. Transformations into pathwise problems are considered in [7, 38].

Optimal control problems for systems governed by Schrödinger equations arise in quantum mechanics, nuclear physics, nonlinear optics and various fields of modern physics and engineering [13, 62].

Optimality conditions for control problems with linear Schrödinger equations are discussed for example in [61], [36]. The nonlinear case is considered for example in [3], [1].

If optimality conditions of maximum principle type are used for control problems involving stochastic partial differential equations, then usually stochastic backward equations occur. This holds especially, when the state equation is a stochastic Schrödinger equation, see [63], [8], [39].

In recent years, the theory of fractional quantum mechanics was developed, the monograph [41] reflects the current state of research in this area. In particular, fractional Schröddinger equations are also considered (see [42], [43], [65], [64]). Fractional Schrödinger equations are characterized by the fractional Laplacian operator and/or a fractional time derivative. In addition to path integrals of a Brownian motion, other path integrals of a fractional Brownian motion [14] or of a fractional Lévy process [15] are investigated. Since the fractional Brownian motion is much more difficult a specific stochastic analysis is required. Especially, the fractional Brownian motion is not a semimartingale.

Fractional Gaussian fields can be represented by a fractional Laplacian and a vector-valued Gaussian white noise [46]. In the scalar case, a representation of the fractional Brownian motion is obtained with respect to a scalar Gaussian white noise and a fractional order of the second derivative.

There are also stochastic Schrödinger equations driven by fractional Brownian motion, see for example [28, 31, 37, 54, 60].

This contribution has the following structure: Section 2 contains some foundations from stochastic analysis and results from functional analysis. In Section 3 is studied an optimal control problem for a stochastic linear homogeneous Schrödinger equation in a bounded domain driven by a linear multiplicative cylindrical fractional Brownian motion with Hurst index $1/2 < h < 1$. The state equation has such a structure that the method of diagonalisable operators from [21] and [47, Chapter 6] can be used and a sequence of scalar problems arises. These problems can be split into a pathwise equation and a process constructed by using the noise term.

An optimal control problem for a nonlinear stochastic Schrödinger equation is considered in Section 4, where the noise process is a linear multiplicative cylindrical fractional Brownian motion with Hurst index $1/2 < h < 1$. The noise structure is chosen such that the transformation into a pathwise equation and a stochastic stochastic process can be used, as in the case of Wiener processes in [7, 38].

Optimality conditions of maximum principle type are proven for both optimal control problems mentioned above. The transformation into pathwise problems has the advantage that for backward equations with a stochastic final condition, despite the fractal noise, a martingale representation theorem can be applied, which assures the adaptedness condition of the solutions. This is possible, because the fractal Brownian motion can be represented by a stochastic integral with a deterministic kernel with respect to a Wiener process, and the filtrations generated by the Wiener process and the fractal Brownian motion are equal, see [52].

In Section 5 we discuss an optimal control problem for a linear stochastic Schrödinger equation, where the multiplicative noise from Section 4 is replaced by an additive noise. This stochastic equation can be easily transformed into a pathwise problem. We also get an optimality condition of maximum principle type.

Section 6 is devoted to a stochastic Schrödinger equation with Lipschitz continuous nonlinearities driven by multiplicative fractional Brownian noise. The equation is controlled by a time dependent disturbed real potential. The solution process can be represented as the product of two stochastic processes. The first process is the solution of a linear homogeneous complex valued fractional stochastic differential equation. The second process solves a controlled pathwise nonlinear Schrödinger equation.

Section 7 contains an optimal control problem for the stochastic Schrödinger equation discussed in Section 6, which can be reduced to an optimal control problem for a pathwise Schrödinger equation. By applying the theory of Itô-Volterra backward equations and a duality principle, a maximum principle is proved. The case of an additive noise is covered in Section 8.

Finally, in Section 9, the optimal control is calculated for a special controlled stochastic Ginzburg-Landau equation. Further, we discuss the optimal control problem from Section 6 in the linear case for Markovian controls. Then, we have to solve a pure deterministic control problem.

2. Foundations

2.1. *Fractional Brownian Motion*

In this contribution we assume that:
- all stochastic processes are defined on a given complete probability space $(\Omega, \mathcal{F}, \mathbb{P})$;
- $\mathbb{E}$ denotes the mathematical expectation with respect to $\mathbb{P}$.

Definition 25. A real-valued Gaussian process $\{B^h(t) : t \geq 0\}$ is called a **fractional Brownian motion** (fBm) with Hurst index $0 < h < 1$, if

(1) $\mathbb{P}(B^h(0) = 0) = 1$;
(2) $\mathbb{E}(B^h(t)) = 0$ for all $t \geq 0$;
(3) $\mathbb{E}(B^h(t)B^h(s)) = \frac{1}{2}(t^{2h} + s^{2h} - |t - s|^{2h})$ for all $s, t \geq 0$. $\qquad \diamond$

The fBm is not a semimartingale and is not a Markovian process for

$h \neq 1/2$. In this contribution we consider $1/2 < h < 1$. Then, the fBm has the so-called long range dependence property. Let $\{B_k^h(t) : t \geq 0\}$ be independent fractional Brownian motions for $k \geq 1$. If $(h_k)_{k\geq 1}$ is a complete orthonormal system in a separable Hilbert space, then $B^h(t) = \sum_{k=1}^{\infty} B_k^h(t)h_k$, $t \geq 0$, is formally defined as a cylindrical fBm.

Mandelbrot and van Ness have given in [49] the stochastic integral representation

$$B_k^h(t) = c_h \int_0^t (t-u)^{h-1/2} dW_k(u) +$$

$$+ c_h \int_{-\infty}^0 \left((t-u)^{h-1/2} - (-u)^{h-1/2} \right) dW_k(u), \qquad (2)$$

where $(W_k)_{k\geq 1}$ are independent two sided Wiener processes and c_h is a constant dependent on h.

Further we denote:
• by $\mathcal{F}_t^X = \sigma\{X(s) : s \leq t\}$, with $t \geq 0$, the natural filtration of the stochastic process $\{X(t) : t \geq 0\}$; we assume that the natural filtration is complete, that is $\mathcal{F}_0^X$ contains all $\mathbb{P}$-null sets, and that it is right continuous, that is $\mathcal{F}_t^X = \mathcal{F}_{t+}^X$, where $\mathcal{F}_{t+}^X = \bigcap_{s>t} \mathcal{F}_s^X$ (these assumptions are not very restrictive, see on p. 2 in [17]);
• by $\left(\mathcal{F}_{k,t}^h\right)_{t\geq 0}$ we denote the natural filtration generated by $\{B_k^h(t) : t \geq 0\}$, $k \geq 1$.

It follows from [52, Theorem 3.2] and its proof that the next result holds:

Theorem 27. *For each $k \geq 1$ and each $t \geq 0$ it holds*

$$\mathcal{F}_t^{W_k} = \mathcal{F}_{k,t}^h.$$

For every $t \geq 0$ we denote the common σ-algebra from Theorem 27 by $\mathcal{F}_t^k$ and we introduce for $t \geq 0$ the σ-algebras

$$\mathcal{F}_t = \sigma\left\{ \bigcup_{k=1}^{\infty} \mathcal{F}_t^k \right\}.$$

It follows from Theorem 27 that the natural filtration of the cylindrical fBm B^h and the natural filtration of the cylindrical Wiener process W are the same, where W is defined by the Wiener processes $(W_k)_{k\geq 1}$ from (2).

In the following we choose for fixed $T > 0$ in $(\Omega, \mathcal{F}, \mathbb{P})$ the σ-algebra $\mathcal{F} = \mathcal{F}_T$. By [29, Theorem 4.3.4] the following martingale representation theorem holds:

Theorem 28. *Let the process $\{M(t) : t \in [0,T]\}$ be a continuous $(\mathcal{F}_t)_{t \in [0,T]}$-adapted martingale with values in the separable Hilbert space H and $\mathbb{E}\|M(t)\|_H^2 < \infty$ for all $t \in [0,T]$. Then there exists a unique predictable process $\{\Phi(t) : t \in [0,T]\}$ with values in the space of Hilbert-Schmidt operators on H and $\mathbb{E} \int_0^T \|\Phi(t)\|_{HS}^2 dt < \infty$ such that*

$$M(t) = \mathbb{E}(M(0)) + \int_0^t \Phi(s)dW(s), t \in [0,T], \; a.s. \,,$$

where $\|\Phi(t)\|_{HS}$ is the Hilbert-Schmidt norm of $\Phi(t)$.

Remark 15. Let $(h_k)_{k \geq 1}$ be a complete orthonormal system in H. In Theorem 28 the stochastic integral in H

$$\int_0^t \Phi(s)dW(s) = \sum_{k=1}^{\infty} \int_0^t \Phi(s)h_k dW_k(s) \,, t \in [0,T] \,,$$

is defined by the Itô integral as in [23]. ◇

The stochastic integrals in this contribution are Wick-Itô-Skorohod (WIS) integrals with respect to a real-valued fractional Brownian motion B^h with $1/2 < h < 1$ (see [9, Chapter 4, p. 99]).

The following Itô formula for functionals of B^h is given in [9, Theorem 4.2.6, p. 107]:

Theorem 29. *Suppose that $f : \mathbb{R} \times \mathbb{R} \to \mathbb{R}$ is a function which belongs to $C^{1,2}(\mathbb{R}_+ \times \mathbb{R})$. Assume that the random variables $f(t, B^h(t))$, $\int_0^t \frac{\partial f}{\partial s}(s, B^h(s))ds$, and $\int_0^t \frac{\partial^2 f}{\partial x}(s, B^h(s))s^{2h-1}ds$ are square integrable for every $t \geq 0$. Then, it holds*

$$f(t, B^h(t)) = f(0,0) + \int_0^t \frac{\partial f}{\partial s}(s, B^h(s))ds + \int_0^t \frac{\partial f}{\partial x}(s, B^h(s))dB^h(s)$$

$$+ h \int_0^t \frac{\partial^2 f}{\partial x} f(s, B^h(s))s^{2h-1}ds \tag{3}$$

for all $t \geq 0$

It follows from page 161 in [9] that the Itô formula (3) holds for all $0 < h < 1$, if the stochastic integral is defined by the fractional Wick-Itô-Skorohod integral. This formula holds for $1/2 < h < 1$ too, if the stochastic integral is defined by the fractional forward Wick-Itô-Skorohod integral or by a fractional divergence-type integral.

Let $\phi : \mathbb{R}_+ \times \mathbb{R}_+ \to \mathbb{R}_+$ be given by

$$\phi(s,t) = h(2h-1)|s-t|^{2h-2}$$

and recall that for $s, t > 0$

$$\int_0^t \int_0^s \phi(u,v)dudv = \frac{1}{2}\left(|s|^{2h} + |t|^{2h} - |t-s|^{2h}\right).$$

Example 11. **(1)** For $t \geq 0$ define the real-valued stochastic process

$$x(t) = x_0 \exp\{aB^h(t) + b(t)\}$$

for $x_0, a \in \mathbb{R}$ and $b : \mathbb{R} \to \mathbb{R}$ is derivable with bounded derivative. This process solves the linear equation

$$x(t) = x_0 + \int_0^t b'(s)x(s)ds + a^2h \int_0^t x(s)s^{2h-1}ds + a \int_0^t x(s)dB^h(s), t \geq 0,$$

by Theorem 29 (or see the computations on p. 108 in [9]).
(2) If we consider for $t \geq 0$ the complex-valued stochastic process

$$z(t) = \exp\{iaB^h(t) + ib(t)\} = \cos\left(aB^h(t) + b(t)\right) + i\sin\left(aB^h(t) + b(t)\right)$$

for $a \in \mathbb{R}$ and for $b : [0,\infty) \to \mathbb{R}$ derivable function with bounded derivative, then this process solves the linear equation

$$z(t) = 1 + i\int_0^t b'(s)z(s)ds - a^2h \int_0^t z(s)s^{2h-1}ds + ia \int_0^t z(s)dB^h(s), t \geq 0,$$

since we can apply Theorem 29 for $\text{Re}z(t) = \cos\left(aB^h(t) + b(t)\right)$ and for $\text{Im}z(t) = \sin\left(aB^h(t) + b(t)\right)$.

Especially, we obtain that

$$z(t) = z_0 \exp\left\{iaB^h(t) + \frac{1}{2}a^2t^{2h}\right\}, t \geq 0,$$

for $z_0 \in \mathbb{C}$ and $a \in \mathbb{R}$ solves the linear equation

$$z(t) = z_0 + ia \int_0^t z(s)dB^h(s), t \geq 0.$$

(3) If we consider for $t \geq 0$ the complex-valued stochastic process

$$z(t) = z_0 \exp\{iaB^h(t) + b_1(t) + ib_2(t)\}$$

for $z_0 \in \mathbb{C}$, $a \in \mathbb{R}$ and $b_1, b_2 : [0,\infty) \to \mathbb{R}$ derivable functions with bounded derivatives. This process solves the linear equation

$$z(t) = z_0 + \int_0^t \left(b_1'(s) + ib_2'(s)\right)z(s)ds - a^2h \int_0^t z(s)s^{2h-1}ds + ia \int_0^t z(s)dB^h(s)$$

for $t \geq 0$. $\diamond$

Example 12. Consider $b : \mathbb{R} \to \mathbb{C}$ to be a function, which is bounded, i.e. $|b(s)| \leq c_b$ for each $s \in [0, T]$. Since it is a deterministic function, we have by [9, Lemma 3.1.3, p. 49]

$$\mathbb{E}\left| \int_0^T b(s)dB^h(s) \right|^2 = \mathbb{E}\left(\int_0^T \operatorname{Re}b(s)dB^h(s) \right)^2 + \mathbb{E}\left(\int_0^T \operatorname{Im}b(s)dB^h(s) \right)^2$$

$$= h(2h-1)\int_0^T\!\!\int_0^T \left(\operatorname{Re}b(u)\operatorname{Re}b(v) + \operatorname{Im}b(u)\operatorname{Im}b(v) \right)|u-v|^{2h-2}dudv \leq 2c_b^2 T^{2h}.$$

Especially, in the case $b(s) = \exp\{i\lambda s\}, s \in [0, T]$ with $\lambda \in \mathbb{R}$ we obtain

$$\mathbb{E}\left| \int_0^T e^{i\lambda s}dB^h(s) \right|^2 \leq T^{2h}.$$

$\Diamond$

Example 13. Let $\{a(t) : t \in [0, T]\}$ be a complex-valued process with a.s. continuous trajectories such that $\mathbb{E} \sup_{s \in [0,T]} |a(s)|^2 < \infty$ and $\{b(t) : t \in [0, T]\}$ be a complex-valued process with $b \in \mathcal{L}[0, T]$ (for the definition of this space see on p. 591 in [25]) such that the following processes exist a.s. and have a.s. continuous versions (for this more assumptions on b are needed as seen in [25, Theorem 3.9 -(2)]: $\mathbb{E} \sup_{s \in [0,T]} |b(s)|^2 + \sup_{s \in [0,T]} \mathbb{E}|D_s^\phi b(s)|^2 < \infty$)

$$u(t) = x_0 + \int_0^t a(s)ds, \quad t \in [0, T],$$

$$v(t) = y_0 + \int_0^t b(s)dB^h(s), \quad t \in [0, T].$$

If $u \cdot b \in \mathcal{L}[0, T]$, then the following product rule (Itô-formula) holds a.s.

$$u(t)v(t) = x_0 y_0 + \int_0^t a(s)v(s)ds + \int_0^t b(s)u(s)dB^h(s), t \in [0, T].$$

The proof can be done by following step by step the ideas from the proofs of [25, Theorem 4.3, Theorem 4.6]. $\Diamond$

Notation 1. Let $(a_k)_{k \geq 1}$ be a sequence of complex-valued functions and let $(b_k)_{k \geq 1}$ be a sequence of real-valued functions.

If the sequence $\left(\sum_{k=1}^n a_k(t) \int_0^t b_k(s)h_k dB_k^h(s) \right)_{n \geq 1}$ is convergent in $L^2(\Omega; H)$ for each $t \in [0, T]$, then we denote its limit (in $L^2(\Omega; H)$) by

$$\sum_{k=1}^\infty a_k(t) \int_0^t b_k(s)h_k dB_k^h(s) = \lim_{n \to \infty} \sum_{k=1}^n a_k(t) \int_0^t b_k(s)h_k dB_k^h(s)$$

for each $t \in [0, T]$. $\Diamond$

2.2. *Some Facts from Functional Analysis*

Throughout this paper we consider the following assumptions:
• Let be (V, H, V^*) a triplet of complex rigged Hilbert spaces. In the following we denote $\|\cdot\|_H$ by $\|\cdot\|$ and the scalar product in H by $(\cdot, \cdot)$.
• Suppose $A : V \to V^*$ to be a linear and continuous operator such that

$$\|Av\|_{V^*} \le c_A \|v\|_V \text{ for all } v \in V,\tag{4}$$

and

$$\langle Au, v \rangle = \overline{\langle Av, u \rangle} \text{ for all } u, v \in V,$$

and there exist constants $\alpha_1 \in \mathbb{R}$ and $\alpha_2 > 0$ with

$$\langle Av, v \rangle + \alpha_2 \|v\|_V^2 \le \alpha_1 \|v\|^2 \text{ for all } v \in V.\tag{5}$$

We consider $D(A) = \{v \in V : Av \in H\}$ and assume that $-A$ has eigenfunctions $(h_j)_{j \ge 1}$ in H, which define a complete orthonormal system in H and the corresponding eigenvalues satisfy $\lambda_j < \lambda_{j+1}$ for $j \ge 1$ and $\lim_{n \to \infty} \lambda_n = \infty$.
• Define

$$\mathcal{S}_t(y) = \sum_{j=1}^{\infty} e^{-\lambda_j t}(y, h_j) h_j \text{ for all } y \in H, \ t \ge 0.$$

$(\mathcal{S}_t)_{t \ge 0}$ is the analytic semigroup of A in H. We introduce

$$\mathcal{T}_t(y) = \sum_{j=1}^{\infty} e^{-i\lambda_j t}(y, h_j) h_j \text{ for all } y \in H, \ t \in \mathbb{R}.$$

It is easy to show, that $(\mathcal{T}_t)_{t \in \mathbb{R}}$ defines a C_0-group of unitary operators in H, that is:

(1) $\mathcal{T}_0 = I$;
(2) $\mathcal{T}_{t+s(y)} = \mathcal{T}_t(\mathcal{T}_s(y))$ for all $y \in H$ and $t \in \mathbb{R}$;
(3) $\|\mathcal{T}_t(y)\| = \|y\|$ for all $y \in H$ and $t \in \mathbb{R}$;
(4) $\lim_{s \to t} \|(\mathcal{T}_s - \mathcal{T}_t)y\| = 0$ for all $y \in H$ and $t \in \mathbb{R}$.

It follows from Stone's theorem that the generator of $(\mathcal{T}_t)_{t \in \mathbb{R}}$ is given by iA, see [58, Theorem VIII.8, p. 266].

Remark 16. If the embedding $V \hookrightarrow H$ is compact and the operator A is linear and self-adjoint satisfying (4) and (5), then A admits eigenfunctions $(h_j)_{j \ge 1}$ in H, which form a complete orthonormal system in H and the corresponding eigenvalues satisfy $\lambda_j < \lambda_{j+1}$ for $j \ge 1$ and $\lim_{n \to \infty} \lambda_n = \infty$ (see [45], page 110). $\diamond$

Example 14. We introduce the Laplacian operator A with homogenous Dirichlet conditions, respectively homogenous Neumann conditions, in the weak sense. Let $G \subset \mathbb{R}^d$ be a bounded set with C^∞ boundary, we consider the complex spaces $V = H_0^{1,2}(G)$ in the homogenous Dirichlet case and

$$V = \left\{ v \in H^{1,2}(G) : \int_G v(x)dx = 0 \right\}$$

in the homogenous Neumann case. In both cases we take $H = L^2(G)$.

We introduce the eigenvalue-eigenfunction problem: we seek $h \in V$ (the eigenfunction) and $\lambda \in \mathbb{R}$ (the eigenvalue) such that

$$\int_G \nabla h(x) \nabla \overline{v}(x)dx - \lambda \int_G h(x)\overline{v}(x)dx = 0$$

for all $v \in V$. This problem admits as solutions eigenfunctions $\left(h_n \right)_{n \geq 1}$ and corresponding eigenvalues $\left(\lambda_n \right)_{n \geq 1}$ such that the eigenfunctions $\left(h_n \right)_{n \geq 1}$ define a complete orthonormal system in H, which is orthogonal in V, with

$$\int_G \nabla v(x) \nabla \overline{v}(x)dx = \sum_{k=1}^{\infty} \lambda_k \int_G |v(x)h_k(x)|^2 dx \text{ for all } v \in V.$$

It is well known that in the Dirichlet case we get eigenvalues with $0 < \lambda_1 < \lambda_2 < \dots$ and in the Neumann case we have $\lambda_1 = 0 < \lambda_2 < \dots$. In both cases it holds $\lim_{n \to \infty} \lambda_n = \infty$.

Choosing $d = 1$, $G =]0, l[\subset \mathbb{R}$ (where $l > 0$ is given), $H = L^2(G)$,
$\rhd$ $V = H_0^{1,2}(G)$ (homogeneous Dirichlet conditions) and the operator

$$\langle -Au, v \rangle = \int_0^l u'(x) \overline{v}'(x)dx, \quad u, v \in V,$$

we have the eigenvalues $\lambda_k = \dfrac{(k\pi)^2}{l^2}$ and the eigenfunctions

$$h_k(x) = \sqrt{\frac{2}{l}} \sin\left(\frac{k\pi x}{l} \right) \text{ for } k \geq 1, x \in G;$$

$\rhd$ $V = \left\{ v \in H^{1,2}(G) : \int_G v(x)dx = 0 \right\}$ (homogeneous Neumann conditions) and the operator

$$\langle -Au, v \rangle = \int_0^l u'(x) \overline{v}'(x)dx, \quad u, v \in V,$$

we have $\lambda_1 = 0$, $h_1(x) = \frac{1}{\sqrt{l}}$ and

$$\lambda_k = \frac{(k-1)^2 \pi^2}{l^2}, \quad h_k(x) = \sqrt{\frac{2}{l}} \cos\left(\frac{(k-1)\pi x}{l} \right) \text{ for } k \geq 2, x \in G.$$

Example 15. We choose the complex Hilbert spaces $V = H^1(\mathbb{R})$, $H = L^2(\mathbb{R})$. Then $\lambda_{k+1} = 2k + 1$, $k \geq 0$ are the eigenvalues and the Hermitian functions are the eigenfunctions of the operator

$$\langle -Au, v \rangle = \int_{\mathbb{R}} \left(u'(x)\overline{v}'(x) + |x|^2 u(x)\overline{v}(x) \right) dx, \ u, v \in V,$$

see [24, Chapter XV, Section 5.2, p. 177]. Note, that the embedding $V \hookrightarrow H$ is in this case not compact.   ◊

3. A Linear Stochastic Schrödinger Equation Driven by Multiplicative Fractional Noise

At first we introduce the set $\mathcal{U}$ of admissible controls. Let $c > 0$ be fixed. The **set of admissible controls** is

$$\mathcal{U} = \{ U = (U_1, U_2, ...) : \ U_k : [0, T] \to \mathbb{R}, \ |U_k(t)| \leq c \text{ for all } t \geq 0, k \geq 1 \}.$$

Further, let $(\mu_k)_{k \geq 1}$ be a sequence of real numbers with $\sum_{k=1}^{\infty} \mu_k^2 < \infty$ and let $X_0 \in H$. We denote for $y \in H$ and for $U = (U_1, U_2, \dots) \in \mathcal{U}$

$$L_1(U)(y) := \sum_{k=1}^{\infty} U_k(y, h_k) h_k, \ L_2(dB^h)(y) = \sum_{k=1}^{\infty} \mu_k(y, h_k) h_k dB_k^h.$$

We consider the controlled equation in symbolic form

$$dX^U(t) = iAX^U(t)dt + iL_1(U)(X^U)(t)dt + iL_2(dB^h)(X^U)(t), \ X^U(0) = X_0.$$
$$(6)$$

Definition 26. The $(\mathcal{F}_t)_{t \geq 0}$-adapted process $\{X^U(t) : t \in [0, T]\}$ with $\mathrm{Re}X^U(t)$, $\mathrm{Im}X^U(t) \in H$ for all $t \in [0, T]$ a.s. and $\sup_{t \in [0, T]} \mathbb{E} \| X^U(t) \|^2 < \infty$ is called **solution** of (6), if

$$X^U(t) = \sum_{k=1}^{\infty} e^{-i\lambda_k t}(X_0, h_k) h_k + i\sum_{k=1}^{\infty} \int_0^t e^{-i\lambda_k(t-s)} U_k(s)(X^U(s), h_k) h_k ds$$

$$+ i\sum_{k=1}^{\infty} \mu_k \int_0^t e^{-i\lambda_k(t-s)}(X^U(s), h_k) h_k dB_k^h(s) \qquad (7)$$

for all $t \in [0, T]$ a.s.

The last term in (7) (containing the stochastic integrals) is understood in the sense of Notation 1.   ◊

Let $U = (U_1, U_2, \ldots) \in \mathcal{U}$. For each $k \geq 1$ we introduce the complex-valued problem

$$X_k^U(t) = (X_0, h_k) - \mathrm{i}\lambda_k \int_0^t X_k^U(s)ds$$

$$+ \mathrm{i} \int_0^t U_k(s)X_k^U(s)ds + \mathrm{i}\mu_k \int_0^t X_k^U(s)dB_k^h(s) \, , \, t \in [0, T]. \quad (8)$$

By Example 11-(2) the following result holds:

Theorem 30. *For each $k \geq 1$ the process*

$$X_k^U(t) = (X_0, h_k) \exp\left\{ -\mathrm{i}\lambda_k t + \mathrm{i}\int_0^t U_k(s)ds + \frac{1}{2}\mu_k^2 t^{2h} + \mathrm{i}\mu_k B_k^h(t) \right\}, t \in [0, T],$$

$$(9)$$

solves (8).

We introduce the complex-valued process

$$Y_k^{(1)}(t) = \exp\left\{ \frac{1}{2}\mu_k^2 t^{2h} + \mathrm{i}\mu_k B_k^h(t), \; t \in [0, T] \right\} \quad (10)$$

which by Example 11-(2) is solution of

$$Y_k^{(1)}(t) = 1 + \mathrm{i}\mu_k \int_0^t Y_k^{(1)}(s)dB_k^h(s) \, , \, t \in [0, T] \, , \quad (11)$$

and the complex-valued deterministic process

$$Y_k^{(2),U}(t) = (X_0, h_k) \exp\left\{ -\mathrm{i}\lambda_k t + \mathrm{i}\int_0^t U_k(s)ds \right\}, t \in [0, T] \, , \quad (12)$$

which is the unique solution of

$$Y_k^{(2),U}(t) = (X_0, h_k) + \int_0^t \left(-\mathrm{i}\lambda_k + \mathrm{i}U_k(s) \right) Y_k^{(2),U}(s)ds \, , t \in [0, T]. \quad (13)$$

Obviously, we can write that

$$X_k^U(t) = Y_k^{(2),U}(t)Y_k^{(1)}(t) \, , t \in [0, T] \, , k \geq 1 \, ,$$

is a solution of (8).

Remark 17. Note, that the formulae (10) and (12) have a nice structure: (10) contains a fractional Brownian motion, but no control term, while (12) contains a control term, but no fractional Brownian motion. $\diamond$

In this contribution C denotes a universal constant.

Theorem 31. *With probability 1 there exists the H-valued process*

$$X^U(t) := \sum_{k=1}^{\infty} X_k^U(t) h_k \, , t \in [0,T] \, ,$$

and

$$\sup_{t \in [0,T]} \|X^U(t)\|^2 \leq C \|X_0\|^2,$$

where $C > 0$ is a constant. Moreover,

$$\mathbb{E} \sup_{t \in [0,T]} \|X^U(t)\|^2 \leq C \|X_0\|^2 \, . \tag{14}$$

Proof. Since $(h_k)_{k \geq 1}$ is a complete orthonormal system, we get by (9)

$$\left\| \sum_{k=1}^{\infty} X_k^U(t) h_k \right\|^2 = \sum_{k=1}^{\infty} |X_k^U(t)|^2 = \sum_{k=1}^{\infty} |(X_0, h_k)|^2 \exp\{\mu_k^2 t^{2h}\} \text{ for all } t \in [0,T].$$

It follows from the assumptions that the sequence $(\mu_k^2)_{k \geq 1}$ is bounded. Consequently, it holds

$$\sup_{t \in [0,T]} \left\| \sum_{k=1}^{\infty} X_k^U(t) h_k \right\|^2 \leq C \|X_0\|^2 \, .$$

Taking expectation in the above inequality we obviously obtain (14). $\quad\square$

By (9) and Example 11-(2) we obtain that

$$e^{i\lambda_k t} X_k^U(t) = (X_0, h_k) + i \int_0^t e^{i\lambda_k s} U_k(s) X_k^U(s) ds + i\mu_k \int_0^t e^{i\lambda_k s} X_k^U(s) dB_k^h(s) \, ,$$

$t \in [0,T]$ and $k \geq 1$, which is equivalent to

$$X_k^U(t) = e^{-i\lambda_k t}(X_0, h_k) + i \int_0^t e^{-i\lambda_k(t-s)} U_k(s) X_k^U(s) ds$$

$$+ i\mu_k \int_0^t e^{-i\lambda_k(t-s)} X_k^U(s) dB_k^h(s) \, , t \in [0,T] \, , k \geq 1 \, . \tag{15}$$

This implies

$$\sum_{k=1}^{n} X_k^U(t) h_k = \sum_{k=1}^{n} \left(e^{-i\lambda_k t} X_k(0) + i \int_0^t e^{-i\lambda_k(t-s)} U_k(s) X_k^U(s) ds \right) h_k$$

$$+ i \sum_{k=1}^{n} \mu_k \int_0^t e^{-i\lambda_k(t-s)} X_k^U(s) dB_k^h(s) h_k \tag{16}$$

for $n \geq 1$. We know from Theorem 31 and (14) that the left hand side is convergent to X^U in mean square in H uniformly with respect to $t \in [0, T]$. Obviously, the following convergence holds in H

$$\lim_{n \to \infty} \sum_{k=1}^{n} e^{-i\lambda_k t} X_k(0) h_k = \sum_{k=1}^{\infty} e^{-i\lambda_k t} (X_0, h_k) h_k.$$

Let $m, n \in \mathbb{N}$ be numbers with $1 \leq m < n$. Then we get with the boundedness of admissible controls and Theorem 31

$$\left\| \sum_{k=m}^{n} \left(\int_0^t e^{-i\lambda_k(t-s)} U_k(s) X_k^U(s) ds \right) h_k \right\|^2 \leq C^2 \sum_{k=m}^{n} \int_0^T |X_k^U(s)|^2 ds \to 0$$

for $m, n \to \infty$. Consequently, the sequence of the partial sums

$$\left(i \sum_{k=1}^{n} \int_0^t e^{-i\lambda_k(t-s)} U_k(s) X_k^U(s) ds h_k \right)_{n \geq 1}$$

is convergent uniformly with respect to $t \in [0, T]$ in mean square in H to

$$i \sum_{k=1}^{\infty} \int_0^t e^{-i\lambda_k(t-s)} U_k(s) X_k^U(s) h_k ds.$$

Then the last sum in (16) must be convergent to

$$i \sum_{k=1}^{\infty} \left(\mu_k \int_0^t e^{-i\lambda_k(t-s)} X_k^U(s) dB_k^h(s) \right) h_k$$

in the same sense (see also Notation 1). Since $X^U(t) \in H$ we have also $X^U(t) = \sum_{k=1}^{\infty} (X^U(t), h_k) h_k$ for all $t \in [0, T]$ with probability 1. Then

$$0 = \left\| \sum_{k=1}^{\infty} X_k^U(t) h_k - \sum_{k=1}^{\infty} (X^U(t), h_k) h_k \right\|^2 = \sum_{k=1}^{\infty} |X_k^U(t) - (X^U(t), h_k)|^2$$

and consequently, $X_k^U(t) = (X^U(t), h_k)$ for all k, $t \in [0, T]$ with probability 1. Hence, we have proven the following result:

Theorem 32. *The process $\{X^U(t) : t \in [0, T]\}$ defined by Theorem 31 is solution of (7) in the sense of Definition 26.*

Remark 18. We consider $A = \Delta$ with $d = 1$ and homogenous Dirichlet or Neuman conditions and the associated triple (V, H, V^*) of rigged Hilbert spaces as in Example 14 and choose $X_0 \in V$. We get for arbitrary $y \in V$

$$\left\| \sum_{k=1}^{\infty} U_k(t)(y, h_k) h_k \right\|_V^2 \leq C \left(\|y\|^2 + \sum_{k=1}^{\infty} \lambda_k |(y, h_k)|^2 \right) = C\|y\|_V^2$$

and

$$\left\| \sum_{k=1}^{\infty} \mu_k(y, h_k) h_k \right\|_V^2 \leq C \left(\|y\|^2 + \sum_{k=1}^{\infty} \lambda_k |(y, h_k)|^2 \right) = C\|y\|_V^2.$$

There exists a unique variational solution of (6), that is: It holds for all $v \in V$ and $t \in [0, T]$ with probability 1

$$(X^U(t), v) = (X_0, v) + i \int_0^t \langle AX^U(s), v \rangle ds$$

$$+ i \sum_{k=1}^{\infty} \int_0^t U_k(s)(X^U(s), h_k)(h_k, v) ds$$

$$+ i \sum_{k=1}^{\infty} \mu_k \int_0^t (X^U(s), h_k)(h_k, v) dB_k^h(s) \qquad (17)$$

and $\sup_{t \in [0,T]} \mathbb{E}\|X^U(t)\|^2 + \mathbb{E} \int_0^T \|X^U(s)\|_V^2 ds < \infty.$

This statemant we can prove if we consider instead of Equation (16) for $X_n(t) = \sum_{k=1}^{n} X_k^n(t) h_k$ the problem

$$X_n(t) = \sum_{k=1}^{n} (X_0, h_k) h_k + i \int_0^t AX_n(s) ds$$

$$+ i \int_0^t U(s) X_n(s) ds + i \sum_{k=1}^{n} \mu_k \int_0^t X_k^n h_k dB_k^h(s), t \in [0, T]$$

with $X_k^n(t) = Y_k^{(2),U}(t) Y_k^{(1)}(t)$, where $Y_k^{(2),U}$ solves (13) and $Y_k^{(1)}$ solves (10). $\qquad \diamond$

Remark 19. The above results are also true, if we replace μ_k by deterministic functions $\mu_k : [0, T] \to \mathbb{R}$ with

$$\sup_{t \in [0,T]} |\mu_k(t)| \leq \nu_k \text{ and } \sum_{k=1}^{\infty} \nu_k^2 < \infty.$$

$\diamond$

4. An Optimal Control Problem for a Linear Stochastic Schrödinger Equation Driven by Multiplicative Fractional Noise

Let $y \in H$ and $\gamma_k : [0, T] \to \mathbb{R}$, $k \geq 1$, be given such that

$$\sum_{k=1}^{\infty} (U_k(t) - \gamma_k(t))^2 dt < \infty$$

for all admissible controls $U = (U_1, U_2, \ldots) \in \mathcal{U}$ ($\mathcal{U}$ is given in Section 3). Now we introduce the objective function

$$J(U, X^U) = \mathbb{E}\|X^U(T) - y\|^2 + \sum_{k=1}^{\infty} \int_0^T (U_k(t) - \gamma_k(t))^2 dt$$

$$= \sum_{k=1}^{\infty} \mathbb{E}|(X^U(T) - y, h_k)|^2 + \sum_{k=1}^{\infty} \int_0^T (U_k(t) - \gamma_k(t))^2 dt, \quad (18)$$

where X^U, constructed in Theorem 31, is the solution of Equation (7) (see Theorem 32). We consider the optimal control problem

$$\min \left\{ J(U, X^U) : U = (U_1, U_2, \ldots) \in \mathcal{U} \right\}. \quad (19)$$

By applying a general result from optimization theory concerning uniformly convex Banach spaces (see [10, Théoréme 4.2, page 23]) the following theorem holds:

Theorem 33. *There exists a unique optimal control* $U^* = (U_1^*, U_2^*, \ldots) \in \mathcal{U}$ *of the problem* (19).

If $\eta_k(t) := \exp\left\{ -\frac{1}{2}\mu_k^2 t^{2h} - \mathrm{i}\mu_k B_k^h(t) \right\} = \frac{1}{Y_k^{(1)}(t)}$, then we can write

$$J(U, X^U) = \sum_{k=1}^{\infty} \mathbb{E} \left| \frac{Y_k^{(2),U}(T)}{\eta_k(T)} - y_k \right|^2 + \sum_{k=1}^{\infty} \int_0^T \left(U_k(t) - \gamma_k(t) \right)^2 dt, \quad (20)$$

where for each $k \geq 1$ and $t \in [0, T]$

$$dY_k^{(2),U}(t) = (-\mathrm{i}\lambda_k + \mathrm{i}U_k(t))Y_k^{(2),U}(t)dt, \quad Y_k^{(2),U}(0) = (X_0, h_k) \quad (21)$$

which is equivalent to

$$Y_k^{(2),U}(t) = e^{-\mathrm{i}\lambda_k t}(X_0, h_k) + \mathrm{i} \int_0^t e^{-\mathrm{i}\lambda_k(t-s)} U_k(s) Y_k^{(2),U}(s) ds. \quad (22)$$

Hence, we have transformed the controlled equation into a pathwise equation.

We want to derive a variational inequality as a necessary optimality condition by using the gradient computed with respect to U of the objective function.

Remark 20. We assume that for at least one index $k \geq 1$ the scalar product (y, h_k) is not zero. If there is k with $y_k = (y, h_k) = 0$, then we get by (10) and by (12)

$$\mathbb{E} \left| \frac{Y_k^{(2),U}(T)}{\eta_k(T)} \right|^2 = \mathbb{E}|Y_k^{(2),U}(T)Y_k^{(1)}(T)|^2 = |(X_0, h_k)|^2 \exp\{\mu_k^2 T^{2h}\},$$

which obviously does not depend on U_k. Consequently, we get $U_k^*(t) = \gamma_k(t)$. If $y = 0$, then $U_k^*(t) = \gamma_k(t)$ for $k \geq 1$ and all $t \in [0, T]$. $\diamond$

Let $U^* = (U_1^*, U_2^*, \ldots) \in \mathcal{U}$ be the optimal control for problem (19). For $k \geq 1$ the process $\{Y_k^{(2),U^*}(t) : t \in [0,T]\}$ denotes the solution of (21) for $U_k = U_k^*$. We know it also solves (22) for $U_k = U_k^*$.

Let $U = (U_1, U_2 \ldots) \in \mathcal{U}$ be an arbitrary admissible control. We introduce for $k \geq 1$ and $\varepsilon > 0$

$$U_{k,\varepsilon} := U_k^* + \varepsilon(U_k - U_k^*).$$

Let $Y_{k,\varepsilon}^{(2)}$ be the solution of (21) for the control $U_{k,\varepsilon}$. By standard arguments one can prove the following result:

Theorem 34. *If* $\xi_{k,\varepsilon}(t) := \dfrac{1}{\varepsilon}\big(Y_{k,\varepsilon}^{(2)}(t) - Y_k^{(2),U^*}(t)\big)$, $t \in [0,T]$, *then*

$$\lim_{\varepsilon \to 0} \sup_{t \in [0,T]} \mathbb{E}\|\xi_{k,\varepsilon}(t) - \xi_k(t)\|^2 = 0,$$

where

$$\xi_k(t) = -\mathrm{i}\lambda_k \int_0^t \xi_k(s)ds + \mathrm{i} \int_0^t U_k^*(s)\xi_k(s)ds$$

$$+ \mathrm{i} \int_0^t \big(U_k(s) - U_k^*(s)\big)Y_k^{(2),U^*}(s)ds, \ t \in [0,T]. \qquad (23)$$

Next we calculate the Gâteaux derivative of (20) in $U = U_k^*$ applied on $U_k - U_k^*$. For each $\varepsilon > 0$ denote by L_ε the following difference

$$\frac{1}{\varepsilon}J\Big(U_1^*, \ldots, U_{k-1}^*, U_{k,\varepsilon}, U_{k+1}^*, \ldots; \frac{Y_1^{(2),U^*}}{\eta_1(T)}, \ldots, \frac{Y_{k-1}^{(2),U^*}}{\eta_{k-1}(T)}, \frac{Y_{k,\varepsilon}^{(2)}}{\eta_k(T)}, \frac{Y_{k+1}^{(2),U^*}}{\eta_{k+1}(T)}, \ldots\Big)$$

$$- \frac{1}{\varepsilon}J\Big(U^*; \frac{Y_1^{(2),U^*}}{\eta_1(T)}, \ldots, \frac{Y_k^{(2),U^*}}{\eta_k(T)}, \ldots\Big).$$

Since U^* is the optimal control we get

$$0 \leq L_\varepsilon = \frac{1}{\varepsilon}\mathbb{E}\Big(\Big|\frac{Y_{k,\varepsilon}^{(2)}(T)}{\eta_k(T)} - (y,h_k)\Big|^2 - \Big|\frac{Y_k^{(2),U^*}(T)}{\eta_k(T)} - (y,h_k)\Big|^2\Big)$$

$$+ \frac{1}{\varepsilon}\Big(\int_0^T |U_{k,\varepsilon}(t) - \gamma_k(t)|^2 dt - \int_0^T |U_k^*(t) - \gamma_k(t)|^2 dt\Big).$$

Theorem 35. *It holds*

$$0 \leq \lim_{\varepsilon \to 0} L_\varepsilon = 2\mathbb{E}\left(\mathrm{Re}\left[\Big(\frac{Y_k^{(2),U^*}(T)}{\eta_k(T)} - y_k\Big)\frac{\overline{\xi_k(T)}}{\overline{\eta_k(T)}}\right]\right)$$

$$+ 2\int_0^T \big(U_k^*(t) - \gamma_k(t)\big)\big(U_k(t) - U_k^*(t)\big)dt$$

for all $k \geq 1$ *and all admissible controls* $U = (U_1, U_2, \ldots) \in \mathcal{U}$.

Proof. We recall the formula $|z_1 + z_2|^2 = |z_1|^2 + 2\text{Re}[z_1\bar{z}_2]$ for $z_1, z_2 \in \mathbb{C}$. Obviously, we get

$$
L_\varepsilon = \frac{1}{\varepsilon}\mathbb{E}\left(\left|\frac{Y_{k,\varepsilon}^{(2)}(T) - Y_k^{(2),U^*}(T)}{\eta_k(T)} + \frac{Y_k^{(2),U^*}(T)}{\eta_k(T)} - (y, h_k)\right|^2\right.
$$

$$
\left. - \left|\frac{Y_k^{(2),U^*}(T)}{\eta_k(T)} - (y, h_k)\right|^2\right)
$$

$$
+ \frac{1}{\varepsilon}\left(\int_0^T |U_{k,\varepsilon}(t) - \gamma_k(t)|^2 dt - \int_0^T |U_k^*(t) - \gamma_k(t)|^2 dt\right)
$$

$$
= \frac{1}{\varepsilon}\mathbb{E}\left(\left|\frac{Y_{k,\varepsilon}^{(2)}(T) - Y_k^{(2),U^*}(T)}{\eta_k(T)}\right|^2\right.
$$

$$
\left. + \frac{2}{\varepsilon}\text{Re}\left[\left(\frac{Y_k^{(2),U^*}(T)}{\eta_k(T)} - (y, h_k)\right)\frac{\overline{Y}_{k,\varepsilon}^{(2)}(T) - \overline{Y}_k^{(2),U^*}(T)}{\overline{\eta}_k(T)}\right]\right)
$$

$$
+ \varepsilon\int_0^T |U_k(t) - U_k^*(t)|^2 dt + 2\int_0^T \left(U_k^*(t) - \gamma_k(t)\right)\left(U_k(t) - U_k^*(t)\right) dt.
$$

It follows from Theorem 34 that the right hand side of the last equation tends to the term of the statement for $\varepsilon \to 0$. $\qquad\square$

For $k \geq 1$ we introduce

$$
\Phi_{k,T} := -2i\frac{1}{\overline{\eta}_k(T)}\left(\frac{Y_k^{(2),U^*}(T)}{\eta_k(T)} - y_k\right)
$$

and the stochastic backward equation

$$
V_k^*(t) = \Phi_{k,T} + i\lambda_k\int_t^T V_k^*(s)ds - i\int_t^T U_k^*(s)V_k^*(s)ds - \int_t^T Z_k^*(s)dW_k(s) \quad (24)
$$

for $t \in [0, T]$. By [48, Theorem 4.2] the following result holds:

Theorem 36. *With probability 1 there exist unique solutions* V_k^*, Z_k^* *of* (24), *which are* $(\mathcal{F}_{k,t})_{t\in[0,T]}$-*adapted, and*

$$
\mathbb{E}\sup_{t\in[0,T]} |V_k^*(t)|^2 < \infty, \quad \mathbb{E}\int_0^T |Z_k^*(s)|^2 ds < \infty
$$

for $k \geq 1$.

We now can prove an optimality condition of maximum principle type.

Theorem 37. *If* $U^* = (U_1^*, U_2^*, \ldots) \in \mathcal{U}$ *is the optimal control, then the following variational inequality holds*

$$0 \leq \mathbb{E} \int_0^T \left(\mathrm{Re}\left[V_k^*(t) \overline{Y}_k^{(2),U^*}(t) \right] + 2(U_k^*(t) - \gamma_k(t)) \right) (U_k(t) - U_k^*(t)) \, dt$$
(25)

for each $k \geq 1$ *and all admissible controls* $U = (U_1, U_2, \ldots) \in \mathcal{U}$.

Proof. We introduce the random variables

$$\Psi_{k,T} := \Phi_{k,T} + \mathrm{i}\lambda_k \int_0^T V_k^*(s) \, ds - \mathrm{i} \int_0^T U_k^*(s) V_k^*(s) \, ds - \int_0^T Z_k^*(s) \, dW_k(s).$$

Then, for $t \in [0, T]$

$$V_k^*(t) = \Psi_{k,T} - \mathrm{i}\lambda_k \int_0^t V_k^*(s) \, ds + \mathrm{i} \int_0^t U_k^*(s) V_k^*(s) \, ds + \int_0^t Z_k^*(s) \, dW_k(s).$$

It follows from the martingale representation Theorem 28, that there are $(\mathcal{F}_{k,t})_{t \in [0,T]}$-adapted processes $Z_k^{(1)}$ with $\mathbb{E} \int_0^T |Z_k^{(1)}(t)|^2 \, dt < \infty$ such that

$$\mathbb{E}(\Psi_{k,T} \mid \mathcal{F}_t) = \mathbb{E}(\Psi_{k,T}) + \int_0^t Z_k^{(1)}(s) \, dW_k(s).$$

Consequently, we get for $t \in [0, T]$

$$V_k^*(t) = \mathbb{E}(\Psi_{k,T}) - \mathrm{i}\lambda_k \int_0^t V_k^*(s) \, ds + \mathrm{i} \int_0^t U_k^*(s) V_k^*(s) \, ds$$
$$+ \int_0^t (Z_k^*(s) + Z_k^{(1)}(s)) \, dW_k(s).$$
(26)

Therefore, by the Itô formula we obtain

$$\mathbb{E}(\Phi_{k,T} \overline{\xi}_k(T)) = \mathbb{E}(V_k^*(T) \overline{\xi}_k(T))$$
$$= \mathbb{E} \int_0^T V_k^*(t) \, d\overline{\xi}_k(t) \, dt + \mathbb{E} \int_0^T \overline{\xi}_k(t) \, dV_k^*(t),$$

since $\xi_k(0) = 0$. By using (23) and (26), we obtain by Theorem 35 the required statement. □

Remark 21.

(1) We see that the right hand side of the inequality in Theorem 35 is the Gâteaux derivative of J in U_k^*.

(2) Analogously, we can calculate the Gâteaux derivative of J in U_k' for every admissible control U'. We get an analogous term, where U_k^* is substituted by U_k'.

(3) We see immediately from Theorem 35 that the Gâteaux derivative depends continuously on U_k'. Theorem 37 shows that, if the Gâteaux derivative is applied to $U_k' - U_k$, then we have a linear and continuous term with respect to $U_k' - U_k$. Consequently, the Gâteaux derivative is the Fréchet derivative too.

(4) It is possible to consider the optimization problem for controls with

$$\int_0^T U_k^2(t)dt < \infty \text{ for each } k \geq 1.$$

The set of all admissible controls is

$$\mathcal{U} = \{U = (U_1, U_2, ...) : U_k \in L^2[0, T] \text{ for each } k \geq 1\}.$$

Fix $k \geq 1$. Then, the optimality condition (25) becomes

$$0 \leq \int_0^T \mathbb{E}\Big(\mathrm{Re}\Big[V_k^*(t)\overline{Y}_k^{(2),U^*}(t) \Big]\Big) + 2(U_k^*(t) - \gamma_k(t))\Big) (u(t) - U_k^*(t)) \, dt$$

for each $u \in L^2[0, T]$. This implies

$$0 \leq \int_0^T \mathbb{E}\Big(\mathrm{Re}\Big[V_k^*(t)\overline{Y}_k^{(2),U^*}(t) \Big]\Big) + 2(U_k^*(t) - \gamma_k(t))\Big) u(t) dt$$

for each $u \in L^2[0, T]$. Hence, for each $k \geq 1$ it holds

$$0 = \mathrm{Re}\Big[- V_k^*(t)\overline{Y}_k^{(2),U^*}(t) \Big] + 2(U_k^*(t) - \gamma_k(t))$$

for a.e. $(t, \omega) \in [0, T] \times \Omega$. $\Diamond$

5. An Optimal Control Problem for a Linear Stochastic Schrödinger Equation Driven by Additive Fractional Noise

Let $\beta^h(t) := \sum_{k=1}^{\infty} \mu_k \int_0^t e^{-i\lambda_k(t-s)} h_k dB_k^h(s)$ with $\sum_{k=1}^{\infty} \mu_k^2 < \infty$. Then, for $U = (U_1, U_2, ...) \in \mathcal{U}$ ($\mathcal{U}$ is given in Section 3), the mild solution of

$$\begin{cases} dX^U(t) = iAX^U(t)dt + i\sum_{k=1}^{\infty} U_k(t)(X^U(t), h_k)h_k dt + i\sum_{k=1}^{\infty} \mu_k h_k dB_k^h(t), \\ X^U(0) = X_0 \end{cases}$$

is the $(\mathcal{F}_t)_{t \geq 0}$-adapted process $\{X^U(t) : t \in [0, T]\}$ such that

$$X^U(t) = \sum_{k=1}^{\infty} e^{-i\lambda_k t}(X_0, h_k)h_k + i\sum_{k=1}^{\infty} \int_0^t e^{-i\lambda_k(t-s)} U_k(s)(X^U(s), h_k)h_k ds$$

$$+ i \sum_{k=1}^{\infty} \mu_k \int_0^t e^{-i\lambda_k(t-s)} h_k dB_k^h(s) \tag{27}$$

for all $t \in [0,T]$ a.s. and $\displaystyle\sup_{t\in[0,T]} \mathbb{E}\|X^U(t)\|^2 < \infty$.

Remark 22. Note, that term $\displaystyle\sum_{k=1}^{\infty} \mu_k \int_0^t e^{-i\lambda_k(t-s)} h_k dB_k^h(s)$ in (27) is understood in the sense of Notation 1. Using computations in the Hilbert space H and Example 12 we have

$$\mathbb{E}\left\| \sum_{k=1}^{n} \mu_k \int_0^t e^{-i\lambda_k(t-s)} h_k dB_k^h(s) \right\|^2 \leq \sum_{k=1}^{n} \mu_k^2 \mathbb{E}\left| \int_0^t e^{i\lambda_k s} dB_k^h(s) \right|^2$$

$$\leq T^{2h} \sum_{k=1}^{n} \mu_k^2 < \infty$$

for each $t \in [0,T]$. Then we obtain that the sequence $\left(\displaystyle\sum_{k=1}^{n} \mu_k \int_0^t e^{-i\lambda_k(t-s)} h_k dB_k^h(s) \right)_{n\geq 1}$ converges in $L^2(\Omega; H)$ and

$$\sup_{t\in[0,T]} \mathbb{E}\left\| \sum_{k=1}^{\infty} \mu_k \int_0^t e^{-i\lambda_k(t-s)} h_k dB_k^h(s) \right\|^2 < \infty.$$

$\diamond$

By the substitution $Y^U(t) := X^U(t) - i\beta^h(t)$, $t \in [0,T,]$ we can transform the above problem into the following pathwise inhomogenous problem

$$Y^U(t) = \sum_{k=1}^{\infty} e^{-i\lambda_k t}(X_0, h_k) h_k$$

$$+ i \sum_{k=1}^{\infty} \int_0^t e^{-i\lambda_k(t-s)} U_k(s) \left((Y^U(s), h_k) + \lambda_k \mu_k B_k^h(s) \right) h_k ds \tag{28}$$

for all $t \in [0,T]$ a.s. and $\displaystyle\sup_{t\in[0,T]} \mathbb{E}\|Y^U(t)\|^2 < \infty$. It follows from the end of the proof of [30, Theorem 6], if we set $g = 0$, that there exists a unique $(\mathcal{F}_t)_{t\in[0,T]}$-adapted solution process $\{Y^U(t) : t \in [0,T]\}$ of (28) with $\displaystyle\sup_{t\in[0,T]} \mathbb{E}\|Y(t)\|^2 < \infty$. By the definition of Y^U and by Remark 22 we see, that X^U has also these properties.

If we introduce for $k \geq 1$

$$dY_k^U(t) = -i\lambda_k Y_k^U(t) dt + i U_k(t)(Y_k^U(t) + i\mu_k B_k^h(t)) dt, \quad Y_k^U(0) = (X_0, h_k) \tag{29}$$

then the solution process is for $t \in [0, T]$

$$Y_k^U(t) = e^{-\mathrm{i}\lambda_k t + \mathrm{i}\int_0^t U_k(s)ds} \left((X_0, h_k) + \int_0^t \mathrm{i}\mu_k B_k^h(s) e^{\mathrm{i}\lambda_k s - \mathrm{i}\int_0^s U_k(r)dr} ds \right). \tag{30}$$

Obviously, we get

$$Y^U(t) = \sum_{k=1}^{\infty} Y_k^U(t) h_k.$$

If the objective function J is given as in (18), then we get

$$J(U, X) = \mathbb{E}\|Y^U(T) + \mathrm{i}\beta^h(T) - y\|^2 + \sum_{k=1}^{\infty} \mathbb{E} \int_0^T (U_k(t) - \gamma_k(t))^2 dt.$$

As in the last section it follows the existence of a unique optimal control U^*. We introduce the stochastic backward equations

$$V_k^*(t) = \Phi_{k,T} + \mathrm{i}\lambda_k \int_t^T V_k^*(s)ds - \mathrm{i} \int_t^T U_k^*(s)V_k^*(s)ds - \int_t^T Z_k^*(s)dW_k(s) \tag{31}$$

for $t \in [0, T]$, where $\Phi_{k,T} = -2i(Y_k^{U^*}(T) + \mathrm{i}\mu_k B_k^h(T) - (y, h_k))$.

Then we can prove with similar proof steps as in the last section the following result:

Theorem 38. *If $U^* = (U_1^*, U_2^*, \dots)$ is the optimal control, then the following variational inequality holds*

$$0 \leq \mathbb{E} \int_0^T \left(\mathrm{Re}\left[V_k^*(t)\overline{Y}_k^{U^*}(t) \right] + 2(U_k^*(t) - \gamma_k(t)) \right) (U_k(t) - U_k^*(t)) \, dt \tag{32}$$

for each $k \geq 1$ and all admissible controls $U = (U_1, U_2, \dots) \in \mathcal{U}$.

6. A Nonlinear Stochastic Schrödinger Equation Driven by Multiplicative Fractional Noise

Let μ be a real parameter and $\{B^h(t) : t \geq 0\}$ be a real-valued fractional Brownian motion. Consider the **set of admissible controls** to be

$$\hat{\mathcal{U}} = \{U : [0, T] \to \mathbb{R}, \ |U(t)| \leq c \text{ for all } t \geq 0\},$$

where $c > 0$ is given.

Let $f : [0, T] \times H \to H$ be a measurable function such that $f(t, 0) = 0$ for each $t \in [0, T]$ and the Fréchet derivative of $x \in H \mapsto f(t, x) \in H$, which we denote by $\nabla f(t, \cdot)$, is bounded for each $t \in [0, T]$ by a constant which is independent of t. This implies that there exists a positive constant K such that

$$\|f(t,x) - f(t,y)\| \leq K\|x - y\|$$

for all $x, y \in H$, $t \in [0,T]$.

The state equation is

$$dX^U(t) = i\Big(AX^U(t) + f(t, X^U(t)) + U(t)X^U(t)\Big)dt + i\mu X^U(t)dB^h(t) \quad (33)$$

with $X(0) = X_0 \in H$ and $U \in \hat{\mathcal{U}}$ being an admissible control.

Definition 27. X^U is a **solution** of (33), if

$$X^U(t) = \sum_{k=1}^{\infty} e^{-i\lambda_k t}(X_0, h_k)h_k + i\sum_{k=1}^{\infty} \int_0^t e^{-i\lambda_k(t-s)}(f(s, X^U(s)), h_k)h_k ds$$

$$+ i\sum_{k=1}^{\infty} \int_0^t e^{-i\lambda_k(t-s)}U(s)(X^U(s), h_k)h_k ds$$

$$+ i\sum_{k=1}^{\infty} \int_0^t e^{-i\lambda_k(t-s)}\mu(X^U(s), h_k)h_k dB^h(s) \quad (34)$$

for all $t \in [0,T]$ a.s.

The last term in (34) (containing the stochastic integrals) is understood in the sense of Notation 1. $\diamond$

We will prove that the process

$$X^U(t) = Y^{(2),U}(t)Y^{(1)}(t), \, t \in [0,T], \quad (35)$$

is a solution of (34), where

$$Y^{(1)}(t) = 1 + i\mu \int_0^t Y^{(1)}(s)dB^h(s), \, t \in [0,T], \quad (36)$$

and $Y^{(2),U}$ is the solution of the pathwise stochastic integral equation

$$Y^{(2),U}(t) = \sum_{k=1}^{\infty} e^{-i\lambda_k t}(X_0, h_k)h_k$$

$$+ i\sum_{k=1}^{\infty} \int_0^t e^{-i\lambda_k(t-s)}\frac{1}{Y^{(1)}(s)}(f(s, Y^{(2),U}(s)Y^{(1)}(s)), h_k)h_k ds$$

$$+ i\sum_{k=1}^{\infty} \int_0^t e^{-i\lambda_k(t-s)}U(s)(Y^{(2),U}(s), h_k)h_k ds, \, t \in [0,T]. \quad (37)$$

Since $\{B^h(t) : t \in [0,T]\}$ is a real-valued fractional Brownian motion, we get by Example 11-(2) that

$$Y^{(1)}(t) = \exp\left\{\frac{1}{2}\mu^2 t^{2h} + i\mu B^h(t)\right\}, \quad t \in [0,T], \tag{38}$$

is a solution of (36). Moreover, by the properties of the ϕ-derivative (see [25] on p. 588) we have

$$D_r^\phi Y^{(1)}(s) = i\mu \exp\left\{\frac{1}{2}\mu^2 s^{2h} + i\mu B^h(s)\right\} \int_0^s \phi(u,r)du, \tag{39}$$

for $r \in [0,s], s \in [0,T]$, otherwise it is equal to 0.

Example 16. Let H be the complex Hilbert space $L^2(G)$ (see Example 14). We consider the measurable function $\hat{f} : [0,T] \times \mathbb{C} \times \mathbb{C} \to \mathbb{C}$ with

$$\hat{f}(t,0,0) = 0,$$

$$\left|\frac{\partial}{\partial\zeta_2}\hat{f}(t,\zeta_2,\overline{\zeta}_2)\right| = \left|\lim_{\zeta_1\to\zeta_2}\frac{\hat{f}(t,\zeta_1,\overline{\zeta}_2) - \hat{f}(t,\zeta_2,\overline{\zeta}_2)}{\zeta_1 - \zeta_2}\right| \leq C$$

and

$$\left|\frac{\partial}{\partial\overline{\zeta}_2}\hat{f}(t,\zeta_2,\overline{\zeta}_2)\right| = \left|\lim_{\overline{\zeta}_1\to\overline{\zeta}_2}\frac{\hat{f}(t,\zeta_2,\overline{\zeta}_1) - \hat{f}(t,\zeta_2,\overline{\zeta}_2)}{\overline{\zeta}_1 - \overline{\zeta}_2}\right| \leq C$$

for all $\zeta_2 \in \mathbb{C}$ and $t \in [0,T]$.

We denote $f(t,z) = \hat{f}(t,z,\overline{z})$ for all $t \in [0,T]$ and $z \in \mathbb{C}$. Then, we get for all $t \in [0,T]$ and $z_1, z_2, z \in \mathbb{C}$

$$|f(t,z_1)-f(t,z_2)|^2 \leq 2|\hat{f}(t,z_1,\overline{z}_1)-\hat{f}(t,z_2,\overline{z}_1)|^2 + 2|\hat{f}(t,z_2,\overline{z}_1)-\hat{f}(t,z_2,\overline{z}_2)|^2$$

$$\leq 2C^2|z_1 - z_2|^2 + 2C^2|\overline{z}_1 - \overline{z}_2|^2 = 4C^2|z_1 - z_2|$$

and

$$|f(t,z)|^2 \leq C^2|z|^2.$$

Therefore, for all $t \in [0,T]$ and $y_1, y_2, y \in L^2(G)$

$$\|f(t,y_1) - f(t,y_2)\|^2_{L^2(G)} = \int_G |\hat{f}(t,y_1(x),\overline{y}_1(x)) - \hat{f}(t,y_2(x),\overline{y}_2(x))|^2 dx$$

$$\leq 4C^2\int_G |y_1(x) - y_2(x)|^2 dx = 4C^2\|y_1 - y_2\|^2_{L^2(G)}$$

and

$$\|f(t,y)\|^2_{L^2(G)} = \int_G |\hat{f}(t,y(x),\overline{y}(x))|^2 dx \leq C^2\|y\|^2_{L^2(G)}.$$

Consequently, the nonlinearity fulfills the conditions for f from the beginning of this section. $\diamondsuit$

We now give two concrete examples for the function f from Example 16.

Example 17. Let λ be a positive parameter. The following two examples can be found in [11].

(1) Let $f(t,z) = \hat{f}(t,z,\overline{z}) = (1 - e^{-\lambda|z|^2})z$. The exponential nonlinearity serves as a useful model in homogenous, unmagnetized plasmas and laser-produced plasmas.

(2) Let $f(t,z) = \hat{f}(t,z,\overline{z}) = \frac{\lambda|z|^2}{1+\lambda|z|^2}z$. The saturating law describes the variation of the dielectric constant of gas vapors through which a laser beam propagates.

Moreover, each of these functions f admits a bounded Fréchet derivative $\nabla f(t,x) : L^2(G) \to L^2(G)$ for $x \in L^2(G)$, which can be computed by

$$\nabla f(t,x)y = \frac{\partial \hat{f}(t,z,\overline{z})}{\partial z}\bigg|_{z=x,\overline{z}=\overline{x}} y + \frac{\partial \hat{f}(t,z,\overline{z})}{\partial \overline{z}}\bigg|_{z=x,\overline{z}=\overline{x}} \overline{y}$$

for $x,y \in L^2(G)$. $\diamondsuit$

Theorem 39. *There exists a unique* $(\mathcal{F}_t)_{t\in[0,T]}$-*adapted solution process* $\{Y^{(2),U}(t) : t \in [0,T]\}$ *for* (37) *with* $\mathbb{E} \sup\limits_{t\in[0,T]} \|Y^{(2),U}(t)\|^2 < \infty$.

Proof. Let $\mathbb{B} = \mathbb{B}[0,T]$ be the Banach space of all $(\mathcal{F}_t)_{t\in[0,T]}$-adapted processes $\{Y(t) : t \in [0,T]\}$ with $\mathbb{E} \sup\limits_{t\in[0,T]} \|Y(t)\|^2 < \infty$. We introduce for $Y \in \mathbb{B}$ and for fixed $U \in \mathcal{U}$ the operator

$$R(Y)(t) = \sum_{k=1}^{\infty} e^{-i\lambda_k t}(X_0, h_k)h_k$$

$$+ i\sum_{k=1}^{\infty} \int_0^t e^{-i\lambda_k(t-s)}(f(s,Y(s)Y^{(1)}(s)), h_k)h_k \frac{1}{Y^{(1)}(s)} ds$$

$$+ i\sum_{k=1}^{\infty} \int_0^t e^{-i\lambda_k(t-s)}U(s)(Y(s), h_k)h_k ds, \ t \in [0,T].$$

By the assumptions for f and U, by the boundedness of $Y^{(1)}$ and by the Schwarz inequality it follows

$$\|R(Y)(t)\|^2 \leq C\|X_0\|^2$$

$$+ C\left\| \sum_{k=1}^{\infty} \int_0^t e^{-i\lambda_k(t-s)}(f(s,Y(s)Y^{(1)}(s)), h_k)h_k \frac{1}{Y^{(1)}(s)} ds \right\|^2$$

$$+ C \left\| \sum_{k=1}^{\infty} \int_0^t e^{-\mathrm{i}\lambda_k(t-s)} U(s)(Y(s), h_k) h_k ds \right\|^2$$

$$\leq C \|X_0\|^2 + C \int_0^t \|Y(s)\|^2 ds \leq C \|X_0\|^2 + C \int_0^T \|Y(s)\|^2 ds$$

for all $t \in [0, T]$. Consequently,

$$\sup_{t \in [0,T]} \|R(Y)(t)\|^2 \leq C \|X_0\|^2 + C \int_0^T \sup_{r \in [0,s]} \|Y(r)\|^2 ds$$

and

$$\mathbb{E} \sup_{t \in [0,T]} \|R(Y)(t)\|^2 \leq C \|X_0\|^2 + C \int_0^T \mathbb{E} \sup_{r \in [0,s]} \|Y(r)\|^2 ds.$$

Then, it follows that $R : \mathbb{B} \to \mathbb{B}$.

We get for $t_1 \in \,]0, T]$ by the Lipschitz continuity of $Y_1, Y_2 \in \mathbb{B}$ with similar considerations

$$\mathbb{E} \sup_{t \in [0,t_1]} \|R_1(Y)(t) - R_2(t)\|^2 \leq C \int_0^{t_1} \mathbb{E} \sup_{r \in [0,s]} \|Y_1(r) - Y_2(r)\|^2 ds$$

$$\leq C t_1 \mathbb{E} \sup_{t \in [0,t_1]} \|Y_1(t) - Y_2(t)\|^2.$$

We choose $t_1 > 0$ such that $C t_1 < 1$. Consequently, it follows from the Banach fixed point theorem the existence of a unique fixed point in $\mathbb{B}[0, t_1]$. The contraction property holds also for R on $\mathbb{B}([kt_1, (k+1)t_1] \cap [0, T])$, $k \geq 1$. So, we get the claim of the theorem. $\qquad \square$

For $n \geq 1$ we consider the process $\{Y_n^{(2),U} : t \in [0, T]\}$, which takes values in span$\{h_1, ..., h_n\}$ and is the solution of the Galerkin equation

$$Y_n^{(2),U}(t) = \sum_{k=1}^{n} e^{-\mathrm{i}\lambda_k t} (X_0, h_k) h_k$$

$$+ \mathrm{i} \sum_{k=1}^{n} \int_0^t e^{-\mathrm{i}\lambda_k(t-s)} (f(s, Y_n^{(2),U}(s) Y^{(1)}(s)), h_k) h_k \frac{1}{Y^{(1)}(s)} ds$$

$$+ \mathrm{i} \sum_{k=1}^{n} \int_0^t e^{-\mathrm{i}\lambda_k(t-s)} U(s)(Y_n^{(2),U}(s), h_k) h_k ds \,, t \in [0, T]. \quad (40)$$

The n-dimensional Equation (40) has a unique $(\mathcal{F}_t)_{t \in [0,T]}$-adapted mild solution $Y_n^{(2),U}(t)$, $t \in [0, T]$ with $\mathbb{E} \sup_{t \in [0,T]} \|Y_n^{(2),U}(t)\|^2 < \infty$ for every $n \geq 1$.

The proof is similar to the proof of Theorem 39.

We write

$$\|Y^{(2),U}(t) - Y_n^{(2),U}(t)\|^2$$

$$\leq S_n(t)$$

$$+ C \left\| \sum_{k=1}^{n} \int_0^t e^{-i\lambda_k(t-s)} (f(s, Y^{(2),U}(s)Y^{(1)}(s)) - f(s, Y_n^{(2),U}(s)Y^{(1)}(s)), h_k) h_k \right.$$

$$\left. \cdot \frac{1}{Y^{(1)}(s)} ds \right\|^2$$

$$+ C \left\| \sum_{k=1}^{n} \int_0^t e^{-i\lambda_k(t-s)} U(s)(Y^{(2),U}(s) - Y_n^{(2),U}(s), h_k) h_k ds \right\|^2 \tag{41}$$

with

$$S_n(t) = C \sum_{k=n+1}^{\infty} |(X_0, h_k)|^2$$

$$+ C \left\| \sum_{k=n+1}^{\infty} \int_0^t e^{-i\lambda_k(t-s)} (f(s, Y^{(2),U}(s)Y^{(1)}(s)), h_k) h_k \frac{1}{Y^{(1)}(s)} ds \right\|^2$$

$$+ C \left\| \sum_{k=n+1}^{\infty} \int_0^t e^{-i\lambda_k(t-s)} U(s)(Y^{(2),U}(s), h_k) h_k ds \right\|^2 .$$

Further we get with the assumptions for f and the boundedness of U

$$S_n(t) \leq C \sum_{k=n+1}^{\infty} |(X_0, h_k)|^2 + C \sum_{k=n+1}^{\infty} \int_0^t |(Y^{(2),U}(s), h_k)|^2 ds.$$

Since

$$\sum_{k=n+1}^{\infty} |(X_0, h_k)|^2 \leq \|X_0\|^2$$

and

$$\sup_{t \in [0,T]} \sum_{k=n+1}^{\infty} \int_0^t |(Y^{(2),U}(s), h_k)|^2 ds \leq T \sup_{t \in [0,T]} \|Y^{(2),U}(t)\|^2 ,$$

we get

$$\lim_{n \to \infty} \mathbb{E}\left(\sup_{t \in [0,T]} S_n(t) \right) = 0.$$

If we denote the second sum in the right hand side of inequality (41) by $S^{(1),n}(t)$, we get with the properties of h_k, $Y^{(2),U}$, $Y_n^{(2),U}$, the Schwarz inequality and the Lipschitz continuity of f for all $t \in [0,T]$ that

$$\mathbb{E}\big(S^{(1),n}(t)\big) \leq C \int_0^T \mathbb{E} \sup_{r \in [0,s]} \|Y^{(2),U}(r) - Y_n^{(2),U}(r)\|^2 ds.$$

Analogously, for the third sum in the right hand side of (41), denoted by $S^{(2),n}(t)$, we have

$$\mathbb{E}\big(S^{(2),n}(t)\big) \leq C \int_0^T \mathbb{E} \sup_{r \in [0,s]} \|Y^{(2),U}(r) - Y_n^{(2),U}(r)\|^2 ds \text{ for all } t \in [0,T].$$

By using Gronwall's lemma, we obtain the following result:

Theorem 40. *The following convergence holds*

$$\lim_{n \to \infty} \mathbb{E} \sup_{t \in [0,T]} \|Y^{(2),U}(t) - Y_n^{(2),U}(t)\|^2 = 0.$$

We introduce $X_n^U = Y_n^{(2),U} Y^{(1)}$ for each $n \geq 1$ and recall from (35) that $X^U = Y^{(2),U} Y^{(1)}$. Obviously by Theorem 40 it holds

$$\lim_{n \to \infty} \mathbb{E} \sup_{t \in [0,T]} \|X^U(t) - X_n^U(t)\|^2 = 0. \tag{42}$$

Theorem 41. *The process X^U constructed above (see (35) and (42)) is a solution of (34) in the sense of Definition 27.*

Proof. We recall the expression of the ϕ-derivative of $Y^{(1)}$, see (39), and the ϕ-derivative of $Y_n^{(2),U}$ which satisfies

$$D_r^\phi Y_n^{(2),U}(t) = \mathrm{i} \sum_{k=1}^n \int_0^t e^{-\mathrm{i}\lambda_k(t-s)} \Big(\frac{1}{Y^{(1)}(s)} D_r^\phi \big(f(s, Y_n^{(2),U}(s) Y^{(1)}(s))\big)$$

$$+ f(s, Y_n^{(2),U}(s) Y^{(1)}(s)) D_r^\phi \Big(\frac{1}{Y^{(1)}(s)} \Big), h_k \Big) h_k ds$$

$$+ \mathrm{i} \sum_{k=1}^n \int_0^t e^{-\mathrm{i}\lambda_k(t-s)} U(s) \Big(D_r^\phi Y_n^{(2),U}(s), h_k \Big) h_k ds,$$

$r \in [0,t], t \in [0,T]$. Moreover, for $r \in [0,s]$

$$D_r^\phi \big(f(s, Y_n^{(2),U}(s) Y^{(1)}(s))\big)$$

$$= \nabla f(s,x) \Big|_{x=Y_n^{(2),U}(s) Y^{(1)}(s)} \Big(D_r^\phi Y_n^{(2),U}(s) Y^{(1)}(s) + Y_n^{(2),U}(s) D_s^\phi Y^{(1)}(s) \Big).$$

Here we used the properties of the ϕ-derivative (see [25] on p. 588). One can easily prove by the properties of $Y^{(1)}$, $Y_n^{(2),U}$ and by the assumptions on f and U that

$$\sup_{t \in [0,T]} \sup_{r \in [0,t]} \mathbb{E}\|D_r^\phi Y_n^{(2),U}(t)\|^2 \leq C \text{ for each } n \geq 1.$$

From (40) we have the finite dimensional process $Y_n^{(2),U}$, which is the unique solution of

$$e^{i\lambda_k t}Y_n^{(2),U}(t) = \sum_{k=1}^{n}(X_0,h_k)h_k$$
$$+ i\sum_{k=1}^{n}\int_0^t e^{i\lambda_k s}(f(s,Y_n^{(2),U}(s)Y^{(1)}(s)),h_k)h_k\frac{1}{Y^{(1)}(s)}ds$$
$$+ i\sum_{k=1}^{n}\int_0^t e^{i\lambda_k s}U(s)(Y_n^{(2),U}(s),h_k)h_k ds\,,\, t\in[0,T].$$

By the product rule (see Example 13, where we use $Y_n^{(2),U}Y^{(1)}\in\mathcal{L}[0,T]$) we obtain

$$d\big(e^{i\lambda_k t}X_n^U(t)\big) = Y^{(1)}(t)d\big(e^{i\lambda_k t}Y_n^{(2),U}(t)\big) + e^{i\lambda_k t}Y_n^{(2),U}(t)dY^{(1)}(t),$$

where $X_n^U(0) = \sum_{k=1}^{n}(X_0,h_k)h_k$.

Therefore, we obtain

$$X_n^U(t) = \sum_{k=1}^{n}e^{-i\lambda_k t}(X_0,h_k)h_k + i\sum_{k=1}^{n}\int_0^t e^{-i\lambda_k(t-s)}(f(s,X_n^U(s)),h_k)h_k ds$$
$$+ i\sum_{k=1}^{n}\int_0^t e^{-i\lambda_k(t-s)}U(s)(X_n^U(s),h_k)h_k ds$$
$$+ i\sum_{k=1}^{n}\int_0^t e^{-i\lambda_k(t-s)}\mu(X_n^U(s),h_k)h_k dB^h(s)\,,\, t\in[0,T]. \quad (43)$$

It follows from (42) that the left hand side of (43) is convergent in mean square in H to $X^U(t)$ for all $t\in[0,T]$. Obviously, we have the convergence in H for the first sum of the right hand side of (43) to the first sum of the right hand side of (34) for all $t\in[0,T]$. It follows also from (42) the convergence in mean square in H of the second and third sums of the right hand side of (43) to the second and third sums of the right hand side of (34). Consequently, the last sums of the right hand side in (43) must be convergent in the same sense to the last term on the right hand side of (34). $\square$

7. An Optimal Control Problem for a Nonlinear Stochastic Schrödinger Equation Driven by Multiplicative Fractional Noise

Let $y : [0,T] \to H$ and $\gamma : [0,T] \to \mathbb{R}$ with $\displaystyle\int_0^T \|y(t)\|^2 dt < \infty$ and $\displaystyle\int_0^T |\gamma(t)|^2 dt < \infty$. We introduce for $U \in \hat{\mathcal{U}}$ (see Section 6) the objective function

$$J(U, X^U) = \mathbb{E} \int_0^T \left(\|X^U(t) - y(t)\|^2 + |U(t) - \gamma(t)|^2 \right) dt$$

and consider the optimal control problem

$$\min \left\{ J(U, X^U) : U \in \hat{\mathcal{U}} \right\}, \tag{44}$$

where X^U is the process given by (35) and is a solution of (34), where we assume that f satisfies the assumptions given at the beginning of Section 6. Problem (44) is equivalent to

$$\min \left\{ J(U, Y^{(2),U} Y^{(1)}) : U \in \hat{\mathcal{U}} \right\}, \tag{45}$$

where $Y^{(2),U}$ is the solution process of (37) and $Y^{(1)}$ is defined in (38). By [10, Théoréme 4.2, page 23] follows the existence of a unique optimal control U^*. In what follows we prove an optimality condition of maximum principle type.

We denote $U_\varepsilon = U^* + \varepsilon(U - U^*)$, where $U^* \in \hat{\mathcal{U}}$ is the optimal control and $U \in \hat{\mathcal{U}}$ is an admissible control and by (37) we get

$$Y^{(2),U_\varepsilon}(t) - Y^{(2),U^*}(t) \tag{46}$$

$$= \mathrm{i} \sum_{k=1}^{\infty} \int_0^t \frac{e^{-\mathrm{i}\lambda_k(t-s)}}{Y^{(1)}(s)} \left(f(s, Y^{(2),U_\varepsilon}(s) Y^{(1)}(s)), h_k \right) h_k ds$$

$$- \mathrm{i} \sum_{k=1}^{\infty} \int_0^t \frac{e^{-\mathrm{i}\lambda_k(t-s)}}{Y^{(1)}(s)} \left(f(s, Y^{(2),U^*}(s) Y^{(1)}(s)), h_k \right) h_k ds$$

$$+ \mathrm{i} \sum_{k=1}^{\infty} \int_0^t e^{-\mathrm{i}\lambda_k(t-s)} \left(U^*(s) + \varepsilon(U(s) - U^*(s)) \right)$$

$$\cdot (Y^{(2),U_\varepsilon}(s) - Y^{(2),U^*}(s), h_k) h_k ds$$

$$+ \varepsilon \mathrm{i} \sum_{k=1}^{\infty} \int_0^t e^{-\mathrm{i}\lambda_k(t-s)} \left(U(s) - U^*(s) \right) \left(Y^{(2),U^*}(s), h_k \right) h_k ds \tag{47}$$

and it holds

$$\mathbb{E}\sup_{t\in[0,T]}\|Y^{(2),U_\varepsilon}(t)-Y^{(2),U^*}(t)\|^2<\infty.$$

Notation 2. For $U\in\hat{\mathcal{U}}, y\in H$ and $t\in[0,T]$ we denote

$$\nabla f\big(t,Y^{(2),U}(t)Y^1(t)\big)y=\nabla f(t,x)\Big|_{x=Y^{(2),U}(t)Y^1(t)}Y^{(1)}(t)y.$$

$\diamond$

Example 18. In the concrete Example 17 we have

$$\nabla f\big(t,Y^{(2),U}(t)Y^{(1)}(t)\big)y=\nabla f(t,x)\Big|_{x=Y^{(2),U}(t)Y^1(t)}Y^{(1)}(t)y$$

$$=\frac{\partial}{\partial z}\hat{f}(t,z,\overline{z})\Big|_{z=Y^{(2),U}(t)Y^{(1)}(t),\overline{z}=\overline{Y}^{(2),U}(t)\overline{Y}^{(1)}(t)}Y^{(1)}(t)y$$

$$+\frac{\partial}{\partial\overline{z}}\hat{f}(t,z,\overline{z})\Big|_{z=Y^{(2),U}(t)Y^{(1)}(t),\overline{z}=\overline{Y}^{(2),U}(t)\overline{Y}^{(1)}(t)}\overline{Y}^{(1)}(t)\overline{y}$$

for $y\in H=L^2(G)$.

$\diamond$

Theorem 42. *The equation*

$$\xi(t)=i\sum_{k=1}^{\infty}\int_0^t\frac{e^{-i\lambda_k(t-s)}}{Y^{(1)}(s)}(\nabla f\big(s,Y^{(2),U^*}(s)Y^{(1)}(s)\big)\xi(s),h_k)h_kds$$

$$+i\sum_{k=1}^{\infty}\int_0^t e^{-i\lambda_k(t-s)}U^*(s)(\xi(s),h_k)h_kds$$

$$+i\sum_{k=1}^{\infty}\int_0^t e^{-i\lambda_k(t-s)}(U(s)-U^*(s))(Y^{(2),U^*}(s),h_k)h_kds\quad(48)$$

has a unique $(\mathcal{F}_t)_{t\in[0,T]}$-*adapted solution* $\{\xi(t):t\in[0,T]\}$ *with*

$$\mathbb{E}\sup_{t\in[0,T]}\|\xi(t)\|^2<\infty$$

and

$$\lim_{\varepsilon\to0}\mathbb{E}\sup_{t\in[0,T]}\left\|\frac{1}{\varepsilon}\big(Y^{(2),U_\varepsilon}(t)-Y^{(2),U^*}(t)\big)-\xi(t)\right\|^2=0.$$

Proof. The existence and uniqueness proof for ξ can be carried out analogously to Theorem 39. We denote

$$\delta_\varepsilon(t)=\frac{1}{\varepsilon}\big(Y^{(2),U_\varepsilon}(t)-Y^{(2),U^*}(t)\big)-\xi(t).$$

Then,

$$\|\delta_\varepsilon(t)\|^2$$

$$= \left\| \frac{i}{\varepsilon} \sum_{k=1}^{\infty} \int_0^t \frac{e^{-i\lambda_k(t-s)}}{Y^{(1)}(s)} (f(s, Y^{(2),U_\varepsilon}(s)Y^{(1)}(s)), h_k) h_k ds \right.$$

$$- \frac{i}{\varepsilon} \sum_{k=1}^{\infty} \int_0^t \frac{e^{-i\lambda_k(t-s)}}{Y^{(1)}(s)} (f(s, Y^{(2),U^*}(s)Y^{(1)}(s)), h_k) h_k ds$$

$$- i \sum_{k=1}^{\infty} \int_0^t \frac{e^{-i\lambda_k(t-s)}}{Y^{(1)}(s)} (\nabla f(s, Y^{(2),U^*}(s)Y^{(1)}(s)) \xi(s), h_k) h_k ds$$

$$+ \frac{i}{\varepsilon} \sum_{k=1}^{\infty} \int_0^t e^{-i\lambda_k(t-s)} (U^*(s) + \varepsilon(U(s) - U^*(s))) (Y^{(2),U_\varepsilon}(s) - Y^{(2),U^*}(s)), h_k) h_k$$

$$- i \sum_{k=1}^{\infty} \int_0^t e^{-i\lambda_k(t-s)} U^*(s) (\xi(s), h_k) h_k ds \left. \right\|^2.$$

From the properties of the Fréchet derivative of f and by Notation 2 we have (in H, pointwise in $s \in [0, T]$)

$$\frac{1}{\varepsilon} \Big(f(s, Y^{(2),U^*}(s) Y^1(s) + \varepsilon Y^1(s) \xi(s)) - f(s, Y^{(2),U^*}(s) Y^1(s)) \Big)$$

$$- \nabla f(s, Y^{(2),U}(s) Y^1(s)) \xi(s) \to 0 \text{ when } \varepsilon \to 0$$

and by the assumptions on f we obtain

$$\left\| \frac{1}{\varepsilon} \Big(f(s, Y^{(2),U_\varepsilon}(s) Y^1(s)) - f(s, Y^{(2),U^*}(s) Y^1(s) + \varepsilon Y^1(s) \xi(s)) \Big) \right\|$$

$$\leq C \left\| \frac{1}{\varepsilon} (Y^{(2),U_\varepsilon}(s) - Y^{(2),U^*}(s)) - \xi(s) \right\|.$$

We get with elementary estimates the convergence property. $\square$

By the above theorem we obtain the following result:

Proposition 1.

$$\lim_{\varepsilon \to 0} \mathbb{E} \sup_{t \in [0,T]} \|Y^{(2),U_\varepsilon}(t) - Y^{(2),U^*}(t)\|^2 = 0.$$

Lemma 7. *It holds*

$$0 \leq \lim_{\varepsilon \to 0} \frac{1}{\varepsilon} \Big(J(U_\varepsilon, Y^{(2),U_\varepsilon} Y^{(1)}) - J(U^*, Y^{(2),U^*} Y^{(1)}) \Big)$$

$$= 2\mathbb{E} \int_0^T \text{Re} \Big[\Big(\xi(t), Y^{(2),U^*}(t) |Y^{(1)}(t)|^2 - y(t) \overline{Y}^{(1)}(t) \Big) \Big] dt$$

$$+ 2\mathbb{E} \int_0^T (U(t) - U^*(t))(U(t) - \gamma(t)) dt.$$

Proof. Consider

$$\frac{1}{\varepsilon}\left(J(U_\varepsilon, Y^{(2),U_\varepsilon}Y^{(1)}) - J(U^*, Y^{(2),U^*}Y^{(1)})\right) = S_1(\varepsilon) + S_2(\varepsilon),$$

where

$$S_1(\varepsilon) = \frac{1}{\varepsilon}\mathbb{E}\int_0^T \left(\|Y^{(2),U_\varepsilon}(t)Y^{(1)}(t) - y(t)\|^2 - \|Y^{(2),U^*}(t)Y^{(1)}(t) - y(t)\|^2\right)dt,$$

$$S_2(\varepsilon) = \frac{1}{\varepsilon}\mathbb{E}\int_0^T \left(|U_\varepsilon(t) - \gamma(t)|^2 - |U^*(t) - \gamma(t)|^2\right)dt.$$

We use the formula $\|x\|^2 - \|y\|^2 = \mathrm{Re}\left[(x - y, x + y)\right]$, where $x, y \in H$, then

$$S_1(\varepsilon) = \mathbb{E}\int_0^T \mathrm{Re}\left[\left(\frac{1}{\varepsilon}(Y^{(2),U_\varepsilon}Y^{(1)}(t) - Y^{(2),U^*}Y^{(1)}(t)),\right.\right.$$
$$\left.\left. Y^{(2),U_\varepsilon}Y^{(1)}(t) + Y^{(2),U^*}Y^{(1)}(t) - 2y(t)\right)\right]dt.$$

By Theorem 42 and Proposition 1 it follows for $\varepsilon \to 0$ that

$$S_1(\varepsilon) \to 2\mathbb{E}\int_0^T \mathrm{Re}\left[\left(\xi(t)Y^1(t), Y^{(2),U^*}(t)Y^{(1)}(t) - y(t)\right)\right]dt$$
$$= 2\mathbb{E}\int_0^T \mathrm{Re}\left[\left(\xi(t), |Y^{(1)}(t)|^2 Y^{(2),U^*}(t) - y(t)\overline{Y}^{(1)}(t)\right)\right]dt.$$

Obviously,

$$S_2(\varepsilon) \to 2\mathbb{E}\int_0^T (U(t) - U^*(t))(U^*(t) - \gamma(t))\,dt$$

for $\varepsilon \to 0$. $\qquad\square$

By [2, Theorem 3.2] we have the following result:

Theorem 43. *For each $t \in [0, T]$ let $\left(\nabla f(t, Y^{(2),U^*}(t)Y^{(1)}(t))\right)^*$ be the adjoint operator of $\nabla f(t, Y^{(2),U^*}(t)Y^{(1)}(t))$. There exist a unique H-valued adapted process $\{V^*(t) : t \in [0,T]\}$ and a unique H-valued process $\{Z(t,s) : s \in [0,T], t \in [0,T]\}$ such that $Z(t,s)$ is $\mathcal{F}_s$-measurable for all $s \in [0,T]$ for every $t \in [0,T]$ with*

$$\mathbb{E}\int_0^T \|V^*(t)\|^2 dt + \mathbb{E}\int_0^T\int_0^T \|Z(t,s)\|^2 ds\,dt < \infty$$

and (V^*, Z) solves the stochastic backward Itô-Volterra integral equation

$$V^*(t) = 2\left(Y^{(2),U^*}(t)|Y^{(1)}(t)|^2 - y(t)\overline{Y}^{(1)}(t)\right)$$

$$- i \sum_{k=1}^{\infty} \int_t^T \frac{e^{i\lambda_k (s-t)}}{\overline{Y}^{(1)}(t)} \left(\nabla f(t, Y^{(2),U^*}(t) Y^{(1)}(t)) \right)^* (V^*(s), h_k) h_k ds$$

$$- i \sum_{k=1}^{\infty} \int_t^T e^{i\lambda_k (s-t)} U^*(t) (V^*(s), h_k) h_k ds - \int_t^T Z(t, s) dW(s) \quad (49)$$

for all $t \in [0, T]$ with probability 1 and

$$V^*(t) = 2\mathbb{E} \left(Y^{(2),U^*}(t) |Y^{(1)}(t)|^2 - y(t) \overline{Y}^{(1)}(t) \right) + \int_0^t Z(t, s) dW(s). \quad (50)$$

We now can prove the optimality condition of maximum principle type.

Theorem 44. *Let U^* be the optimal control for* (44). *Then,*

$$0 \le \left(\mathbb{E} \left(\int_t^T \mathrm{Re} \left[(V^*(s), i \sum_{k=1}^{\infty} e^{-i\lambda_k (s-t)} (Y^{(2),U^*}(t), h_k) h_k ds) \right] \Big| \mathcal{F}_t \right) \right.$$

$$\left. + 2 \big(U^*(t) - \gamma(t) \big) \right) \big(U(t) - U^*(t) \big) \quad (51)$$

for a.e. $(t, \omega) \in [0, T] \times \Omega$ and $U \in \mathcal{U}$, where V^* is the solution of (49).

Proof. We introduce

$$L = \mathbb{E} \mathrm{Re} \left[\int_0^T (V^*(t), i \sum_{k=1}^{\infty} \int_0^t e^{-i\lambda_k (t-s)} (U(s) - U^*(s)) (Y^{(2),U^*}(s), h_k) h_k ds) dt \right].$$

Then by Equation (48) we obtain

$$L = \mathbb{E} \mathrm{Re} \Big[\int_0^T (V^*(t), \xi(t)) dt$$

$$+ \mathbb{E} \sum_{k=1}^{\infty} \int_0^T \int_0^t (V^*(t), -i \frac{e^{-i\lambda_k (t-s)}}{Y^{(1)}(s)}$$

$$\cdot (\nabla f(s, Y^{(2),U^*}(s) Y^{(1)}(s)) \xi(s), h_k) h_k) ds dt$$

$$+ \mathbb{E} \sum_{k=1}^{\infty} \int_0^T \int_0^t (V^*(t), -i e^{-i\lambda_k (t-s)} U^*(s) (\xi(s), h_k) h_k) ds dt \Big].$$

For $k \ge 1$ let $\Pi_k(x)$ be the orthogonal projection of $x \in H$ on the one dimensional subspace generated by h_k, i.e. $\Pi_k(x) = (x, h_k) h_k$. Then we get with $V^*(t) = \sum_{k=1}^{\infty} (V^*(t), h_k) h_k$ and the properties of the orthogonal projection

$$\mathbb{E} \sum_{k=1}^{\infty} \mathrm{Re} \Big[\int_0^T \int_0^t (V^*(t), -i \frac{e^{-i\lambda_k (t-s)}}{Y^{(1)}(s)}$$

$$\cdot (\nabla f(s, Y^{(2),U^*}(s) Y^{(1)}(s)) \xi(s), h_k) h_k) ds dt \Big]$$

$$= \mathbb{E} \sum_{k=1}^{\infty} \mathrm{Re} \Big[\int_0^T \int_0^t (\Pi_k(V^*(t)), -\mathrm{i} \frac{e^{-\mathrm{i}\lambda_k(t-s)}}{Y^{(1)}(s)}$$

$$\cdot \Pi_k(\nabla f(s, Y^{(2),U^*}(s) Y^{(1)}(s)) \xi(s))) ds dt \Big]$$

$$= \mathbb{E} \sum_{k=1}^{\infty} \mathrm{Re} \Big[\int_0^T \int_0^t (\Pi_k(V^*(t)), -\mathrm{i} \frac{e^{-\mathrm{i}\lambda_k(t-s)}}{Y^{(1)}(s)}$$

$$\cdot \nabla f(s, Y^{(2),U^*}(s) Y^{(1)}(s)) \xi(s)) ds dt \Big]$$

$$= \mathbb{E} \sum_{k=1}^{\infty} \mathrm{Re} \Big[\int_0^T \int_0^t ((\nabla f(s, Y^{(2),U^*}(s) Y^{(1)}(s)))^* (V^*(t), h_k) h_k,$$

$$- \mathrm{i} \frac{e^{-\mathrm{i}\lambda_k(t-s)}}{Y^{(1)}(s)} \xi(s)) ds dt \Big].$$

Then we get by exchanging the order of integration and by the properties of the complex scalar product and by substituting s by t and t by s

$$L = \mathbb{E} \sum_{k=1}^{\infty} \mathrm{Re} \Big[\int_0^T (V^*(t)$$

$$+ \mathrm{i} \int_t^T \frac{e^{\mathrm{i}\lambda_k(s-t)}}{\overline{Y}^{(1)}(t)} ((\nabla f(t, Y^{(2),U^*}(t) Y^{(1)}(t)))^* (V^*(s), h_k) h_k ds$$

$$+ \mathrm{i} \int_t^T e^{\mathrm{i}\lambda_k(s-t)} U^*(t) (V^*(s), h_k) h_k ds, \xi(t)) dt \Big].$$

If we substitute $V^*(t)$ by the right side of Equation (49), then

$$L = \mathbb{E} \mathrm{Re} \Big[\int_0^T (2(Y^{(2),U^*}(t) |Y^{(1)}(t)|^2 - y(t) \overline{Y}^{(1)}(t)) - \int_t^T Z(t,s) dW(s), \xi(t)) dt \Big].$$

Since $\xi(t)$ and $Y^{(2),U^*}(t) |Y^{(1)}(t)|^2 - y(t) \overline{Y}^{(1)}(t)$ are $\mathcal{F}_t$-adapted and

$$\mathbb{E} \Big(\int_t^T Z(t,s) dW(s) \,|\, \mathcal{F}_t \Big) = 0,$$

we get

$$L = 2 \mathbb{E} \mathrm{Re} \Big[\int_0^T \Big(Y^{(2),U^*}(t) |Y^{(1)}(t)|^2 - y(t) \overline{Y}^{(1)}(t), \xi(t) \Big) dt \Big].$$

Consequently,

$$\mathbb{E} \mathrm{Re} \Big[\int_0^T (V^*(t), \mathrm{i} \sum_{k=1}^{\infty} \int_0^t e^{-\mathrm{i}\lambda_k(t-s)} (U(s) - U^*(s)) (Y^{(2),U^*}(s), h_k) h_k) ds dt \Big]$$

$$= 2\mathbb{E}\mathrm{Re}\Big[\int_0^T (Y^{(2),U^*}(t)|Y^{(1)}(t)|^2 - y(t)\overline{Y}^{(1)}(t), \xi(t))dt\Big].$$

Then Lemma 7 shows

$$0 \le 2\mathbb{E}\mathrm{Re}\int_0^T \Big(Y^{(2),U^*}(t)|Y^{(1)}(t)|^2 - y(t)\overline{Y}^{(1)}(t), \xi(t)\Big) dt$$

$$+ 2\mathbb{E}\int_0^T (U^*(t) - \gamma(t))(U(t) - U^*(t))dt$$

$$= \mathbb{E}\mathrm{Re}\Big[\int_0^T (V^*(t), \mathrm{i}\sum_{k=1}^\infty \int_0^t e^{-\mathrm{i}\lambda_k(t-s)}(U(s)-U^*(s))(Y^{(2),U^*}(s), h_k)h_k)dsdt\Big]$$

$$+ 2\mathbb{E}\int_0^T (U^*(t) - \gamma(t))(U(t) - U^*(t))dt.$$

We exchange the order of integration and after that we substitute s by t and t by s, therefore

$$0 \le \mathbb{E}\mathrm{Re}\Big[\int_0^T (V^*(t), \mathrm{i}\sum_{k=1}^\infty \int_0^t e^{-\mathrm{i}\lambda_k(t-s)}(U(s) - U^*(s))(Y^{(2),U^*}(s), h_k)h_k)dsdt\Big]$$

$$+ 2\mathbb{E}\int_0^T (U^*(t) - \gamma(t))(U(t) - U^*(t))dt$$

$$= \mathbb{E}\int_0^T \Big(\int_t^T \mathrm{Re}\Big[(V^*(s), \mathrm{i}\sum_{k=1}^\infty e^{-\mathrm{i}\lambda_k(s-t)}(Y^{(2),U^*}(t), h_k)h_k)\Big]ds$$

$$+ 2(U^*(t) - \gamma(t))\Big)(U(t) - U^*(t))dt \qquad (52)$$

for all admissible controls U. It follows immediately with an indirect proof

$$0 \le \Big(\mathbb{E}\Big(\int_t^T \mathrm{Re}\Big[(V^*(s), \mathrm{i}\sum_{k=1}^\infty e^{-\mathrm{i}\lambda_k(s-t)}(Y^{(2),U^*}(t), h_k)h_k)\Big]ds\Big| \mathcal{F}_t\Big)$$

$$+ 2(U^*(t) - \gamma(t))\Big)(U(t) - U^*(t))$$

for all admissible controls U and a.e. $(t, \omega) \in [0, T] \times \Omega$. $\qquad\square$

8. A Nonlinear Stochastic Schrödinger Equation Driven by Additive Fractional Noise

We consider the set of admissible controls $\hat{\mathcal{U}}$, the objective function J and f as in Section 7. Consider the sequence $(\mu_k)_{k\ge 1}$ such that $\sum_{k=1}^\infty \mu_k^2 < \infty$.

We consider the following SPDE written in symbolic form as

$$
\begin{cases}
dX^U(t) = \mathrm{i}\big(AX^U(t) + f(t, X^U(t)) + U(t)X^U(t)\big)dt + \mathrm{i}\sum_{k=1}^{\infty}\mu_k h_k dB_k^h(s) \\
X^U(0) = X_0.
\end{cases}
$$

(53)

We denote

$$
\beta^h(t) = \mathrm{i}\sum_{k=1}^{\infty}\int_0^t e^{-\mathrm{i}\lambda_k(t-s)}\mu_k h_k dB_k^h(t).
$$

We introduce the optimal control problem

$$
\min\{J(U, Y^U + \beta^h) : U \in \hat{\mathcal{U}}\}
$$

(54)

and the pathwise state equation

$$
Y^U(t) = \sum_{k=1}^{\infty} e^{-\mathrm{i}\lambda_k t}(X_0, h_k)h_k
$$

$$
+ \mathrm{i}\sum_{k=1}^{\infty}\int_0^t e^{-\mathrm{i}\lambda_k(t-s)}(f(s, Y^U(s) + \beta^h(s)), h_k)h_k ds
$$

$$
+ \mathrm{i}\sum_{k=1}^{\infty}\int_0^t e^{-\mathrm{i}\lambda_k(t-s)}U(s)(Y^U(s) + \beta^h(s), h_k)h_k ds,
$$

(55)

$t \in [0, T]$. We can prove by using the fixed point theorem of Banach that a unique solution process Y^U exists, where $E\sup_{t\in[0,T]}\|Y^U(t)\|^2 < \infty$. We define $X = Y^U - \beta^h$ and we get from (55)

$$
X^U(t) = \sum_{k=1}^{\infty} e^{-\mathrm{i}\lambda_k t}(X_0, h_k)h_k + \mathrm{i}\sum_{k=1}^{\infty}\int_0^t e^{-\mathrm{i}\lambda_k(t-s)}(f(s, X^U(s)), h_k)h_k ds
$$

$$
+ \mathrm{i}\sum_{k=1}^{\infty}\int_0^t e^{-\mathrm{i}\lambda_k(t-s)}U(s)(X^U(s), h_k)h_k ds
$$

$$
+ \mathrm{i}\sum_{k=1}^{\infty}\int_0^t e^{-\mathrm{i}\lambda_k(t-s)}\mu_k h_k dB_k^h(s).
$$

(56)

Denote

$$
\nabla f(s, Y^U(s) + \beta^h(s)) = \nabla f(s, x)\big|_{x = Y^U(s) + \beta^h(s)}.
$$

We get for

$$
\xi(t) = \lim_{\varepsilon \to 0}\frac{1}{\varepsilon}\big(Y^{U_\varepsilon}(t) - Y^{U^*}(t)\big), t \in [0, T]
$$

the equation

$$\xi(t) = i \sum_{k=1}^{\infty} \int_0^t e^{-i\lambda_k(t-s)} (\nabla f(s, Y^{U^*}(s) + \beta^h(s))\xi(s), h_k)h_k ds$$

$$+ i \sum_{k=1}^{\infty} \int_0^t e^{-i\lambda_k(t-s)} U^*(s)(\xi(s), h_k)h_k ds$$

$$+ i \sum_{k=1}^{\infty} \int_0^t e^{-i\lambda_k(t-s)} (U(s) - U^*(s))(Y^{U^*}(s) + \beta^h(s), h_k)h_k ds.$$

We consider the stochastic Itô-Volterra backward equation

$$V^*(t) = 2(Y^{U^*}(t) + \beta^h(t) - y(t))$$

$$- i \sum_{k=1}^{\infty} \int_t^T e^{-i\lambda_k(s-t)} (\nabla f(t, Y^{U^*}(t) + \beta^h(t)))^* (V^*(s), h_k)h_k ds$$

$$- i \sum_{k=1}^{\infty} \int_t^T e^{-i\lambda_k(s-t)} U^*(s)(V^*(s), h_k)h_k ds - \int_t^T Z(t,s)dW(s).$$

Then we get with similar considerations as in the last section the following necessary optimality condition:

Theorem 45. *The following variational inequality holds*

$$0 \leq \Big(\mathbb{E}\Big(\int_t^T \mathrm{Re}\Big[(V^*(s), i \sum_{k=1}^{\infty} e^{-i\lambda_k(s-t)}(Y^{U^*}, h_k)h_k)\Big] ds | \mathcal{F}_t \Big)$$

$$+ 2(U^*(t) - \gamma(t))\Big)(U(t) - U^*(t)),$$

for all admissible controls U and a.e. $(t, \omega) \in [0, T] \times \Omega$.

9. Further Optimal Control Problems

First, we consider the following stochastic equation of Ginzburg-Landau type in the context of the assumptions of Section 3 for $U \in \mathcal{U}$

$$X^U(t) = \sum_{k=1}^{\infty} e^{-i\lambda_k t}(X_0, h_k)h_k + \sum_{k=1}^{\infty} \int_0^t e^{-i\lambda_k(t-s)} U_k(s)(X^U(s), h_k)h_k ds$$

$$+ i \sum_{k=1}^{\infty} \mu_k \int_0^t e^{-i\lambda_k(t-s)}(X^U(s), h_k)h_k dB_k^h(s), \ t \in [0, T]. \quad (57)$$

All proofs for the existence of solution work for this case similarly to (7), but we use the results from Example 11-(3).

For $k \geq 1$ recall the process $\{Y_k^{(1)}(t), t \in [0,T]\}$ given by (10). We modify the process defined in (12) by

$$Y_k^{(2),U}(t) = (X_0, h_k) \exp\left\{ -\mathrm{i}\lambda_k t + \int_0^t U_k(s)ds \right\}, \tag{58}$$

which is a deterministic process. Obviously, for each $k \geq 1$ and $t \in [0,T]$ it satisfies the equation

$$Y_k^{(2),U}(t) = (X_0, h_k) + \int_0^t \left(-\mathrm{i}\lambda_k + U_k(s) \right) Y_k^{(2),U}(s)ds. \tag{59}$$

We introduce a discounted objective function of the type

$$J(U, X^U) = \sum_{k=1}^{\infty} \left(\mathbb{E}\left| (X^U(T), h_k) \right|^2 + \int_0^T \exp\{-\mu_k^2 t^{2h}\} U_k^2(t) \mathbb{E}\left| (X^U(t), h_k) \right|^2 dt \right) \tag{60}$$

and we introduce the optimal control problem

$$\min\left\{ J(U, X^U) : U \in \mathcal{U} \right\}, \tag{61}$$

where X^U solves (57) and, in fact,

$$X^U(t) = \sum_{i=1}^{\infty} Y_k^{(2),U}(t) Y_k^{(1)}(t) h_k, \quad t \in [0,T].$$

Analogous to Section 4, we transform (61) equivalently into a problem of optimal control for the state equation (59) with the objective function

$$J(U, X^U) = \sum_{k=1}^{\infty} |Y_k^{(2),U}(T)|^2 \exp\{\mu_k^2 T^{2h}\} + \sum_{k=1}^{\infty} \int_0^T U_k^2(t) |Y_k^{(2),U}(t)|^2 dt. \tag{62}$$

Thus, we have got a pure deterministic problem of optimal control. Consequently, it is not necessary to solve stochastic backward equations.

For $k \geq 1$ we introduce the special backward Riccati differential equation

$$\frac{d}{dt} P_k(t) = P_k^2(t) \quad \text{with} \quad P_k(T) = \exp\{\mu_k^2 T^{2h}\}. \tag{63}$$

The solutions are the positive functions

$$P_k(t) = -\frac{1}{t - \exp\{-\mu_k^2 T^{2h}\} - T}, \quad t \in [0,T], \ k \geq 1. \tag{64}$$

Moreover, (59) implies

$$d(\mathrm{Re}[Y_k^{(2),U}(t)]) = \left(\lambda_k \mathrm{Im}[Y_k^{(2),U}(t)] + U_k(t) \mathrm{Re}[Y_k^{(2),U}(t)] \right) dt,$$

$$d(\text{Im}[Y_k^{(2),U}(t)]) = \big(-\lambda_k \text{Re}[Y_k^{(2),U}(t)] + U_k(t)\text{Im}[Y_k^{(2),U}(t)]\big)dt.$$

By (63) we finally obtain

$$d\left(|Y_k^{(2),U}(t)|^2 P_k(t)\right) = P_k'(t)|Y_k^{(2),U}(t)|^2 dt$$

$$+ 2P_k(t)\left(\text{Re}[Y_k^{(2),U}(t)]d(\text{Re}[Y_k^{(2),U}(t)]) + \text{Im}[Y_k^{(2),U}(t)]d(\text{Im}[Y_k^{(2),U}(t)])\right)$$

$$= P_k^2(t)|Y_k^{(2),U}(t)|^2 + 2U_k(t)P_k(t)|Y_k^{(2),U}(t)|^2.$$

Then

$$|Y_k^{(2),U}(T)|^2 \exp\{\mu_k^2 T^{2h}\} = |Y_k^{(2),U}(T)|^2 P_k(T)$$

$$= |Y_k^{(2),U}(0)|^2 P_k(0) + \int_0^T \left(2U_k(t)P_k(t)|Y_k^{(2),U}(t)|^2 + P_k^2(t)|Y_k^{(2),U}(t)|^2\right)dt.$$

Consequently we get for (62) with $Y_k^{(2),U}(0) = (X_0, h_k)$,

$$J(U) = \sum_{k=1}^{\infty} |(X_0, h_k)|^2 P_k(0)$$

$$+ \sum_{k=1}^{\infty} \int_0^T |Y_k^{(2),U}(t)|^2 \left(P_k^2(t) + 2P_k(t)U_k(t) + U_k^2(t)\right)dt.$$

The last term will be minimal, if

$$P_k^2(t) + 2P_k(t)U_k(t) + U_k^2(t) = 0 \text{ for a.e. } t \in [0, T]$$

holds. Consequently, we have $U_k^*(t) = -P_k(t)$.

Theorem 46. *The optimal control U^* for (61) is given by the sequence*

$$U^*(t) = (-P_1(t), -P_2(t), \ldots) \text{ for a.e. } t \in [0, T]$$

and $J(U^*) = \sum_{k=1}^{\infty} |(X_0, h_k)|^2 P_k(0)$, where P_k is defined by (64) and solves (63).

Finally, we discuss the optimal control problem from Section 6 in the linear case $(f = 0)$ for admissible control $U \in \hat{\mathcal{U}}$. Then, Equation (37) is deterministic and we get from (52)

$$\mathbb{E}\int_0^T \left(\int_t^T \text{Re}\left[(V^*(s), i\sum_{k=1}^{\infty} e^{-i\lambda_k(s-t)}(Y^{(2),U^*}(t), h_k)h_k)\right](U(t) - U^*(t))ds\right.$$

$$\left. + 2(U^*(t) - \gamma(t))(U(t) - U^*(t))\right)dt$$

$$= \int_0^T \left(\int_t^T \mathrm{Re}\left[(\mathbb{E}(V^*(s)), \mathrm{i} \sum_{k=1}^{\infty} e^{-\mathrm{i}\lambda_k(s-t)} (Y^{(2),U^*}(t), h_k)h_k) \right] (U(t) - U^*(t)) ds \right.$$

$$\left. + 2(U^*(t) - \gamma(t))(U(t) - U^*(t)) \right) dt \geq 0. \tag{65}$$

It follows from (38) and (36) that

$$E\big(\overline{Y}^{(1)}(s)\big) = 1 \text{ and } E|Y^{(1)}(s)|^2 = \exp\{\mu^2 s^{2h}\}$$

and from (50) that

$$E\big(V^*(s)\big) = 2Y^{(2),U^*}(s)\exp\{\mu^2 s^{2h}\} - y(s).$$

Consequently, the variational inequality (65) is a pure deterministic problem. We get again with an indirect proof

$$\mathrm{Re}\left[\int_t^T (\mathbb{E}(V^*(s)), \mathrm{i} \sum_{k=1}^{\infty} e^{-\mathrm{i}\lambda_k(s-t)} (Y^{(2),U^*}(t), h_k)h_k) \right] (U(t) - U^*(t)) ds$$

$$+ 2(U^*(t) - \gamma(t))(U(t) - U^*(t)) \geq 0$$

for all admissible controls $U \in \hat{\mathcal{U}}$ and for a.e. $t \in [0, T]$.

References

[1] N. U. Ahmed, Nonlinear Schrödinger Equation and its Optimal Control Using Laser Induced Dynamical Potential, *Pure Appl. Funct. Anal.* **4**, 1–19 (2019).

[2] Vo. V. Ahn, W. Grecksch and J. Yong, Regularity of Backward Stochastic Volterra Integral Equations in Hilbert Spaces. *Stoch. Anal. Appl.* **29**, 146–168 (2011).

[3] N. Y. Aksoy, E. Aksoy and Y. Kosac, An Optimal Control Problem with Final Observation for Systems Governed by Nonlinear Schrödinger Equation. *Filomat* **30**, 649–665 (2016).

[4] D. Anderson, M. Bonnedal, Variational approach to nonlinear self-focusing of Gaussian laser beams. *Phys. Fluids* **22**, 105–109 (1979).

[5] N. W. Ashcroft, N. D. Mermin, *Solid state physics*. Holt, Rinehart and Winstor, New York (1976).

[6] O. Bang, P. L. Christiansen, F. If, K. O. Rasmussen, Y. B. Gaididei, Temperature effects in a nonlinear model of monolayer Scheibe aggregates. *Phys. Rev.* E **49**, 4627–4636 (1994).

[7] V. Barbu, M. Röckner and D. Zhang, Stochastic Nonlinear Schrödinger Equations with Linear Multiplicative Noise: Rescaling Approach, *J. Nonlinear Sci.* **24**, 383–409 (2014).

[8] V. Barbu, M. Röckner and D. Zhang, Optimal Bilinear Control of Stochastic Schrödinger Equations Driven by Linear Multiplicative Noise, *Ann. Probab*, **4**, 1957–1999 (2018).

[9] F. Biagini, Y. Hu, B. Øksendal and T. Zhang, *Stochastic Caculus for Fractional Brownian Motion and Applications.* Springer-Verlag (2008).

[10] M. F. Bidaut, *Théorème d'existence et d'existence "en général" d'un contrôle optimal pour des systèmes régis par dés équations aux dérivées partielles non linéaires*, PhD Thesis. Université Paris VI (1973).

[11] A. Biswas and K. Swapan, *Introduction to non-Kerr Law Optical Solutions*, Chapman and Hall/CRC (2006).

[12] M. Blencowe, Quantum electromechanical systems. *Phys. Rep.* **39**, 5, 159–222 (2004).

[13] A. G. Butkovskiy and Yu. I. Samoilenko, *Control of Quantum-Mechanical Processes and Systems.* Kluwer Academic Publishers Group, Dordrecht (1990).

[14] I. Calvo and R. Sánchez, The Path Integral Formulation of Fractional Brownian Motion for the General Hurst Index. *J. Phys A: Math. Theor.* **41**, 282002, 1–5 (2008).

[15] I. Calvo, R. Saánchez and B. A. Carreras, Fractional Lévy Motion through Paths Integrals. *J. Phys. A: Math. Theor.* **42**, 05503, 1–8 (2009).

[16] H. T. Chu, Eigen energies and eigen states of conduction electrons in pure bismuth under size and magnetic field quantizations. *J. Phys. Chem. Solids* **50**, 319–324 (1989).

[17] D. Coculescu and A. Nikeghbali, Filtrations. *arXiv:0712.0622v1 [math.PR]* (2007).

[18] A. De Bouard and A. Debussche, A stochastic nonlinear Schrödinger equation with multiplicative noise, *Commun. Math. Phys.* **2050**, 161–181 (1999).

[19] A. De Bouard and A. Debussche, A semi-discrete scheme for the stochastic nonlinear Schrödinger equation. *Numer. Math.* **6**, 733–770 (2004).

[20] H. Brezis, *Functional Analysis Sobolev Spaces and Partial Differential Equations.* Springer, New York (2011).

[21] I. Cialenko, Parameter Estimation for SPDE's with Multiplicative Fractional Noise, *Stoch. Dyn.* **10**, 561–576 (2010).

[22] A. Debussche and C. Odasso, Ergodicity for a weakly damped stochastic non-linear Schrödinger equation, *J. Evol. Equ.* **5**, 317–356 (2005).

[23] G. Da Prato and J. Zabczyk, *Stochastic Equations in Infinite Dimensions.* University Press, Cambridge (1992).

[24] R. Dautray and J.-L. Lions, *Mathematical Analysis And Numerical Methods for Science and Technology*, Volume 5: Evolution Problems I. Springer Verlag, Berlin (2000).

[25] T. E. Duncan, Y. Hu and B. Pasik-Duncan, Stochastic calculus for fractional Brownian motion. I. Theory. *SIAM J. Control Optim.* **38**, 582–612 (2000).

[26] T. E. Duncan, B. Pasik-Duncan and J. Jakubowski, Stochastic Integration for Fractional Brownian Motion in a Hilbert Space. *Stoch. Dyn.* **06**, 53–75 (2006).

[27] G. E. Falkovich, I. Kolokolov, V. Lebedev and S. K. Turitsyn, Statistics of soliton-bearing systems with additive noise, *Phys. Rev. E 63*: 025601(R) (2001).

[28] E. Gautier, Stochastic Nonlinear Schrödinger Equations Driven by a Fractional Noise: Well-posedness, Large Deviations and Support, *Electronic Journal of Probability*, **12**, 848–861 (2007).

[29] L. Gawarecki and V. Mandrekar, *Stochastic Differential Equations in Infinite Dimensions, with Applications to Stochastic Partial Differential Equations*. Springer Verlag, Berlin (2011).

[30] W. Grecksch and H. Lisei, Stochastic Nonlinear Equations of Schrödinger Type. *Stoch. Anal. Appl.* **29**, 631–651 (2011).

[31] W. Grecksch and H. Lisei, Stochastic Schrödinger Equation Driven by Cylindrical Wiener Process and Fractional Brownian Motion. *Stud. Univ. Babes-Bolyai Math.* **56**, 2, 381–391 (2011).

[32] W. Grecksch and H. Lisei, Approximation of Stochastic Nonlinear Equations of Schrödinger Type by the Splitting Method. *Stoch. Anal. Appl.* **31**, 314–335 (2013).

[33] W. Grecksch, H. Lisei, and J. Lueddeckens, Parameter Estimations for Linear Parabolic Fractional SPDEs with Jumps. *Stud. Univ. Babes-Bolyai Math.* **64**, 279–289 (2019).

[34] E. P. Gross, Structure of a quantized vortex in boson systems. *Nuovo Cimento* **20**, 454–477 (1961).

[35] Y. Hu, Itô Stochastic Differential Equations Driven by Fractional Brownian Motions of Hurst Index $H > 1/2$. *arXiv:1610.01137v2 [math.PR]*, 1–39 (2016).

[36] K. Ito and K. Kunisch, Optimal Bilinear Control of an Abstract Schrödinger Equation, *SIAM J. Control Optim.* **46**, 274–287 (2007).

[37] G. Jumarie, Schrödinger Equations for Quantum Fractal Space-Time of Order n via the Complex-Valued Fractional Brownian Motion. *Internat. J. Modern Phys. A* **16** 31, 5061–5084 (2001).

[38] D. Keller and H. Lisei, Variational Solution of Stochastic Schrödinger Equations with Power-Type Nonlinearity. *Stoch. Anal. Appl.* **33**, 653–672 (2015).

[39] D. Keller, Optimal Control of a Nonlinear Stochastic Schrödinger Equation. *J. Optim. Theory Appl.* **167**, 862–873 (2015).

[40] P. I. Kelley, Self-focusing of optical beams. *Phys.Rev.Lett.* **15**, 1005–1008 (1965).

[41] N. Laskin, *Fractional Quantum Mechanics*. World Scientific Publ. Co. Pte. Ltd., Hackensack (2018).

[42] N. Laskin, Time Fractional Quantum Mechanics, Future Directions In Fractional Research and Applications. *Chaos, Solitons and Fractals: The International Journal of Nonlinear Science and Nonequilibrium and Complex Phenomena* **102**, 16–28 (2017).

[43] N. Laskin, Fractional Schrödinger Equation. *Physical Review E* **66**, 056108-1–056108-7 (2002).

[44] H. Lisei and D. Keller, A Stochastic Nonlinear Schrödinger Problem in Variational Formulation. *Nonlinear Differ. Equ. Appl* **23**, 1–27 (2016).

[45] J. L. Lions, *Optimal Control of Systems Governed by Partial Differential Equations*. Springer Verlag, Berlin - New York (1971).

[46] A. Lodhia, S. Sheffield, X. Sun and S. S. Watson, Fractional Gaussian Fields: A Survey. *arXiv:1407.5598v2 [math.PR]* (5 Feb. 2016).

[47] S. V. Lototsky and B. L. Rozovsky, *Stochastic Partial Differential Equations.* Springer International Publishing, Basel (2017).

[48] J. Ma and J. Yong, *Forward-Backward Stochastic Differential Equations and Their Applications.* Lecture Notes in Mathematics 1702, Springer Verlag, Berlin (1999).

[49] B. Mandelbrot and J. van Ness, Fractional Brownian Motions, Fractional Noises and Appilcations. *SIAM Rev.* **10**, 422–437 (1968).

[50] C. M. Mora and R. Rebolledo, Regularity of solutions to linear stochastic Schrödinger equations. *Infin. Dimens. Anal. Quantum Probab. Relat. Top.* **10**, 237–259 (2007).

[51] C. M. Mora and R. Rebolledo, Basic properties of nonlinear stochastic Schrödinger equations driven by Brownian motions. *Ann. Appl. Probab.* **18**, 591–619 (2008).

[52] I. Norros, E. Valkeila and J. Virtamo, An Elementary Approach to a Girsanov Formula and Other Analytical Results on Fractional Brownian Motions. *Bernoulli* **5**, 571–587 (1999).

[53] D. Nualart, Stochastic Analysis with respect to Fractional Brownian Motion. *Ann. Fac. Sci. Toulouse Math. (6)* **15**, 63–78 (2006).

[54] O. Pinaud, A Note on Stochastic Schrödinger Equations with Fractional Multiplicative Noise, *J. Differential Equations* **256**, 1467–1491 (2014).

[55] L. P. Pitaevskii, Vortex lines in an imperfect Bose gas. *Sov.Phys. JETP* **13**, 451–454 (1961).

[56] C. Prévôt and M. Röckner, *A Concise Course on Stochastic Partial Differential Equations.* Lecture Notes in Mathematics Vol. 1905, Springer Verlag, Berlin (2007).

[57] K. Ø. Rasmussen,Y. B. Gaididei, O. Bang and P. L. Christiansen, The influence of noise on critical collapse in the nonlinear Schrödinger equation. *Phys. Rev. A* **204**, 121–127 (1995).

[58] M. Reed and B. Simon, *Methods on modern mathematical physics. IV: Analysis of Operators.* Academic Press, New York (1972).

[59] B. L. Rozovskii, *Stochastic Evolution Systems. Linear Theory and Applications to Non-Linear Filtering.* Kluwer Academic Publishers, Dordrecht (1990).

[60] J. Shu, Random Attractors for Disctree Klein-Gordon-Schrödinger Equations Driven by Fractional Brownian Motion. *Discrete Contin. Dyn. Syst. Ser. B* **22**, 1587–1599 (2017).

[61] M. Subasi, An Optimal Control Problem governed by the Potential of a Linear Schrödinger Equation, *Appl. Math. Comput.* **131**, 414–422 (2002).

[62] M. A. Vorontsov and V. I. Shmalgauzen, *The Principles of Adaptive Optics.* Nauka, Moscow (1985). (in Russian)

[63] D. Zhang, Recent Progress on Nonlinear Stochastic Schrödinger Equations, *Springer Proc. Math. Stat.* **229**, Springer Cham, 279–289 (2018).

[64] J. Zhang and D. Li, Efficient Numerical Computation of Time-Fractional Nonlinear Schrödinger Equations in Unbounded Domain. *Commun. Comput. Phys.* **25**, 218–243 (2019).

[65] H. Zhang, X. Jiang, C. Wang and W. Fan, Galerkin-Legendre Spectral Schemes for Nonlinear Space Fractional Schrödinger Equation. *Numer. Algorithms* **79**, 337–356 (2018).

Chapter 4

Optimal Control of the Stochastic Navier-Stokes Equations

Peter Benner*

*Max Planck Institute for Dynamics of Complex Technical Systems
Sandtorstrasse 1, 39106 Magdeburg, Germany*

Christoph Trautwein†

*Friedrich Schiller University Jena,
School of Mathematics and Computer Science, Institute for Mathematics,
Ernst-Abbe-Platz 2, 07743 Jena, Germany*

We consider a control problem of uncertain fluid flows, where the velocity field is described by the stochastic Navier-Stokes equations in multidimensional domains. We distinguish between velocity fields driven by additive noise and multiplicative noise. Using a stochastic maximum principle, we derive a necessary optimality condition to obtain an explicit formula the optimal control has to satisfy. Moreover, we show that the optimal control satisfies a sufficient optimality condition based on the higher order Fréchet derivative of the cost functional related to the control problem. As a consequence, we are able to solve uniquely control problems of the stochastic Navier-Stokes equations especially for two-dimensional as well as for three-dimensional domains.

Contents

*E-mail: benner@mpi-magdeburg.mpg.de
†E-mail (corresponding author): christoph.trautwein@uni-jena.de

1. Introduction

We consider the optimal control problem for flow problems described by the incompressible time-dependent Navier-Stokes equations perturbed by dynamical noise processes. In fluid dynamics, noise may enter the system due to structural vibration, wind gusts, and other environmental effects. As a consequence, the velocity field of the fluid may show an undesired behavior. Thus, the aim is to control these velocity fields affected by noise in a desired way, where we incorporate physical requirements, such as drag minimization, lift enhancement, mixing enhancement, turbulence minimization, and stabilization, see [43] and the references therein. Here, we focus on control terms appearing as distributed controls, which are defined inside the domain.

Throughout this chapter, let $\mathcal{D} \subset \mathbb{R}^n$, $n \geq 2$, be a bounded and connected domain with sufficiently smooth boundary $\partial\mathcal{D}$. Moreover, we assume that $(\Omega, \mathcal{F}, \mathbb{P})$ is a complete probability space endowed with a filtration $(\mathcal{F}_t)_{t \geq 0}$ satisfying $\mathcal{F}_t = \bigcap_{s > t} \mathcal{F}_s$ for all $t > 0$ and $\mathcal{F}_0$ contains all sets of $\mathcal{F}$ with $\mathbb{P}$-measure 0. Let $T > 0$. We consider the Navier-Stokes equations with homogeneous Dirichlet boundary condition:

$$\begin{cases} \dfrac{\partial}{\partial t} y + (y \cdot \nabla)y + \nabla p - \nu \Delta y = f & \text{in } (0, T) \times \mathcal{D} \times \Omega, \\[2mm] \operatorname{div} y = 0 & \text{in } (0, T) \times \mathcal{D} \times \Omega, \\[2mm] y = 0 & \text{on } (0, T) \times \partial\mathcal{D}, \\[2mm] y(0, x, \omega) = \xi(x, \omega) & \text{in } \mathcal{D} \times \Omega, \end{cases} \quad (1)$$

where $y = y(t, x, \omega) \in \mathbb{R}^n$ denotes the velocity field with $\mathcal{F}_0$-measurable initial value $\xi(x, \omega) \in \mathbb{R}^n$. The term $p = p(t, x, \omega) \in \mathbb{R}$ describes the pressure of the fluid and $\nu > 0$ is the viscosity parameter (for the sake of simplicity, we assume $\nu = 1$). For the external random force $f \in \mathbb{R}^n$, we distinguish between two cases:

(i) independent of the velocity field, i.e. $f = f(t, x, \omega)$;
(ii) dependent on the velocity field, i.e. $f = f(t, x, \omega, y)$.

In the last decades, existence and uniqueness results of solutions to the stochastic Navier-Stokes equations have been studied extensively. Unique weak solutions of the stochastic Navier-Stokes equations exist only for two-dimensional domains. In [38, 44], weak solutions are considered with noise terms given by Wiener processes. Weak solutions with Lévy noise are considered in [9, 18]. For three-dimensional domains, uniqueness is still an open problem and weak solutions are introduced as martingale solutions, see [8, 11, 19, 20, 40]. Another approach uses the theory of semigroups leading to solutions in a mild sense. The existence and uniqueness of a mild solution over an arbitrary time interval can be obtained under certain additional assumptions, see [13, 15]. In general, a unique mild solution of the stochastic Navier-Stokes equations does not exist. Thus, stopping times are required to define local mild solutions. For the local mild solution with additive noise given by Wiener processes, we refer to [5]. In [17, 17], the stochastic Navier-Stokes equations with additive Lévy noise are considered. A generalization to multiplicative Lévy noise can be found in [3]. In [27], an existence and uniqueness result for strong pathwise solutions is given. For further definitions of solutions to the fractional stochastic Navier-Stokes equations, we refer to [15]. Here, we treat both cases (i) and (ii) separately, where we assume that the system is driven by a Wiener noise. Case (i) will lead us to a SPDE with additive noise studied in Section 2 and case (ii) results in a SPDE with multiplicative noise considered in Section 3. Using suitable stopping times, we show existence and uniqueness of local mild solutions especially for two-dimensional as well as three-dimensional domains. This approach is mainly based on [3].

The control problem is given by

$$\inf_{u} \mathbb{E} \int_0^T \int_{\mathcal{D}} g(y(t,x,\omega)) + h(u(t,x,\omega)) \, dx \, dt$$

subject to (1). The solution of this problem is called an optimal control. As discussed above, we can not ensure the existence of a unique solution to system (1) in general. Therefore, a stopping time $\tau_m = \tau_m(\omega)$ has to be incorporated. Mainly two cases for the function $g \colon \mathbb{R} \to \mathbb{R}$ are analyzed:

(a) $g(y(t,x,\omega)) = \mathbb{1}_{[0,\tau_m)}(t) \, |y(t,x,\omega) - y_d(t,x)|$;
(b) $g(y(t,x,\omega)) = \mathbb{1}_{[0,\tau_m)}(t) \, |\nabla \times y(t,x,\omega)|$

with $\mathbb{1}$ the indicator function and $y_d(t,x) \in \mathbb{R}$ a given (deterministic) desired velocity field. Furthermore, the function $h \colon \mathbb{R} \to \mathbb{R}$ characterizes the costs of the control. The control problem considered here is motivated by common control strategies in fluid dynamics. In [32, 37, 41, 48], the problem is formulated as a tracking type problem arising in data assimilation, which is in line with case (a). For a tracking type problem of velocity fields desribed by the stochastic Stokes equations, we refer to [4]. Approaches that minimize the enstrophy corresponding to (b) can be found in [12, 16, 30, 43]. In [51], the cost functional combines both strategies by introducing weights. The shortcoming of these papers is the restriction to two-dimensional domains such that $\tau_m = T$. Here, we will consider a generalized control problem involving both cases (a) and (b). Using a stochastic maximum principle, an explicit formula of the optimal control based on the corresponding adjoint equation is derived in [2, 46]. This adjoint equation is given by a backward SPDE, which requires a martingale representation theorem in order to obtain a well defined solution. Moreover, additional restrictions are needed such that the control problem for case (b) cannot be solved explicitly. The results are stated in Section 3. To overcome this problem, we first consider the stochastic Navier-Stokes equations driven by an additive noise in Section 2. We show that these additional restrictions vanishes, which results mainly from the fact that the corresponding adjoint equation is given by a deterministic backward equation.

2. Additive Noise

2.1. *The State Equation*

In this section, we analyze the evolution equation of the controlled stochastic Navier-Stokes equations with distributed control driven by additive Wiener noise. Using suitable stopping times, we introduce a local mild solution of the corresponding system and we show an existence and uniqueness result. Here, we closely follow [15]. As a consequence, we obtain a well defined solution especially for two-dimensional as well as three-dimensional domains. Finally, we show some useful properties, which we use in the following sections.

For $s \geq 0$, let $H^s(\mathcal{D})$ denote the usual Sobolev space and for $s \geq \frac{1}{2}$, let $H_0^s(\mathcal{D}) = \{y \in H^s(\mathcal{D}) : y = 0 \text{ on } \partial\mathcal{D}\}$. We introduce the following spaces:
$$H = \left\{y \in (L^2(\mathcal{D}))^n : \text{div } y = 0 \text{ in } \mathcal{D}, y \cdot \eta = 0 \text{ on } \partial\mathcal{D}\right\},$$
$$V = \left\{y \in \left(H_0^1(\mathcal{D})\right)^n : \text{div } y = 0 \text{ in } \mathcal{D}\right\},$$
where η denotes the unit outward normal to $\partial\mathcal{D}$. The spaces H and V equipped with the inner product on $(L^2(\mathcal{D}))^n$ and $\left(H_0^1(\mathcal{D})\right)^n$, respectively, are Hilbert spaces. The norms in H and V are denoted by $\|\cdot\|_H$ and $\|\cdot\|_V$, respectively. We define the Stokes Operator $A: D(A) \subset H \to H$ with $D(A) = \left(H^2(\mathcal{D})\right)^n \cap V$ by
$$Ay = -\Pi\Delta y$$

for every $y \in D(A)$, where $\Pi: (L^2(\mathcal{D}))^n \to H$ denotes the orthogonal Helmholtz projection as introduced in [23]. The Stokes operator A is positive, self-adjoint, and has a bounded inverse. Moreover, the operator $-A$ is the infinitesimal generator of an analytic semigroup $(e^{-At})_{t \geq 0}$ such that $\left\| e^{-At} \right\|_{\mathcal{L}(H)} \leq 1$ for all $t \geq 0$. For more details, see [22, 25, 26, 50]. Hence, we can introduce fractional powers of the Stokes operator denoted by A^α with $\alpha \in \mathbb{R}$, see [42, 49, 50]. The following lemma provides some useful properties of fractional powers of the Stokes operator.

Lemma 8 (cf. Section 2.6, [42]). *Let* $A: D(A) \subset H \to H$ *be the Stokes operator. Then*

(i) for $\alpha, \beta \in \mathbb{R}$, *we have* $A^{\alpha+\beta}y = A^\alpha A^\beta y$ *for every* $y \in D(A^\gamma)$, *where* $\gamma = \max\{\alpha, \beta, \alpha + \beta\}$;

(ii) $e^{-At}: H \to D(A^\alpha)$ *for all* $t > 0$ *and* $\alpha \geq 0$;

(iii) we have $A^\alpha e^{-At}y = e^{-At}A^\alpha y$ *for every* $y \in D(A^\alpha)$ *with* $\alpha \in \mathbb{R}$,

(iv) the operator $A^\alpha e^{-At}$ *is bounded for all* $t > 0$ *and there exist constants* $M_\alpha, \theta > 0$ *such that*

$$\left\| A^\alpha e^{-At} \right\|_{\mathcal{L}(H)} \leq M_\alpha t^{-\alpha} e^{-\theta t};$$

(v) $0 \leq \beta \leq \alpha \leq 1$ *implies* $D(A^\alpha) \subset D(A^\beta)$;

(vi) the space $D(A^\alpha)$ *for all* $\alpha \geq 0$ *equipped with the inner product*

$$\langle y, z \rangle_{D(A^\alpha)} = \langle A^\alpha y, A^\alpha z \rangle_H$$

for every $y, z \in D(A^\alpha)$ *is a Hilbert space;*

(vii) the operator A^α *is self-adjoint for all* $\alpha \in \mathbb{R}$.

We define the bilinear operator $B(y, z) = \Pi(y \cdot \nabla)z$ for certain $y, z \in H$. If $y = z$, then we write $B(y) = B(y, y)$. The following results provide some useful properties.

Lemma 9 (cf. Lemma 2.2, [26]). *Let* $0 \leq \delta < \frac{1}{2} + \frac{n}{4}$. *If* $y \in D(A^{\alpha_1})$ *and* $z \in D(A^{\alpha_2})$, *then we have*

$$\left\| A^{-\delta} B(y, z) \right\|_H \leq \widetilde{M} \left\| A^{\alpha_1} y \right\|_H \left\| A^{\alpha_2} z \right\|_H,$$

with some constant $\widetilde{M} = \widetilde{M}_{\delta, \alpha_1, \alpha_2}$, *provided that* $\alpha_1, \alpha_2 > 0$, $\delta + \alpha_2 > \frac{1}{2}$, *and* $\delta + \alpha_1 + \alpha_2 \geq \frac{n}{4} + \frac{1}{2}$.

Corollary 1. *Let* α_1, α_2 *and* δ *be as in Lemma 9. If* $y, z \in D(A^\beta)$ *with* $\beta = \max\{\alpha_1, \alpha_2\}$, *then we have*

$$\left\| A^{-\delta}(B(y) - B(z)) \right\|_H$$
$$\leq \widetilde{M} \left(\left\| A^{\alpha_1} y \right\|_H \left\| A^{\alpha_2}(y - z) \right\|_H + \left\| A^{\alpha_1}(y - z) \right\|_H \left\| A^{\alpha_2} z \right\|_H \right).$$

We introduce the space $L_{\mathcal{F}}^k(\Omega; L^r([0,T]; D(A^\beta)))$ containing all $\mathcal{F}_t$-adapted processes $(v(t))_{t \in [0,T]}$ with values in $D(A^\beta)$ such that $\mathbb{E}(\int_0^T \|v(t)\|_{D(A^\beta)}^r dt)^{k/r} < \infty$ with $k, r \in [0, \infty)$ and $\beta \in [0, \alpha]$. The space $L_{\mathcal{F}}^k(\Omega; L^r([0,T]; D(A^\beta)))$ equipped with the norm

$$\|v\|_{L_{\mathcal{F}}^k(\Omega; L^r([0,T]; D(A^\beta)))}^k = \mathbb{E}\left(\int_0^T \|v(t)\|_{D(A^\beta)}^r dt\right)^{k/r}$$

for every $v \in L_{\mathcal{F}}^k(\Omega; L^r([0,T]; D(A^\beta)))$ is a Banach space.

Next, we assume that the external random force f in system (1) is independent of the velocity field and can be decomposed into a sum of a control term and a noise term. Using the spaces and the operators introduced above, we obtain the stochastic Navier-Stokes equations with additive noise in $D(A^\alpha)$:

$$\begin{cases} dy(t) = -[Ay(t) + B(y(t)) - Fu(t)] \, dt + G(t) \, dW(t), \\ y(0) = \xi, \end{cases} \tag{2}$$

where $(W(t))_{t \geq 0}$ is a Q-Wiener process with values in H and covariance operator $Q \in \mathcal{L}(H)$. The process $(G(t))_{t \in [0,T]}$ is a predictable process with values in the space of Hilbert-Schmidt operators mapping $Q^{1/2}(H)$ into $D(A^\alpha)$, which we denote by $\mathcal{L}_{(HS)}(Q^{1/2}(H); D(A^\alpha))$. Moreover, the process $(G(t))_{t \in [0,T]}$ satisfies $\mathbb{E} \int_0^T \|G(t)\|_{\mathcal{L}_{(HS)}(Q^{1/2}(H); D(A^\alpha))}^2 \, dt < \infty$. We assume that $u \in L_{\mathcal{F}}^2(\Omega; L^2([0,T]; D(A^\beta)))$ and $F \colon D(A^\beta) \to D(A^\beta)$ is linear and bounded. Since the operator B is only locally Lipschitz continuous, we can not ensure the existence and uniqueness of a mild solution over an arbitrary time interval $[0,T]$ in general. Thus, we need the following definition of a local mild solution.

Definition 28 (cf. Definition 3.2, [15]). Let τ be a stopping time taking values in $(0,T]$ and $(\tau_m)_{m \in \mathbb{N}}$ be an increasing sequence of stopping times taking values in $[0,T]$ satisfying

$$\lim_{m \to \infty} \tau_m = \tau.$$

A predictable process $(y(t))_{t \in [0,\tau)}$ with values in $D(A^\alpha)$ is called a *local mild solution of system* (2) if for fixed $m \in \mathbb{N}$

$$\mathbb{E} \sup_{t \in [0,\tau_m)} \|y(t)\|_{D(A^\alpha)}^2 < \infty,$$

and we have for each $m \in \mathbb{N}$, all $t \in [0, T]$, and $\mathbb{P}$-a.s.

$$y(t \wedge \tau_m) = e^{-A(t \wedge \tau_m)}\xi - \int_0^{t \wedge \tau_m} A^\delta e^{-A(t \wedge \tau_m - s)} A^{-\delta} B(y(s)) \, ds$$

$$+ \int_0^{t \wedge \tau_m} e^{-A(t \wedge \tau_m - s)} Fu(s) \, ds + I_{\tau_m}(G)(t \wedge \tau_m),$$

where $t \wedge \tau_m = \min\{t, \tau_m\}$ and

$$I_{\tau_m}(G)(t) = \int_0^t \mathbb{1}_{[0, \tau_m)}(s) e^{-A(t-s)} G(s) \, dW(s). \tag{3}$$

Remark 23. The stopped stochastic convolution $(I_{\tau_m}(G)(t \wedge \tau_m))_{t \in [0,T]}$ is well defined according to [10, Appendix].

Remark 24. Alternatively, one can define a local mild solution $(y(t))_{t \in [0,\rho]}$ of system (2) such that $\mathbb{E} \sup_{t \in [0,\rho]} \|y(t)\|^2_{D(A^\alpha)} < \infty$ for a certain stopping time ρ with values in $[0, T]$ and independent of $m \in \mathbb{N}$, see [5, 17].

The proof of the existence and uniqueness of a local mild solution to system (2) is done in two steps. First, we consider a modified system to obtain a mild solution well defined over the whole time interval $[0, T]$. Second, we introduce suitable stopping times such that the mild solution of the modified system and the local mild solution of system (2) coincides. Here, we closely follow [15]. We introduce the following modified system in $D(A^\alpha)$:

$$\begin{cases} dy_m(t) = -[Ay_m(t) + B(\pi_m(y_m(t))) - Fu(t)] \, dt + G(t) \, dW(t), \\ y_m(0) = \xi, \end{cases} \tag{4}$$

where $m \in \mathbb{N}$ and $\pi_m \colon D(A^\alpha) \to D(A^\alpha)$ is defined by

$$\pi_m(y) = \begin{cases} y & \|y\|_{D(A^\alpha)} \leq m, \\ m\|y\|^{-1}_{D(A^\alpha)} y & \|y\|_{D(A^\alpha)} > m. \end{cases} \tag{5}$$

Then we get for every $y, z \in D(A^\alpha)$

$$\|\pi_m(y)\|_{D(A^\alpha)} \leq \min\{m, \|y\|_{D(A^\alpha)}\}, \tag{6}$$

$$\|\pi_m(y) - \pi_m(z)\|_{D(A^\alpha)} \leq 2\|y - z\|_{D(A^\alpha)}. \tag{7}$$

Definition 29. A predictable process $(y_m(t))_{t \in [0,T]}$ with values in $D(A^\alpha)$ is called a *mild solution of (4)* if

$$\mathbb{E} \sup_{t \in [0,T]} \|y_m(t)\|^2_{D(A^\alpha)} < \infty,$$

and we have for all $t \in [0,T]$ and $\mathbb{P}$-a.s.

$$y_m(t) = e^{-At}\xi - \int_0^t A^\delta e^{-A(t-s)} A^{-\delta} B(\pi_m(y_m(s))) \, ds + \int_0^t e^{-A(t-s)} Fu(s) \, ds$$

$$+ \int_0^t e^{-A(t-s)} G(s) \, dW(s).$$

In the following lemma, we state some properties of the stochastic convolution, which we use for the proof of the existence and uniqueness for the solutions of system (2) and system (4).

Lemma 10. *We have*

(i) for all $t \in [0,T]$ and $\mathbb{P}$-a.s. $\int_0^t e^{-A(t-s)} G(s) \, dW(s) \in D(A^\alpha)$ with $\alpha \in \mathbb{R}$ and

$$A^\alpha \int_0^t e^{-A(t-s)} G(s) \, dW(s) = \int_0^t e^{-A(t-s)} A^\alpha G(s) \, dW(s);$$

(ii) for all $k \in (0, \infty)$ and all $r \in [0, T]$

$$\mathbb{E} \sup_{t \in [0,r]} \left\| \int_0^t e^{-A(t-s)} G(s) \, dW(s) \right\|^k_{D(A^\alpha)}$$

$$\leq c_k^k \, \mathbb{E} \left(\int_0^r \|G(t)\|^2_{\mathcal{L}_{(HS)}(Q^{1/2}(H); D(A^\alpha))} \, dt \right)^{k/2},$$

where $c_k > 0$ is a constant.

Proof. The claim (i) results from Lemma 8 and [14, Proposition 4.15] and (ii) is a consequence of (i) and [31, Proposition 1.3 (ii)]. □

Theorem 47. *Let $\alpha \in (0,1)$ and $\delta \in [0,1)$ satisfy $1 > \delta + \alpha > \frac{1}{2}$ and $\delta + 2\alpha \geq \frac{n}{4} + \frac{1}{2}$. Furthermore, let $u \in L^2_{\mathcal{F}}(\Omega; L^2([0,T]; D(A^\beta)))$ be fixed for $\beta \in [0,\alpha]$ such that $\alpha - \beta < \frac{1}{2}$. Then for any $\xi \in L^2(\Omega; D(A^\alpha))$, there exists a unique local mild solution $(y(t))_{t \in [0,\tau)}$ of system (2). Moreover, the process $(y(t))_{t \in [0,\tau)}$ has a continuous modification.*

Proof. We give only the idea of the proof. For a detailed version in a more general setting, we refer to [3]. Let $m \in \mathbb{N}$ be fixed and let the space $\mathcal{Y}_{[t_0, t_1]}$ contain all predictable processes $(\tilde{y}(t))_{t \in [t_0, t_1]}$ with values in $D(A^\alpha)$ such that $\mathbb{E} \sup_{t \in [t_0, t_1]} \|\tilde{y}(t)\|^2_{D(A^\alpha)} < \infty$. We define for all $t \in [0, T]$ and $\mathbb{P}$-a.s.

$$\mathcal{J}_m(\tilde{y})(t) = e^{-At}\xi - \int_0^t A^\delta e^{-A(t-s)} A^{-\delta} B(\pi_m(\tilde{y}(s))) \, ds$$

$$+ \int_0^t e^{-A(t-s)} Fu(s) \, ds + \int_0^t e^{-A(t-s)} G(s) \, dW(s).$$

Using Lemma 8, Lemma 10, and Corollary 1, the operator $\mathcal{J}_m$ is a contraction mapping $\mathcal{Y}_{[0, T_1]}$ into itself for sufficient small $T_1 \in (0, T]$. Especially, note that $(\mathcal{J}_m(\tilde{y})(t))_{t \in [0, T_1]}$ is predictable due to [14, Proposition 3.7]. Applying the Banach fixed point theorem, we get a unique element $y_m \in \mathcal{Y}_{[0, T_1]}$ such that $y_m(t) = \mathcal{J}_m(y_m)(t)$ for all $t \in [0, T_1]$ and $\mathbb{P}$-almost surely. By definition, we get for all $t \in [T_1, T]$ and $\mathbb{P}$-a.s.

$$\mathcal{J}_m(\tilde{y})(t) = e^{-A(t-T_1)} y_m(T_1) - \int_{T_1}^t A^\delta e^{-A(t-s)} A^{-\delta} B(\pi_m(\tilde{y}(s))) \, ds$$

$$+ \int_{T_1}^t e^{-A(t-s)} Fu(s) \, ds + \int_{T_1}^t e^{-A(t-s)} G(s) \, dW(s).$$

Again, we find $T_2 \in [T_1, T]$ such that there exists a unique fixed point of $\mathcal{J}_m$ on $\mathcal{Y}_{[T_1, T_2]}$. By continuing the method, we get the existence and uniqueness of a mild solution $(y_m(t))_{t \in [0, T]}$ to system (4). The fact that $(y_m(t))_{t \in [0, T]}$ has a continuous modification is a consequence of [14, Theorem 5.11].

We define a sequence of stopping times $(\tau_m)_{m \in \mathbb{N}}$ by

$$\tau_m = \inf\{t \in (0, T) : \|y_m(t)\|_{D(A^\alpha)} > m\} \wedge T$$

$\mathbb{P}$-a.s. with the usual convention that $\inf\{\emptyset\} = +\infty$. The sequence $(\tau_m)_{m \in \mathbb{N}}$ is increasing and bounded and hence, there exists a stopping time τ with values in $(0, T]$ such that $\tau_m = \tau$ as $m \to \infty$. By definition, we get $\pi_m(y_m(t)) = y_m(t)$ for all $t \in [0, \tau_m)$ and $\mathbb{P}$-almost surely. Thus, we get for all $t \in [0, T]$ and $\mathbb{P}$-a.s.

$$y_m(t \wedge \tau_m) = e^{-A(t \wedge \tau_m)}\xi - \int_0^{t \wedge \tau_m} A^\delta e^{-A(t \wedge \tau_m - s)} A^{-\delta} B(y_m(s)) \, ds$$

$$+ \int\limits_0^{t \wedge \tau_m} e^{-A(t \wedge \tau_m - s)} Fu(s)\, ds + I_{\tau_m}(G)(t \wedge \tau_m),$$

where $I_{\tau_m}(G)(t)$ is given by (3). For each $m \in \mathbb{N}$, we set $y(t) = y_m(t)$ for all $t \in [0, \tau_m)$ and $\mathbb{P}$-almost surely. Then the process $(y(t))_{t \in [0, \tau)}$ is the unique local mild solution of system (2). $\qquad \square$

Remark 25. Throughout this section, we will always assume that the parameters $\alpha \in (0,1)$, $\delta \in [0,1)$ and $\beta \in [0, \alpha]$ satisfy the assumptions of Theorem 47 and the stopping times $(\tau_m)_{m \in \mathbb{N}}$ in Definition 28 are given by

$$\tau_m = \inf\{t \in (0, T) : \|y_m(t)\|_{D(A^\alpha)} > m\} \wedge T \tag{8}$$

$\mathbb{P}$-a.s. with the usual convention that $\inf\{\emptyset\} = +\infty$.

Next, we show some useful properties. Let us denote by $(y_m(t; u))_{t \in [0,T]}$ and $(y(t; u))_{t \in [0, \tau^u)}$ the mild solution of system (4) and the local mild solution of system (2), respectively. Note that the stopping times $(\tau_m^u)_{m \in \mathbb{N}}$ depend on the control as well. Whenever these processes and the stopping times are considered for fixed control, we use the notation introduced above.

Lemma 11. *For fixed $m \in \mathbb{N}$, let $(y_m(t; u))_{t \in [0,T]}$ be the mild solution of system (4) corresponding to the control $u \in L_{\mathcal{F}}^2(\Omega; L^2([0,T]; D(A^\beta)))$. If $u_1, u_2 \in L_{\mathcal{F}}^k(\Omega; L^2([0,T]; D(A^\beta)))$ with $k \geq 2$, then there exists a constant $c > 0$ such that*

$$\mathbb{E} \sup_{t \in [0,T]} \|y_m(t; u_1) - y_m(t; u_2)\|_{D(A^\alpha)}^k \leq c \|u_1 - u_2\|_{L_{\mathcal{F}}^k(\Omega; L^2([0,T]; D(A^\beta)))}^k.$$

Proof. By definition, we have for all $t \in [0, T]$ and $\mathbb{P}$-a.s.

$$y_m(t; u_1) - y_m(t; u_2)$$

$$= - \int\limits_0^t A^\delta e^{-A(t-s)} A^{-\delta} \left[B(\pi_m(y_m(s; u_1))) - B(\pi_m(y_m(s; u_2))) \right] ds$$

$$+ \int\limits_0^t e^{-A(t-s)} F[u_1(s) - u_2(s)]\, ds.$$

Using Lemma 8, Corollary 1, inequality (6) and (7), and the Cauchy-Schwarz inequality, there exists a constant $c_1 > 0$ such that for sufficient small $T_1 \in (0, T]$

$$\mathbb{E} \sup_{t \in [0,T_1]} \|y_m(t; u_1) - y_m(t; u_2)\|_{D(A^\alpha)}^k \leq c_1 \|u_1 - u_2\|_{L_{\mathcal{F}}^k(\Omega; L^2([0,T]; D(A^\beta)))}^k.$$

Next, we consider for all $t \in [T_1, T]$, $\mathbb{P}$-almost surely, and for $i = 1, 2$

$$y_m(t; u_i) = e^{-A(t-T_1)} y_m(T_1; u_i) - \int_{T_1}^{t} A^{\delta} e^{-A(t-s)} A^{-\delta} B(\pi_m(y_m(s; u_i))) \, ds$$

$$+ \int_{T_1}^{t} e^{-A(t-s)} F u_i(s) \, ds.$$

Again, we find $T_2 \in [T_1, T]$ and a constant $c_2 > 0$ such that

$$\mathbb{E} \sup_{t \in [T_1, T_2]} \|y_m(t; u_1) - y_m(t; u_2)\|_{D(A^{\alpha})}^{k} \leq c_2 \|u_1 - u_2\|_{L^{k}_{\mathcal{F}}(\Omega; L^2([0,T]; D(A^{\beta})))}^{k}.$$

By continuing this method, we obtain the result. □

By definition, we have $y(t; u) = y_m(t; u)$ for all $t \in [0, \tau_m^u)$ and $\mathbb{P}$-almost surely. Hence, a result similar to the previous lemma holds for the local mild solution of system (2). In the following lemmas, we show some useful properties of the stopping times.

Lemma 12. *For fixed $m \in \mathbb{N}$, let $(y_m(t; u))_{t \in [0,T]}$ be the mild solution of system (4) corresponding to the control $u \in L^2_{\mathcal{F}}(\Omega; L^2([0,T]; D(A^{\beta})))$ and let the stopping time τ_m^u be given by equation (8). Then we have*

$$\lim_{u_1 \to u_2} \mathbb{P}\left(\tau_m^{u_1} \neq \tau_m^{u_2}\right) = 0.$$

Proof. Using the extended version of Markov's inequality, see [7], and Lemma 11 with $k = 2$, we get for all $\varepsilon > 0$

$$\mathbb{P}\left(\sup_{t \in [0,T]} \|y_m(t; u_1)) - y_m(t; u_2)\|_{D(A^{\alpha})} \geq \varepsilon \right)$$

$$\leq \frac{c}{\varepsilon^2} \mathbb{E} \int_{0}^{T} \|u_1(t) - u_2(t)\|_{D(A^{\beta})}^{2} \, dt. \tag{9}$$

The proof is done by a contradiction. We assume that $\mathbb{P}(\tau_m^{u_1} < \tau_m^{u_2}) > 0$ as $u_1 \to u_2$. Since the stopping times satisfy equation (8), we get $\mathbb{P}\left(\{\|y_m(\tau_m^{u_1}; u_1)\|_{D(A^{\alpha})} > m\} \cap \{\|y_m(\tau_m^{u_1}; u_2)\|_{D(A^{\alpha})} \leq m\}\right) > 0$ as $u_1 \to u_2$. Therefore, there exists $\varepsilon_0 > 0$ such that $\mathbb{P}\left(\|y_m(\tau_m^{u_1}; u_1)\|_{D(A^{\alpha})} - \|y_m(\tau_m^{u_1}; u_2)\|_{D(A^{\alpha})} \geq \varepsilon_0\right) > 0$ as $u_1 \to u_2$. This implies that $\mathbb{P}\left(\|y_m(\tau_m^{u_1}; u_1) - y_m(\tau_m^{u_1}; u_2)\|_{D(A^{\alpha})} \geq \varepsilon_0\right) > 0$ as $u_1 \to u_2$, which is a contradiction to inequality (9). Hence, we get $\mathbb{P}(\tau_m^{u_1} < \tau_m^{u_2}) = 0$. as $u_1 \to u_2$. Similarly, we obtain $\mathbb{P}(\tau_m^{u_1} > \tau_m^{u_2}) = 0$ as $u_1 \to u_2$, which completes the proof. □

Lemma 13. *For fixed $m \in \mathbb{N}$, let $(y_m(t;u))_{t \in [0,T]}$ be the mild solution of system (4) corresponding to the control $u \in L^2_{\mathcal{F}}(\Omega; L^2([0,T]; D(A^\beta)))$ and let the stopping time τ^u_m be given by equation (8). If we assume that $u_1, u_2 \in L^{k+1}_{\mathcal{F}}(\Omega; L^2([0,T]; D(A^\beta)))$ for $k \geq 1$, then*

$$\lim_{\theta \to 0} \frac{\mathbb{P}\left(\tau^{u_1}_m \neq \tau^{u_1 + \theta u_2}_m\right)}{\theta^k} = 0.$$

Proof. The claim follows similarly to Lemma 12. □

In this section, we introduced the stochastic controlled Navier-Stokes equations driven by additive Wiener noise on bounded and connected domains. An existence result of a unique local mild solution as well as some basic properties are provided. We will utilize these results to show the existence of a unique optimal control of the corresponding control problem stated in the following section.

2.2. A Generalized Control Problem

Here, we introduce the control problem, which is a non-convex optimization problem. We show the existence and uniqueness of an optimal control, which is based on results stated in [6, 28].

We define the cost functional $J_m \colon L^2_{\mathcal{F}}(\Omega; L^2([0,T]; D(A^\beta))) \to \mathbb{R}$ by

$$J_m(u) = \frac{1}{2}\mathbb{E} \int\limits_0^{\tau^u_m} \|A^\gamma(y(t;u) - y_d(t))\|^2_H \, dt + \frac{1}{2}\mathbb{E} \int\limits_0^T \|A^\beta u(t)\|^2_H \, dt, \quad (10)$$

where $m \in \mathbb{N}$ is fixed and $\gamma \in [0, \alpha]$. Moreover, the process $(y(t;u))_{t \in [0,\tau^u)}$ is the local mild solution of system (2) corresponding to the control $u \in L^2_{\mathcal{F}}(\Omega; L^2([0,T]; D(A^\beta)))$ and $y_d \in L^2([0,T]; D(A^\gamma))$ is a given desired velocity field. The set of admissible controls U is a nonempty, closed, bounded, and convex subset of the Hilbert space $L^2_{\mathcal{F}}(\Omega; L^2([0,T]; D(A^\beta)))$ such that $0 \in U$. The task is to find a control $\overline{u}_m \in U$ such that

$$J_m(\overline{u}_m) = \inf_{u \in U} J_m(u).$$

The control $\overline{u}_m \in U$ is called an optimal control. For $\gamma = 0$, the formulation coincides with a tracking problem, see [32, 37, 41, 48]. For $\gamma = \frac{1}{2}$ and $y_d = 0$, we minimize the enstrophy, see [12, 30, 43]. Hence, we are dealing with a generalized cost functional, which incorporates common control strategies in fluid dynamics.

Since the velocity field as well as the stopping times are non-convex with respect to the control, we formulated a control problem using a non-convex cost functional. We have the following existence and uniqueness result.

Theorem 48. *Let the functional* $J_m \colon L^2_{\mathcal{F}}(\Omega; L^2([0,T]; D(A^\beta))) \to \mathbb{R}$ *be given by (10). Then there exists a unique optimal control* $\overline{u}_m \in U$.

Proof. First, we show that $f_m \colon L^2_{\mathcal{F}}(\Omega; L^2([0,T]; D(A^\beta))) \to \mathbb{R}$ defined by

$$ f_m(u) = \mathbb{E} \int_0^{\tau_m^u} \| A^\gamma (y(t;u) - y_d(t)) \|_H^2 \, dt $$

is continuous with respect to the control $u \in L^2_{\mathcal{F}}(\Omega; L^2([0,T]; D(A^\beta)))$. Let the process $(y_m(t;u))_{t \in [0,T]}$ be the mild solution of system (4) corresponding to the control $u \in L^2_{\mathcal{F}}(\Omega; L^2([0,T]; D(A^\beta)))$. Moreover, assume that $u_1, u_2 \in L^2_{\mathcal{F}}(\Omega; L^2([0,T]; D(A^\beta)))$. Using Lemma 8 and the Cauchy-Schwarz inequality, there exists a constant $K > 0$ such that

$$ |f_m(u_1) - f_m(u_2)| $$

$$ \leq K \left(\mathbb{E} \sup_{t \in [0,T]} \| y_m(t;u_1) - y_m(t;u_2) \|_{D(A^\alpha)}^2 \right)^{1/2} $$

$$ + K \int_0^T \mathbb{P}(\tau_m^{u_1} \wedge \tau_m^{u_2} \leq t < \tau_m^{u_1} \vee \tau_m^{u_2}) \left(1 + \| y_d(t) \|_{D(A^\gamma)}^2 \right) dt, $$

where $\tau_m^{u_1} \vee \tau_m^{u_2} = \max\{\tau_m^{u_1}, \tau_m^{u_2}\}$. We get $|f_m(u_1) - f_m(u_2)| = 0$ as $u_1 \to u_2$ by Lemma 11 with $k = 2$, Lemma 12, and the dominated convergence theorem. Furthermore, we obtain $f_m(u) \geq 0$ for every control $u \in L^2_{\mathcal{F}}(\Omega; L^2([0,T]; D(A^\beta)))$.

The space $L^2_{\mathcal{F}}(\Omega; L^2([0,T]; D(A^\beta)))$ is a Hilbert space and thus, a uniformly convex Banach space. By definition, the set of admissible controls $U \subset L^2_{\mathcal{F}}(\Omega; L^2([0,T]; D(A^\beta)))$ is bounded and closed such that $0 \in U$. Applying [6, Partie (A), Théorème 4.2] with $p = 2$, we get the existence of a dense subset $V_0 \subset U$ such that for any $v \in V_0$ the functional $f_m(u) + \| u - v \|_{L^2(\Omega; L^2([0,T]; D(A^\beta))}^2$ attains its unique minimum over U denoted by $u_m(v)$, which is continuous with respect to $v \in V_0$. Since V_0 is a dense subset, there exists a sequence $(v_k)_{k \in \mathbb{N}} \subset V_0$ such that $\| v_k \|_{L^2(\Omega; L^2([0,T]; D(A^\beta)))} = 0$ as $k \to \infty$. We obtain

$$ J_m(u_m(0)) = \lim_{k \to \infty} \left(f_m(u_m(v_k)) + \| u_m(v_k) - v_k \|_{L^2(\Omega; L^2([0,T]; D(A^\beta)))}^2 \right) $$

$$= \lim_{k \to \infty} \inf_{u \in U} \left(f_m(u) + \|u - v_k\|^2_{L^2(\Omega; L^2([0,T]; D(A^\beta)))} \right)$$

$$= \inf_{u \in U} \left(f_m(u) + \|u\|^2_{L^2(\Omega; L^2([0,T]; D(A^\beta)))} \right)$$

$$= \inf_{u \in U} J_m(u)$$

and thus, there exists a unique optimal control $\bar{u}_m = u_m(0) \in U$.   □

We introduced the control problem and we showed the existence of a unique optimal control. Next, we state necessary and sufficient optimality conditions the optimal control has to satisfy. Since the control problem is a non-convex optimization problem on a bounded subset, the necessary optimality condition is stated as a variational inequality using the Gâteaux derivative of the cost functional. The sufficient optimality condition is based on the Fréchet derivative of order two of the cost functional. Hence, the aim of the following section is to derive the required derivatives.

2.3. Necessary Optimality Condition

Here, we state a necessary optimality condition the optimal control has to satisfy, where we use results shown in [33, 52]. We first derive that the Gâteaux derivative of the local mild solution to system (2) is given by the local mild solution to the linearized stochastic Navier-Stokes equations. Consequently, we can calculate the Gâteaux derivative of the cost functional in order to obtain a necessary optimality condition. Moreover, we calculate the Fréchet derivative of order two of the cost functional. Thus, we will show that also a sufficient optimality condition holds.

2.3.1. The linearized stochastic Navier-Stokes equations

We introduce the following linearized system in $D(A^\alpha)$:

$$\begin{cases} dz(t) = -[Az(t) + B(z(t), y(t)) + B(y(t), z(t)) - Fv(t)] \, dt, \\ z(0) = 0, \end{cases} \tag{11}$$

where $v \in L^2_{\mathcal{F}}(\Omega; L^2([0,T]; D(A^\beta)))$ and $(y(t))_{t \in [0,\tau)}$ is the local mild solution of system (2). The operators A, B, F are introduced in Section 2.1. The solution of system (11) is defined in a local mild sense analogously to Definition 28 with stopping times $(\tau_m)_{m \in \mathbb{N}}$ given by equation (8). Similarly to Section 2.1, we first consider the following modified system in $D(A^\alpha)$:

$$\begin{cases} dz_m(t) = -[Az_m(t) + B(z_m(t), \pi_m(y_m(t)))] \, dt \\ \qquad\qquad - [B(\pi_m(y_m(t)), z_m(t)) - Fv(t)] \, dt, \\ z_m(0) = 0, \end{cases} \tag{12}$$

where the process $(y_m(t))_{t\in[0,T]}$ is the mild solution of system (4) and π_m is given by (5). The solution of system (12) is given in a mild sense analog to Definition 29. Recall that the initial value $\xi \in L^2(\Omega; D(A^\alpha))$ is fixed. Due to Theorem 47, we get the existence and uniqueness of the local mild solution $(y(t))_{t\in[0,\tau)}$ to system (4) for fixed control $u \in L^2_{\mathcal{F}}(\Omega; L^2([0,T]; D(A^\beta)))$. Thus, there exists a unique local mild solution $(z(t))_{t\in[0,\tau)}$ of system (11) having a continuous modification. The proof can be obtained similarly to Theorem 47.

Next, we state some properties, which we use to calculate the Gâteaux derivative of the cost functional (10). Recall that both the local mild solution of system (2) and the mild solution of system (4) depend on the control $u \in L^2_{\mathcal{F}}(\Omega; L^2([0,T]; D(A^\beta)))$. Hence, the local mild solution of system (11) and the mild solution of system (12) depend on $u \in L^2_{\mathcal{F}}(\Omega; L^2([0,T]; D(A^\beta)))$ as well as on $v \in L^2_{\mathcal{F}}(\Omega; L^2([0,T]; D(A^\beta)))$. Let us denote these solutions by $(z(t; u, v))_{t\in[0,\tau)}$ and $(z_m(t; u, v))_{t\in[0,T]}$. Whenever these processes are considered for fixed controls, we use the notation introduced above.

Lemma 14. *For fixed $m \in \mathbb{N}$, let $(z_m(t; u, v))_{t\in[0,T]}$ be the mild solution of system (12) corresponding to the controls $u, v \in L^2_{\mathcal{F}}(\Omega; L^2([0,T]; D(A^\beta)))$. Then*

(i) for every control $u \in L^2_{\mathcal{F}}(\Omega; L^2([0,T]; D(A^\beta)))$ and every control $v \in L^k_{\mathcal{F}}(\Omega; L^2([0,T]; D(A^\beta)))$ with $k \geq 2$, there exists a constant $\tilde{c} > 0$ such that

$$\mathbb{E} \sup_{t\in[0,T]} \|z_m(t; u, v)\|^k_{D(A^\alpha)} \leq \tilde{c} \, \|v\|^k_{L^k_{\mathcal{F}}(\Omega; L^2([0,T]; D(A^\beta)))};$$

(ii) we have for every $u, v_1, v_2 \in L^2_{\mathcal{F}}(\Omega; L^2([0,T]; D(A^\beta)))$, all $a, b \in \mathbb{R}$, all $t \in [0,T]$, and $\mathbb{P}$-a.s.

$$z_m(t; u, a\, v_1 + b\, v_2) = a\, z_m(t; u, v_1) + b\, z_m(t; u, v_2);$$

(iii) for all controls $u_1, u_2 \in L^2_{\mathcal{F}}(\Omega; L^2([0,T]; D(A^\beta)))$ and for every control $v \in L^4_{\mathcal{F}}(\Omega; L^2([0,T]; D(A^\beta)))$, there exists a constant $\bar{c} > 0$ such that

$$\mathbb{E} \sup_{t\in[0,T]} \|z_m(t; u_1, v) - z_m(t; u_2, v)\|^2_{D(A^\alpha)}$$

$$\leq \bar{c} \, \|v\|^2_{L^4_{\mathcal{F}}(\Omega; L^2([0,T]; D(A^\beta)))} \|u_1 - u_2\|_{L^2_{\mathcal{F}}(\Omega; L^2([0,T]; D(A^\beta)))}.$$

By definition, we have for all $t \in [0, \tau_m^u)$ and $\mathbb{P}$-a.s.

$$z(t; u, v) = z_m(t; u, v).$$

Hence, one can easily obtain similar results for the local mild solution of system (11).

Here, we introduced the linearized stochastic Navier-Stokes equations and provided some basic properties. Next, we will calculate the derivatives of the cost functional and state the necessary optimality condition.

2.3.2. *The Derivatives of the Cost Functional*

Let X, Y, and Z be arbitrary Banach spaces. For $f \colon M \subset X \to Y$ with M nonempty and open, we denote the Gâteaux derivative and the Fréchet derivative at $x \in M$ in direction $h \in X$ by $d^G f(x)[h]$ and $d^F f(x)[h]$, respectively. Derivatives of order $k \in \mathbb{N}$ at $x \in M$ in directions $h_1, ..., h_k \in X$ are represented by $d^G(f(x))^k[h_1, ..., h_k]$ and $d^F(f(x))^k[h_1, ..., h_k]$. For a mapping $f \colon M_X \times M_Y \to Z$ with $M_X \subset X$, $M_Y \subset Y$ nonempty and open, we denote the partial Gâteaux derivative and the partial Fréchet derivative at $x \in M_X$ in direction $h \in X$ for fixed $y \in M_Y$ by $d_x^G f(x, y)[h]$ and $d_x^F f(x, y)[h]$, respectively. Analogously, the partial Gâteaux derivative and the partial Fréchet derivative at $y \in M_y$ in direction $h \in Y$ for fixed $x \in M_X$ are represented by $d_y^G f(x, y)[h]$ and $d_y^F f(x, y)[h]$, respectively.

First, we show that the local mild solution of system (11) is the partial Gâteaux derivative of the local mild solution to system (2) with respect to the control variable.

Theorem 49. *Let the processes $(y(t; u))_{t \in [0, \tau^u)}$ and $(z(t; u, v))_{t \in [0, \tau^u)}$ be the local mild solutions of system (2) and system (11), respectively, corresponding to the controls $u, v \in L_{\mathcal{F}}^2(\Omega; L^2([0, T]; D(A^\beta)))$. Then for fixed $m \in \mathbb{N}$, the Gâteaux derivative of $y(t; u)$ at $u \in L_{\mathcal{F}}^2(\Omega; L^2([0, T]; D(A^\beta)))$ in direction $v \in L_{\mathcal{F}}^2(\Omega; L^2([0, T]; D(A^\beta)))$ satisfies for all $t \in [0, \tau_m^u)$ and $\mathbb{P}$-a.s.*

$$d_u^G y(t; u)[v] = z(t; u, v).$$

Proof. First, we assume that $v \in L_{\mathcal{F}}^4(\Omega; L^2([0, T]; D(A^\beta)))$. Moreover, let $0 = T_0 < T_1 < ... < T_l = T$, which we specify below. Since the stopping time $\tau_m^u \wedge \tau_m^{u+\theta v}$ with $\theta \in \mathbb{R} \backslash \{0\}$ takes values in $[0, T]$, we have $\mathbb{P}$-a.s.

$$\mathbb{1}_{\tau_m^u \wedge \tau_m^{u+\theta v} \in [0, T_1]}(\omega) + \sum_{j=1}^{l-1} \mathbb{1}_{\tau_m^u \wedge \tau_m^{u+\theta v} \in (T_j, T_{j+1}]}(\omega) = 1, \qquad (13)$$

where $\mathbb{1}$ denotes the indicator function. We set $\mathbb{1}_0 = \mathbb{1}_{\tau_m^u \wedge \tau_m^{u+\theta v} \in [0, T_1]}$ and $\mathbb{1}_j = \mathbb{1}_{\tau_m^u \wedge \tau_m^{u+\theta v} \in (T_j, T_{j+1}]}$ for $j = 1, ..., l - 1$. Furthermore, let the processes $(y_m(t; u^*))_{t \in [0,T]}$ and $(z_m(t; u^*, v^*))_{t \in [0,T]}$ be the mild solutions of system (4) and system (12), respectively, corresponding to arbitrary $u^*, v^* \in L^2_{\mathcal{F}}(\Omega; L^2([0, T]; D(A^\beta)))$. By definition, the process $(\frac{1}{\theta}[y(t; u + \theta v) - y(t; u)] - z(t; u, v))_{t \in [0,T]}$ satisfies a deterministic equation. We show that this process tends to zero as $\theta \to 0$ on the intervals $[0, T_1]$ and $(T_j, T_{j+1}]$ for $j = 1, ..., l-1$. Due to Lemma 8, Lemma 9, Lemma 14 (i) with $k = 4$, and the Cauchy-Schwarz inequality, we have for all $\theta \in \mathbb{R} \setminus \{0\}$ and for $j = 1, ..., l - 1$

$$\mathbb{E}\left[\mathbb{1}_j \sup_{t \in [0,T_1]} \left\|\frac{1}{\theta}[y(t; u + \theta v) - y(t; u)] - z(t; u, v)\right\|^2_{D(A^\alpha)}\right]$$

$$\leq C_{T_1} \mathbb{E}\left[\mathbb{1}_j \sup_{t \in [0,T_1]} \left\|\frac{1}{\theta}[y(t; u + \theta v) - y(t; u)] - z(t; u, v)\right\|^2_{D(A^\alpha)}\right]$$

$$+ C\left(\mathbb{E} \sup_{t \in [0,T_1]} \|y_m(t; u + \theta v) - y_m(t; u)\|^2_{D(A^\alpha)}\right)^{1/2}$$

with a constant $C_{T_1} > 0$ dependent on the point of time $T_1 \in (0, T]$ and a constant $C > 0$. We chose $T_1 \in (0, T]$ such that $C_{T_1} < 1$. Lemma 11 with $k = 2$ implies that the second part of the inequality tends to zero, and thus, taking limits on both sides, we can conclude that for $j = 1, ..., l - 1$

$$\lim_{\theta \to 0} \mathbb{E}\left[\mathbb{1}_j \sup_{t \in [0,T_1]} \left\|\frac{1}{\theta}[y(t; u + \theta v) - y(t; u)] - z(t; u, v)\right\|^2_{D(A^\alpha)}\right] = 0.$$

Similarly, we get

$$\lim_{\theta \to 0} \mathbb{E}\left[\mathbb{1}_0 \sup_{t \in [0, \tau_m^u \wedge \tau_m^{u+\theta v})} \left\|\frac{1}{\theta}[y(t; u + \theta v) - y(t; u)] - z(t; u, v)\right\|^2_{D(A^\alpha)}\right] = 0.$$

By continuing the method similarly to Lemma 11, we obtain

$$\lim_{\theta \to 0} \mathbb{E}\left[\mathbb{1}_j \sup_{t \in [0, \tau_m^u \wedge \tau_m^{u+\theta v})} \left\|\frac{1}{\theta}[y(t; u + \theta v) - y(t; u)] - z(t; u, v)\right\|^2_{D(A^\alpha)}\right] = 0$$

for $j = 0, 1, ..., l - 1$. Due to equation (13), we have

$$\lim_{\theta \to 0} \mathbb{E} \sup_{t \in [0, \tau_m^u \wedge \tau_m^{u+\theta v})} \left\|\frac{1}{\theta}[y(t; u + \theta v) - y(t; u)] - z(t; u, v)\right\|^2_{D(A^\alpha)} = 0.$$

Therefore, the Gâteaux derivative of $y(t; u)$ at $u \in L^2_{\mathcal{F}}(\Omega; L^2([0, T]; D(A^\beta)))$ in direction $v \in L^4_{\mathcal{F}}(\Omega; L^2([0, T]; D(A^\beta)))$ satisfies for all $t \in [0, \tau^u_m \wedge \tau^{u+\theta v}_m)$ and $\mathbb{P}$-a.s.

$$d^G_u y(t; u)[v] = z(t; u, v). \tag{14}$$

Note that by Lemma 12, we have $\mathbb{P}(\tau^u_m \neq \tau^{u+\theta v}_m) = 0$ as $\theta \to 0$. Moreover, the operator $d^G_u y(t; u)$ is linear and bounded as a consequence of Lemma 14 (i) with $k = 4$ and Lemma 14 (ii). Since the space $L^4_{\mathcal{F}}(\Omega; L^2([0, T]; D(A^\beta)))$ is dense in $L^2_{\mathcal{F}}(\Omega; L^2([0, T]; D(A^\beta)))$, the equation (14) holds for every direction $v \in L^2_{\mathcal{F}}(\Omega; L^2([0, T]; D(A^\beta)))$. □

This enables us to calculate the Gâteaux derivative of the cost functional.

Theorem 50. *Let the functional* $J_m \colon L^2_{\mathcal{F}}(\Omega; L^2([0, T]; D(A^\beta))) \to \mathbb{R}$ *be defined by (10). The Gâteaux derivative at* $u \in L^2_{\mathcal{F}}(\Omega; L^2([0, T]; D(A^\beta)))$ *in direction* $v \in L^2_{\mathcal{F}}(\Omega; L^2([0, T]; D(A^\beta)))$ *satisfies*

$$d^G J_m(u)[v] = \mathbb{E} \int_0^{\tau^u_m} \langle A^\gamma(y(t; u) - y_d(t)), A^\gamma z(t; u, v) \rangle_H \, dt$$

$$+ \mathbb{E} \int_0^T \langle A^\beta u(t), A^\beta v(t) \rangle_H \, dt,$$

where the process $(z(t; u, v))_{t \in [0, \tau^u)}$ *is the local mild solution of system (11) corresponding to the controls* $u, v \in L^2_{\mathcal{F}}(\Omega; L^2([0, T]; D(A^\beta)))$.

Proof. Define the functionals $\Phi, \Psi \colon L^2_{\mathcal{F}}(\Omega; L^2([0, T]; D(A^\beta))) \to \mathbb{R}$ by

$$\Phi(u) = \frac{1}{2} \mathbb{E} \int_0^{\tau^u_m} \|A^\gamma(y(t; u) - y_d(t))\|^2_H \, dt \quad \text{and} \quad \Psi(u) = \frac{1}{2} \mathbb{E} \int_0^T \|A^\beta u(t)\|^2_H \, dt.$$

First, we calculate the Gâteaux derivative of the functional Φ at the point $u \in L^2_{\mathcal{F}}(\Omega; L^2([0, T]; D(A^\beta)))$ in direction $v \in L^2_{\mathcal{F}}(\Omega; L^2([0, T]; D(A^\beta)))$. We set $\tilde{z}_\theta(t; u, v) = \frac{1}{\theta}[y(t; u + \theta v) - y(t; u)] - z(t; u, v)$ for all $\theta \in \mathbb{R}\backslash\{0\}$, all $t \in [0, T]$, and $\mathbb{P}$-almost surely. We get for all $\theta \in \mathbb{R}\backslash\{0\}$

$$\left| \frac{1}{\theta}[\Phi(u + \theta v) - \Phi(u)] - \mathbb{E} \int_0^{\tau^u_m} \langle A^\gamma(y(t; u) - y_d(t)), A^\gamma z(t; u, v) \rangle_H \, dt \right|$$

$$\leq \mathcal{I}_1(\theta) + \mathcal{I}_2(\theta) + \mathcal{I}_3(\theta),$$

where

$$\mathcal{I}_1(\theta) = \left| \frac{1}{2\theta} \mathbb{E} \int_0^{\tau_m^u \wedge \tau_m^{u+\theta v}} \|A^\gamma(y(t; u + \theta v) - y(t; u))\|_H^2 \, dt \right|,$$

$$\mathcal{I}_2(\theta) = \left| \mathbb{E} \int_0^{\tau_m^u \wedge \tau_m^{u+\theta v}} \langle A^\gamma(y(t; u) - y_d(t)), A^\gamma \tilde{z}_\theta(t; u, v) \rangle_H \, dt \right|,$$

$$\mathcal{I}_3(\theta) = \left| \frac{1}{2\theta} \mathbb{E} \int_{\tau_m^u \wedge \tau_m^{u+\theta v}}^{\tau_m^{u+\theta v}} \|A^\gamma(y(t; u + \theta v) - y_d(t))\|_H^2 \, dt \right|$$

$$+ \left| \frac{1}{2\theta} \mathbb{E} \int_{\tau_m^u \wedge \tau_m^{u+\theta v}}^{\tau_m^u} \|A^\gamma(y(t; u) - y_d(t))\|_H^2 \, dt \right|$$

$$+ \left| \mathbb{E} \int_{\tau_m^u \wedge \tau_m^{u+\theta v}}^{\tau_m^u} \langle A^\gamma(y(t; u) - y_d(t)), A^\gamma z(t; u, v) \rangle_H \, dt \right|.$$

Using Lemma 8 (v) and Lemma 11 with $k = 2$, we can conclude $\mathcal{I}_1(\theta) = 0$ as $\theta \to 0$. Due to the Cauchy-Schwarz inequality, Lemma 8 (v), and Theorem 49, we can infer $\mathcal{I}_2(\theta) = 0$ as $\theta \to 0$. Using Lemma 8 (v), Fubini's theorem, Lemma 12, Lemma 13 with $k = 1$, and Lebesgue's dominated convergence theorem, we obtain $\mathcal{I}_3(\theta) = 0$ as $\theta \to 0$. Thus, we get

$$\lim_{\theta \to 0} \left| \frac{1}{\theta} [\Phi(u + \theta v) - \Phi(u)] - \mathbb{E} \int_0^{\tau_m^u} \langle A^\gamma(y(t; u) - y_d(t)), A^\gamma z(t; u, v) \rangle_H \, dt \right| = 0$$

and the Gâteaux derivative of Φ at $u \in L_{\mathcal{F}}^2(\Omega; L^2([0, T]; D(A^\beta)))$ in direction $v \in L_{\mathcal{F}}^2(\Omega; L^2([0, T]; D(A^\beta)))$ is given by

$$d^G \Phi(u)[v] = \mathbb{E} \int_0^{\tau_m^u} \langle A^\gamma(y(t; u) - y_d(t)), A^\gamma z(t; u, v) \rangle_H \, dt.$$

By Lemma 14 (ii), the functional $d^G \Phi(u)$ is linear. Moreover, the functional $d^G \Phi(u)$ is bounded as a consequence of Lemma 8 (v), Lemma 14 (i) with $k = 2$, and the Cauchy-Schwarz inequality.

Note that the functional Ψ is given by the squared norm on the Hilbert space $L^2(\Omega; L^2([0, T]; D(A^\beta)))$. Therefore, the Gâteaux derivative of Ψ

at $u \in L^2_{\mathcal{F}}(\Omega; L^2([0,T]; D(A^\beta)))$ in direction $v \in L^2_{\mathcal{F}}(\Omega; L^2([0,T]; D(A^\beta)))$ satisfies

$$d^G\Psi(u)[v] = \mathbb{E} \int\limits_0^T \langle A^\beta u(t), A^\beta v(t) \rangle_H \, dt.$$

Obviously, the functional $d^G\Psi(u)$ is linear and bounded.

The Gâteaux derivative of J_m at $u \in L^2_{\mathcal{F}}(\Omega; L^2([0,T]; D(A^\beta)))$ in direction $v \in L^2_{\mathcal{F}}(\Omega; L^2([0,T]; D(A^\beta)))$ is given by

$$d^G J_m(u)[v] = d^G\Phi(u)[v] + d^G\Psi(u)[v].$$

Since $d^G\Phi(u)$ and $d^G\Psi(u)$ are linear and bounded, the functional $d^G J_m(u)$ is linear and bounded as well. $\square$

Since the set of admissible controls U is a closed, bounded, and convex subset of the Hilbert space $L^2_{\mathcal{F}}(\Omega; L^2([0,T]; D(A^\beta)))$ such that $0 \in U$, the optimal control $\overline{u}_m \in U$ satisfies the following necessary optimality condition for fixed $m \in \mathbb{N}$ and every $u \in U$:

$$d^G J_m(\overline{u}_m)[u - \overline{u}_m] \geq 0. \tag{15}$$

Due to the previous theorem, we get for fixed $m \in \mathbb{N}$ and every $u \in U$:

$$\mathbb{E} \int\limits_0^{\tau_m^{\overline{u}_m}} \langle A^\gamma(y(t; \overline{u}_m) - y_d(t)), A^\gamma z(t; \overline{u}_m, u - \overline{u}_m) \rangle_H \, dt$$

$$+ \mathbb{E} \int\limits_0^T \langle A^\beta \overline{u}_m(t), A^\beta(u(t) - \overline{u}_m(t)) \rangle_H \, dt \geq 0. \tag{16}$$

For more details on necessary optimality conditions of general optimization problems, see [33, 52].

In order to obtain a sufficient optimality condition, we calculate the Fréchet derivative of the cost functional (10) of order two. First, we show that the Gâteaux derivative of the cost functional coincides with its Fréchet derivative.

Corollary 2. *Let the functional $J_m \colon L^2_{\mathcal{F}}(\Omega; L^2([0,T]; D(A^\beta))) \to \mathbb{R}$ be defined by (10). The Fréchet derivative at $u \in L^2_{\mathcal{F}}(\Omega; L^2([0,T]; D(A^\beta)))$ in direction $v \in L^2_{\mathcal{F}}(\Omega; L^2([0,T]; D(A^\beta)))$ satisfies*

$$d^F J_m(u)[v] = \mathbb{E} \int\limits_0^{\tau_m^u} \langle A^\gamma(y(t; u) - y_d(t)), A^\gamma z(t; u, v) \rangle_H \, dt$$

$$+\, \mathbb{E} \int_0^T \big\langle A^\beta u(t), A^\beta v(t) \big\rangle_H \, dt,$$

where the process $(z(t; u, v))_{t \in [0, \tau^u)}$ is the local mild solution of system (11) corresponding to the controls $u, v \in L^2_{\mathcal{F}}(\Omega; L^2([0, T]; D(A^\beta)))$. Moreover, the functional $d^F J_m(u)[v]$ is continuous with respect to u.

Proof. By Theorem 50, we get the Gâteaux derivative of the functional J_m. Moreover, the processes $(y(t; u))_{t \in [0, \tau^u)}$ and $(z(t; u, v))_{t \in [0, \tau^u)}$ are continuous with respect to $u \in L^2_{\mathcal{F}}(\Omega; L^2([0, T]; D(A^\beta)))$ as a consequence of Lemma 11 and Lemma 14 (iii). Due to Lemma 12, one can show that $u \mapsto d^G J_m(u)[v]$ is a continuous mapping from $L^2_{\mathcal{F}}(\Omega; L^2([0, T]; D(A^\beta)))$ into $\mathbb{R}$. By the mean value theorem, see [36, Theorem 4.1.2], we get

$$\lim_{\|v\|_{L^2(\Omega; L^2([0,T]; D(A^\beta)))} \to 0} \frac{\big| J_m(u + v) - J_m(u) - d^G J_m(u)[v] \big|}{\|v\|_{L^2(\Omega; L^2([0,T]; D(A^\beta)))}} = 0.$$

Hence, the Fréchet derivative of J_m at $u \in L^2_{\mathcal{F}}(\Omega; L^2([0, T]; D(A^\beta)))$ in direction $v \in L^2_{\mathcal{F}}(\Omega; L^2([0, T]; D(A^\beta)))$ coincides with the Gâteaux derivative, which is linear and bounded by Theorem 50. $\qquad \square$

Similarly to Theorem 50, the second-order Gâteaux derivative of the cost functional given by (10) at $u \in L^2_{\mathcal{F}}(\Omega; L^2([0, T]; D(A^\beta)))$ in directions $v_1, v_2 \in L^2_{\mathcal{F}}(\Omega; L^2([0, T]; D(A^\beta)))$ satisfies

$$d^G (J_m(u))^2 [v_1, v_2] = \mathbb{E} \int_0^{\tau_m^u} \big\langle A^\gamma z(t; u, v_1), A^\gamma z(t; u, v_2) \big\rangle_H \, dt$$

$$+\, \mathbb{E} \int_0^T \big\langle A^\beta v_1(t), A^\beta v_2(t) \big\rangle_H \, dt,$$

where $(z(t; u, v_i))_{t \in [0, \tau^u)}$ are the local mild solutions of system (11) corresponding to the controls $u, v_i \in L^2_{\mathcal{F}}(\Omega; L^2([0, T]; D(A^\beta)))$ for $i = 1, 2$. Moreover, this functional coincides with its Fréchet derivative and is continuous with respect to u, where we can adopt the proof of Corollary 2.

In this section, we calculated the required derivatives of the cost function in order to obtain necessary and sufficient optimality conditions the optimal control has to satisfy. We will use these derivatives to derive an explicit formula of the optimal control such that the sufficient optimality condition holds.

2.4. The Optimal Control

In this section, we use the variational inequality (16) to derive an explicit formula the optimal control $\overline{u}_m \in U$ has to satisfy. Therefore, we need a duality principle, which gives us a relation between the linearized stochastic Navier-Stokes equations and the adjoint equation. Since the control problem considered here is constrained by a SPDE with additive noise, the adjoint equation is specified by a deterministic backward PDE, which has a well defined mild solution with values in the space of fractional powers of the Stokes operator. In general, a duality principle of solutions to forward and backward equations can be obtained using an Itô product formula. This formula is not applicable to solutions in a mild sense. Hence, we need to approximate the mild solutions of the linearized stochastic Navier-Stokes equations and the adjoint equation by strong formulations. One method is given by introducing the Yosida approximation of the operator A, see [14]. For applications regarding duality principles, we refer to [21, 45]. Since this approximation is done only in the underlying Hilbert space H, we do not obtain convergence results in the domain of fractional powers of the Stokes operator and hence, we can not use this approach. Here, we apply the method introduced in [29, 35]. The basic idea is to formulate a mild solution with values in $D(A)$ by using the resolvent operator. As a consequence, we get convergence results in the required spaces and the mild solutions coincide with strong solutions. This allows us to obtain the duality principle and we get an explicit formula for the optimal control. Finally, we show that this optimal control satisfies a sufficient optimality condition.

2.4.1. The Adjoint Equation

We introduce the following backward equation in $D(A^\delta)$:

$$\begin{cases} dz_m^*(t) = -\mathbb{1}_{[0,\tau_m)}(t)[-Az_m^*(t) - A^{2\alpha}B_\delta^* \left(y(t), A^\delta z_m^*(t)\right) \\ \qquad\qquad - A^{2\gamma} \left(y(t) - y_d(t)\right)] \, dt \\ z_m^*(T) = 0, \end{cases} \tag{17}$$

where $m \in \mathbb{N}$ and $(y(t))_{t \in [0,\tau)}$ is the local mild solution of system (2). The stopping times $(\tau_m)_{m \in \mathbb{N}}$ are defined by (8) and $y_d \in L^2([0,T]; D(A^\gamma))$ is the given desired velocity field. The operator A and its fractional powers are introduced in Section 2.1. Moreover, the operator $B_\delta^* \left(y(t), \cdot\right) : H \to D(A^\alpha)$ is linear and bounded. A precise meaning is given in the following remark.

Remark 26. (i) The operator $A^{-\delta}[B(\cdot, y) + B(y, \cdot)] : D(A^\alpha) \to H$ is linear and bounded for every $y \in D(A^\alpha)$ satisfying $\|y\|_{D(A^\alpha)} \leq m$ as a consequence of Lemma 9. Therefore, there exists a linear and bounded operator $B_\delta^*(y, \cdot) : H \to D(A^\alpha)$ such that for every $h \in H$ and every $z \in D(A^\alpha)$

$$\langle A^{-\delta}[B(z, y) + B(y, z)], h \rangle_H = \langle z, B_\delta^*(y, h) \rangle_{D(A^\alpha)}.$$

We can rewrite this equivalently as

$$\langle A^{-\delta}[B(z, y) + B(y, z)], h \rangle_H = \langle A^\alpha z, A^\alpha B_\delta^*(y, h) \rangle_H \qquad (18)$$

for every $h \in H$ and every $z \in D(A^\alpha)$. Moreover, we get that the operator $A^\alpha B_\delta^*(y, \cdot) : H \to H$ is linear and bounded due to the closed graph theorem.

(ii) Recall that $\|y(t)\|_{D(A^\alpha)} \leq m$ for all $t \in [0, \tau_m)$ and $\mathbb{P}$-almost surely.

Definition 30. A stochastic process $(z_m^*(t))_{t \in [0,T]}$ with values in $D(A^\delta)$ is called a *mild solution of system (17)* if

$$\mathbb{E} \sup_{t \in [0,T]} \|z_m^*(t)\|_{D(A^\delta)}^2 < \infty,$$

and we have for all $t \in [0, T]$ and $\mathbb{P}$-a.s.

$$z_m^*(t) = - \int_t^T \mathbb{1}_{[0,\tau_m)}(s) A^\alpha e^{-A(s-t)} A^\alpha B_\delta^* \left(y(s \wedge \tau_m), A^\delta z_m^*(s) \right) ds$$

$$+ \int_t^T \mathbb{1}_{[0,\tau_m)}(s) A^\gamma e^{-A(s-t)} A^\gamma \left(y(s \wedge \tau_m) - y_d(s) \right) ds.$$

The existence and uniqueness of a mild solution to system (17) can be obtained similarly to Theorem 47 if we additionally require $\gamma + \delta < \frac{1}{2}$. If $y_d \in L^\infty([0, T]; D(A^\gamma))$, then this restriction vanishes. Moreover, we have for fixed $m \in \mathbb{N}$

$$\mathbb{E} \sup_{t \in [\tau_m, T]} \|z_m^*(t)\|_{D(A^\delta)}^2 = 0. \qquad (19)$$

Here, we introduced the adjoint equation. Next, we will approximate the local mild solution of the linearized stochastic Navier-Stokes equations as well as the mild solution of the adjoint equation by strong formulations.

2.4.2. *Approximation by a Strong Formulation*

Here, we omit the dependence on the controls for the sake of simplicity.

Let $\lambda \in \mathbb{C}$ such that $\lambda I + A$ is invertible, i.e. $(\lambda I + A)^{-1}$ is a linear and bounded operator. Then $R(\lambda; -A) = (\lambda I + A)^{-1}$ is called the *resolvent operator.* The operator $R(\lambda; -A)$ maps H into $D(A)$ and using the closed graph theorem, we can conclude that the operator $AR(\lambda; -A)$ is linear and bounded on H. Moreover, we have the following representation:

$$R(\lambda; -A) = \int_0^\infty e^{-\lambda r} e^{-Ar} dr. \tag{20}$$

For all $\lambda \in \mathbb{R}$ with $\lambda > 0$, we get $\|R(\lambda; -A)\|_{\mathcal{L}(H)} \leq \frac{1}{\lambda}$ and since the semigroup $(e^{-At})_{t \geq 0}$ is self-adjoint, the operator $R(\lambda; -A)$ is self-adjoint as well. Let the operator $R(\lambda) \colon H \to D(A)$ be defined by $R(\lambda) = \lambda R(\lambda; -A)$. We have for all $\lambda > 0$

$$\|R(\lambda)\|_{\mathcal{L}(H)} \leq 1. \tag{21}$$

By Lemma 8 (iii) and equation (20), we obtain for every $y \in D(A^\alpha)$

$$A^\alpha R(\lambda) y = R(\lambda) A^\alpha y \tag{22}$$

with $\alpha \in \mathbb{R}$. Moreover, we get for every $y \in H$

$$\lim_{\lambda \to \infty} \|R(\lambda) y - y\|_H = 0. \tag{23}$$

For more details on the resolvent operator, we refer to [42].

Next, we give an approximation of the mild solution of system (12). We introduce the following system in $D(A^{1+\alpha})$:

$$\begin{cases} dz_m(t, \lambda) = -[Az_m(t, \lambda) + R(\lambda)B(R(\lambda)z_m(t, \lambda), \pi_m(y_m(t)))]\, dt \\ \qquad\qquad - [R(\lambda)B(\pi_m(y_m(t)), R(\lambda)z_m(t, \lambda)) - R(\lambda)Fv(t)]\, dt, \quad (24) \\ z_m(0, \lambda) = 0, \end{cases}$$

where $m \in \mathbb{N}$, $\lambda > 0$ and $v \in L^2_{\mathcal{F}}(\Omega; L^2([0, T]; D(A^\beta)))$. The operators A, B, F are introduced in Section 2.1. The mapping π_m is given by (5) and the process $(y_m(t))_{t \in [0,T]}$ is the mild solution of system (4).

Remark 27. Note that the approximation scheme provided in [29, 35] differs from the approximation scheme introduced by system (24). Here, the additional operator $R(\lambda)$ is necessary to obtain a duality principle.

The solution of system (24) is defined in a mild sense according to Definition 29. Recall that the operators $R(\lambda)$ and $AR(\lambda)$ are linear and bounded on H. Hence, an existence and uniqueness result of a mild solution $(z_m(t, \lambda))_{t \in [0,T]}$ to system (24) with fixed $m \in \mathbb{N}$ and fixed $\lambda > 0$ can be adopted from the proof of Theorem 47. Moreover, we have the following strong formulation for fixed $m \in \mathbb{N}$, fixed $\lambda > 0$, all $t \in [0, T]$, and $\mathbb{P}$-almost surely:

$$
z_m(t, \lambda)
$$

$$
= - \int_0^t A z_m(s, \lambda) + A^\delta R(\lambda) A^{-\delta} B(R(\lambda) z_m(s, \lambda), \pi_m(y_m(s))) \, ds
$$

$$
- \int_0^t A^\delta R(\lambda) A^{-\delta} B(\pi_m(y_m(s)), R(\lambda) z_m(s, \lambda)) + R(\lambda) F v(s) \, ds, \quad (25)
$$

which is an immediate consequence of [35, Proposition 2.3]. Furthermore, we have the following convergence result.

Lemma 15. *Let $(z_m(t))_{t \in [0,T]}$ and $(z_m(t, \lambda))_{t \in [0,T]}$ be the mild solutions of system (12) and system (24), respectively. Then we have for fixed $m \in \mathbb{N}$*

$$
\lim_{\lambda \to \infty} \mathbb{E} \sup_{t \in [0,T]} \|z_m(t) - z_m(t, \lambda)\|_{D(A^\alpha)}^2 = 0.
$$

Proof. We define $\widetilde{B}(y, z) = B(z, y) + B(y, z)$ for every $y, z \in D(A^\alpha)$. Since B is bilinear on $D(A^\alpha) \times D(A^\alpha)$, the operator $\widetilde{B}$ is bilinear as well and we get for every $y, z \in D(A^\alpha)$

$$
\left\| A^{-\delta} \widetilde{B}(y, z) \right\|_H \leq 2 \widetilde{M} \|y\|_{D(A^\alpha)} \|z\|_{D(A^\alpha)} \quad (26)
$$

as a consequence of Lemma 9. By definition, we have for a point of time $T_1 \in (0, T]$ and all $\lambda > 0$

$$
\mathbb{E} \sup_{t \in [0,T_1]} \|z_m(t) - z_m(t, \lambda)\|_{D(A^\alpha)}^2
$$

$$
\leq 2 \, \mathbb{E} \sup_{t \in [0,T_1]} \|\mathcal{I}_1(t, \lambda)\|_{D(A^\alpha)}^2 + 2 \, \mathbb{E} \sup_{t \in [0,T_1]} \|\mathcal{I}_2(t, \lambda)\|_{D(A^\alpha)}^2, \quad (27)
$$

where

$$
\mathcal{I}_1(t, \lambda) = \int_0^t A^\delta e^{-A(t-s)} R(\lambda) A^{-\delta} \widetilde{B}(\pi_m(y_m(s)), R(\lambda) [z_m(s) - z_m(s, \lambda)]) \, ds,
$$

$$
\mathcal{I}_2(t, \lambda) = \int_0^t A^\delta e^{-A(t-s)} A^{-\delta} \widetilde{B}(\pi_m(y_m(s)), [I - R(\lambda)] z_m(s)) \, ds
$$

$$
+ \int_0^t A^\delta e^{-A(t-s)} [I - R(\lambda)] A^{-\delta} \widetilde{B}(\pi_m(y_m(s)), R(\lambda) z_m(s)) \, ds
$$

$$+ \int_0^t e^{-A(t-s)}[I - R(\lambda)]Fv(s)\,ds$$

with I the identity operator on H. By Lemma 8, equation (22), and inequalities (6), (21), and (26), there exists a constant C_{T_1} dependent on the point of time T_1 such that for all $\lambda > 0$

$$\mathbb{E} \sup_{t \in [0,T_1]} \|\mathcal{I}_1(t,\lambda)\|^2_{D(A^\alpha)} \leq C_{T_1} \mathbb{E} \sup_{t \in [0,T_1]} \|z_m(t) - z_m(t,\lambda)\|^2_{D(A^\alpha)}. \quad (28)$$

Similarly, there exists a constant $C > 0$ such that for all $\lambda > 0$

$$\mathbb{E} \sup_{t \in [0,T_1]} \|\mathcal{I}_2(t,\lambda)\|^2_{D(A^\alpha)}$$

$$\leq C\,\mathbb{E} \sup_{t \in [0,T_1]} \|[I - R(\lambda)]A^\alpha z_m(t)\|^2_H$$

$$+ C\,\mathbb{E} \sup_{t \in [0,T_1]} \left\|[I - R(\lambda)]A^{-\delta}\widetilde{B}(\pi_m(y_m(t)), R(\lambda)z_m(t))\right\|^2_H$$

$$+ C\,\mathbb{E} \int_0^{T_1} \left\|[I - R(\lambda)]A^\beta Fv(t)\right\|^2_H dt.$$

Using equation (23) and Lebesgue's dominated convergence theorem, we can conclude

$$\lim_{\lambda \to \infty} \mathbb{E} \sup_{t \in [0,T_1]} \|\mathcal{I}_2(t,\lambda)\|^2_{D(A^\alpha)} = 0. \quad (29)$$

Due to inequality (27) and inequality (28), we find for all $\lambda > 0$

$$\mathbb{E} \sup_{t \in [0,T_1]} \|z_m(t) - z_m(t,\lambda)\|^2_{D(A^\alpha)} \leq 2C_{T_1} \mathbb{E} \sup_{t \in [0,T_1]} \|z_m(t) - z_m(t,\lambda)\|^2_{D(A^\alpha)}$$

$$+ 2\,\mathbb{E} \sup_{t \in [0,T_1]} \|\mathcal{I}_2(t,\lambda)\|^2_{D(A^\alpha)}.$$

We chose $T_1 \in (0,T]$ such that $C_{T_1} < 1/2$. Then we obtain for all $\lambda > 0$

$$\mathbb{E} \sup_{t \in [0,T_1]} \|z_m(t) - z_m(t,\lambda)\|^2_{D(A^\alpha)} \leq \frac{2}{1 - 2C_{T_1}} \mathbb{E} \sup_{t \in [0,T_1]} \|\mathcal{I}_2(t,\lambda)\|^2_{D(A^\alpha)}.$$

By equation (29), we can conclude $\mathbb{E} \sup_{t \in [0,T_1]} \|z_m(t) - z_m(t,\lambda)\|^2_{D(A^\alpha)} = 0$ as $\lambda \to \infty$. Similarly to Lemma 11, we can conclude that the result holds for the whole time interval $[0,T]$. $\qquad\square$

Next, we give an approximation of the mild solution to system (17). We introduce the following backward equation in $D(A^{1+\delta})$:

$$\begin{cases} dz_m^*(t,\lambda) = -\mathbb{1}_{[0,\tau_m)}(t)\left[-Az_m^*(t,\lambda)\right.\\ \qquad\qquad - A^\alpha R(\lambda)A^\alpha B_\delta^*\left(y(t), R(\lambda)A^\delta z_m^*(t,\lambda)\right)\\ \qquad\qquad \left.+ A^\gamma R(\lambda)A^\gamma\left(y(t) - y_d(t)\right)\right]dt,\\ z_m^*(T,\lambda) = 0, \end{cases} \tag{30}$$

where $m \in \mathbb{N}$, $\lambda > 0$, and the operator B_δ^* is introduced in Section 2.4.2. The process $(y(t))_{t\in[0,\tau)}$ is the local mild solution of system (2) with stopping times $(\tau_m)_{m\in\mathbb{N}}$ defined by (8) and $y_d \in L^2([0,T]; D(A^\gamma))$ is the given desired velocity field. We consider the solution of system (30) in a mild sense according to Definition 30. Recall that the operators $R(\lambda)$ and $AR(\lambda)$ are linear and bounded on H. Hence, a proof of the existence and uniqueness of a mild solution to system (30) with fixed $m \in \mathbb{N}$ and fixed $\lambda > 0$ can be obtained similarly to Theorem 47. Especially, we get for fixed $m \in \mathbb{N}$ and fixed $\lambda > 0$

$$\mathbb{E}\sup_{t\in[\tau_m,T]}\|z_m^*(t,\lambda)\|_{D(A^{1+\delta})}^2 = 0. \tag{31}$$

Moreover, we have the following strong formulation for fixed $m \in \mathbb{N}$, fixed $\lambda > 0$, all $t \in [0,T]$, and $\mathbb{P}$-almost surely:

$$z_m^*(t,\lambda)$$

$$= -\int_t^T \mathbb{1}_{[0,\tau_m)}(s)\left[Az_m^*(s,\lambda) + A^\gamma R(\lambda)A^\gamma\left(y(s\wedge\tau_m) - y_d(s)\right)\right]ds$$

$$+ \int_t^T \mathbb{1}_{[0,\tau_m)}(s)A^\alpha R(\lambda)A^\alpha B_\delta^*\left(y(s\wedge\tau_m), R(\lambda)A^\delta z_m^*(s,\lambda)\right)ds, \tag{32}$$

which is a consequence of [1, Theorem 3.4 and Theorem 4.1]. The proof of the following convergence result can be adopted from Lemma 15.

Lemma 16. *Let $(z_m^*(t))_{t\in[0,T]}$ and $(z_m^*(t,\lambda))_{t\in[0,T]}$ be the mild solutions of system (17) and system (30), respectively. Then we have for fixed $m \in \mathbb{N}$*

$$\lim_{\lambda\to\infty}\mathbb{E}\sup_{t\in[0,T]}\|z_m^*(t) - z_m^*(t,\lambda)\|_{D(A^\delta)}^2 = 0.$$

Here, we provided approximations of the local mild solution of the linearized stochastic Navier-Stokes equations and the mild solution of the adjoint equation by strong formulations. Next, we will show the duality principle and derive an explicit formula for the optimal control.

2.4.3. The Duality Principle and Derivation of the Optimal Control

Based on the results provided in the previous sections, we are able to show a duality principle, which provides us with a relation between the local mild solution of system (11) and the mild solution of system (17).

First, we state a product formula for infinite dimensional stochastic processes. The formula is an immediate consequence of the Itô formula, see [24, Theorem 2.9].

Lemma 17. *Let $\mathcal{H}$ and E be arbitrary separable Hilbert spaces. For $i = 1, 2$, let X_i^0 be $\mathcal{H}$-valued $\mathcal{F}_0$-measurable random variables, $(f_i(t))_{t \in [0,T]}$ be $\mathcal{H}$-valued $\mathcal{F}_t$-adapted processes such that*

$$\mathbb{E} \int_0^T \|f_i(t)\|_{\mathcal{H}} \, dt < \infty,$$

and $(\Phi_i(t))_{t \in [0,T]}$ be $\mathcal{L}_{(HS)}(Q^{1/2}(E); \mathcal{H})$-valued predictable processes such that

$$\mathbb{E} \int_0^T \|\Phi_i(t)\|^2_{\mathcal{L}_{(HS)}(Q^{1/2}(E); \mathcal{H})} \, dt < \infty.$$

For $i = 1, 2$, assume that the processes $(X_i(t))_{t \in [0,T]}$ satisfy for all $t \in [0, T]$ and $\mathbb{P}$-a.s.

$$X_i(t) = X_i^0 + \int_0^t f_i(s) \, ds + \int_0^t \Phi_i(s) \, dW(s).$$

Then we have for all $t \in [0, T]$ and $\mathbb{P}$-a.s.

$$\langle X_1(t), X_2(t) \rangle_{\mathcal{H}} = \langle X_1^0, X_2^0 \rangle_{\mathcal{H}} + \int_0^t [\langle X_1(s), f_2(s) \rangle_{\mathcal{H}} + \langle X_2(s), f_1(s) \rangle_{\mathcal{H}}] \, ds$$

$$+ \int_0^t \langle \Phi_1(s), \Phi_2(s) \rangle_{\mathcal{L}_{(HS)}(Q^{1/2}(E); \mathcal{H})} \, ds$$

$$+ \int_0^t \langle X_1(s), \Phi_2(s) \, dW(s) \rangle_{\mathcal{H}} + \int_0^t \langle X_2(s), \Phi_1(s) \, dW(s) \rangle_{\mathcal{H}}.$$

Note that the local mild solution of system (4) depends on the control $u \in L^2_{\mathcal{F}}(\Omega; L^2([0, T]; D(A^\beta)))$. Hence, the mild solution of system (17) depends on the control $u \in L^2_{\mathcal{F}}(\Omega; L^2([0, T]; D(A^\beta)))$ as well. Let us denote this solution by $(z^*_m(t; u))_{t \in [0,T]}$. With this, we are ready to state the following duality principle for the optimal control problem introduced in Section 2.2.

Theorem 51. *Let $(y(t; u))_{t \in [0,\tau^u)}$ and $(z(t; u, v))_{t \in [0,\tau^u)}$ be the local mild solutions of system (2) and system (11), respectively, corresponding to the controls $u, v \in L^2_{\mathcal{F}}(\Omega; L^2([0, T]; D(A^\beta)))$. Moreover, assume that $(z^*_m(t; u))_{t \in [0,T]}$ is the mild solution of system (17) corresponding to the control $u \in L^2_{\mathcal{F}}(\Omega; L^2([0, T]; D(A^\beta)))$. Then we have for fixed $m \in \mathbb{N}$*

$$\mathbb{E} \int_0^{\tau^u_m} \langle A^\gamma(y(t; u) - y_d(t)), A^\gamma z(t; u, v) \rangle_H \, dt = \mathbb{E} \int_0^{\tau^u_m} \langle z^*_m(t; u), Fv(t) \rangle_H \, dt.$$

Proof. For the sake of simplicity, we omit the dependence on the controls. First, we prove the result for the approximations introduced in Section 2.4.2. Let $(z^*_m(t, \lambda))_{t \in [0,T]}$ be the mild solution of system (30). By equation (32), we get a strong formulation and hence, we find for all $\lambda > 0$, all $t \in [0, T]$, and $\mathbb{P}$-a.s.

$$\mathbb{E}\left[z^*_m(t, \lambda) | \mathcal{F}_t\right]$$

$$= M(t) + \int_0^t \mathbb{1}_{[0,\tau_m)}(s) \left[A z^*_m(s, \lambda) + A^\gamma R(\lambda) A^\gamma \left(y(s \wedge \tau_m) - y_d(s)\right)\right] ds$$

$$- \int_0^t \mathbb{1}_{[0,\tau_m)}(s) A^\alpha R(\lambda) A^\alpha B^*_\delta \left(y(s \wedge \tau_m), R(\lambda) A^\delta z^*_m(s, \lambda)\right) ds,$$

where

$$M(t)$$

$$= -\mathbb{E}\left[\int_0^T \mathbb{1}_{[0,\tau_m)}(s) \left[A z^*_m(s, \lambda) + A^\gamma R(\lambda) A^\gamma \left(y(s \wedge \tau_m) - y_d(s)\right)\right] ds \middle| \mathcal{F}_t\right]$$

$$+ \mathbb{E}\left[\int_0^T \mathbb{1}_{[0,\tau_m)}(s) A^\alpha R(\lambda) A^\alpha B^*_\delta \left(y(s \wedge \tau_m), R(\lambda) A^\delta z^*_m(s, \lambda)\right) ds \middle| \mathcal{F}_t\right].$$

By a martingale representation theorem, see [24, Theorem 2.5], applied to the process $(M(t))_{t \in [0,T]}$, there exists a unique predictable process

$(\Psi_m(t,\lambda))_{t\in[0,T]}$ with values in $\mathcal{L}_{(HS)}(Q^{1/2}(H);H)$ such that for all $\lambda > 0$, all $t \in [0,T]$, and $\mathbb{P}$-a.s.

$$
\mathbb{E}\left[z_m^*(t,\lambda)|\mathcal{F}_t\right]
$$

$$
= -\mathbb{E}\left[\int_0^T \mathbb{1}_{[0,\tau_m)}(s)\left[Az_m^*(s,\lambda) + A^\gamma R(\lambda)A^\gamma\left(y(s\wedge\tau_m) - y_d(s)\right)\right]ds\right]
$$

$$
+ \mathbb{E}\left[\int_0^T \mathbb{1}_{[0,\tau_m)}(s)A^\alpha R(\lambda)A^\alpha B_\delta^*\left(y(s\wedge\tau_m), R(\lambda)A^\delta z_m^*(s,\lambda)\right)ds\right]
$$

$$
+ \int_0^t \mathbb{1}_{[0,\tau_m)}(s)\left[Az_m^*(s,\lambda) + A^\gamma R(\lambda)A^\gamma\left(y(s\wedge\tau_m) - y_d(s)\right)\right]ds
$$

$$
- \int_0^t \mathbb{1}_{[0,\tau_m)}(s)A^\alpha R(\lambda)A^\alpha B_\delta^*\left(y(s\wedge\tau_m), R(\lambda)A^\delta z_m^*(s,\lambda)\right)ds
$$

$$
+ \int_0^t \Psi_m(s,\lambda)\,dW(s). \tag{33}
$$

Assume that $(z_m(t,\lambda))_{t\in[0,T]}$ is the mild solution of system (24). Next, we apply Lemma 17 to the strong formulation of $(z_m(t,\lambda))_{t\in[0,T]}$ given by equation (25) and equation (33). Using additionally equation (31), we obtain for all $\lambda > 0$ and $\mathbb{P}$-a.s.

$$
0 = \mathcal{I}_1(\tau_m,\lambda) + \mathcal{I}_2(\tau_m,\lambda) + \mathcal{I}_3(\tau_m,\lambda) + \mathcal{I}_4(\tau_m,\lambda),
$$

where

$$
\mathcal{I}_1(\tau_m,\lambda) = \int_0^{\tau_m}\langle z_m(s,\lambda), Az_m^*(s,\lambda)\rangle_H\,ds - \int_0^{\tau_m}\langle z_m^*(s,\lambda), Az_m(s,\lambda)\rangle_H\,ds,
$$

$$
\mathcal{I}_2(\tau_m,\lambda) = \int_0^{\tau_m}\langle z_m(s,\lambda), A^\alpha R(\lambda)A^\alpha B_\delta^*\left(y(s\wedge\tau_m), R(\lambda)A^\delta z_m^*(s,\lambda)\right)\rangle_H\,ds
$$

$$
- \int_0^{\tau_m}\langle z_m^*(s,\lambda), A^\delta R(\lambda)A^{-\delta}B(R(\lambda)z_m(s,\lambda), \pi_m(y_m(s)))\rangle_H\,ds
$$

$$
- \int_0^{\tau_m}\langle z_m^*(s,\lambda), A^\delta R(\lambda)A^{-\delta}B(\pi_m(y_m(s)), R(\lambda)z_m(s,\lambda))\rangle_H\,ds,
$$

$$\mathcal{I}_3(\tau_m, \lambda) = \int_0^{\tau_m} \langle z_m^*(s, \lambda), R(\lambda) F v(s) \rangle_H \, ds$$

$$- \int_0^{\tau_m} \langle z_m(s, \lambda), A^\gamma R(\lambda) A^\gamma \left(y(s \wedge \tau_m) - y_d(s) \right) \rangle_H \, ds,$$

$$\mathcal{I}_4(\tau_m, \lambda) = \int_0^{\tau_m} \langle z_m(s, \lambda), \Psi_m(s, \lambda) \, dW(s) \rangle_H \, .$$

Since the operator A is self-adjoint, we have $\mathcal{I}_1(\tau_m, \lambda) = 0$ for all $\lambda > 0$ and $\mathbb{P}$-almost surely. Recall that $R(\lambda)$ is self-adjoint on H and $y(t) = \pi_m(y_m(t))$ for all $t \in [0, \tau_m)$ and $\mathbb{P}$-almost surely. Using Lemma 8, equation (18), and equation (22), we find $\mathcal{I}_2(\tau_m, \lambda) = 0$ for all $\lambda > 0$ and $\mathbb{P}$-almost surely. Using additionally $\mathbb{E} \, \mathcal{I}_4(\tau_m, \lambda) = 0$, we get $0 = \mathbb{E} \, \mathcal{I}_3(\tau_m, \lambda)$ for all $\lambda > 0$. Hence, we have for all $\lambda > 0$

$$\mathbb{E} \int_0^{\tau_m} \langle R(\lambda) A^\gamma z_m(t, \lambda), A^\gamma \left(y(t) - y_d(t) \right) \rangle_H \, dt$$

$$= \mathbb{E} \int_0^{\tau_m} \langle R(\lambda) z_m^*(t, \lambda), F v(t) \rangle_H \, dt.$$

Moreover, the left hand side and the right hand side of the above equation converge as $\lambda \to \infty$ as a consequence of Lemma 15, Lemma 16, and equation (23), which completes the proof. $\qquad\square$

Based on the necessary optimality condition formulated as the variational inequality (16) and the duality principle derived in the previous theorem, we are able to deduce a formula the optimal control has to satisfy. First, we introduce the following projection operator. Note that the set of admissible controls U is a closed subset of the Hilbert space $L^2_{\mathcal{F}}(\Omega; L^2([0,T]; D(A^\beta)))$. By $P_U \colon L^2_{\mathcal{F}}(\Omega; L^2([0,T]; D(A^\beta))) \to U$ we denote the projection onto U, i.e.

$$\|P_U(v) - v\|_{L^2_{\mathcal{F}}(\Omega; L^2([0,T]; D(A^\beta)))} = \min_{u \in U} \|u - v\|_{L^2_{\mathcal{F}}(\Omega; L^2([0,T]; D(A^\beta)))}$$

for every $v \in L^2_{\mathcal{F}}(\Omega; L^2([0,T]; D(A^\beta)))$. We have $u = P_U(v)$ if and only if

$$\langle v - u, \tilde{u} - u \rangle_{L^2_{\mathcal{F}}(\Omega; L^2([0,T]; D(A^\beta)))} \leq 0 \tag{34}$$

for every $\tilde{u} \in U$, see [33, Lemma 1.10 (b)]. We get the following result.

Theorem 52. *Let* $(z_m^*(t; u), \Phi_m(t; u))_{t \in [0,T]}$ *be the mild solution of system (17) corresponding to the control* $u \in L^2_{\mathcal{F}}(\Omega; L^2([0,T]; D(A^\beta)))$. *Then for* *fixed $m \in \mathbb{N}$, the optimal control $\overline{u}_m \in U$ satisfies for almost all $t \in [0, T]$ and $\mathbb{P}$-a.s.*

$$\overline{u}_m(t) = -P_U \left(F^* A^{-2\beta} \mathbb{E}[z_m^*(t; \overline{u}_m) | \mathcal{F}_t] \right), \tag{35}$$

where $P_U \colon L_\mathcal{F}^2(\Omega; L^2([0, T]; D(A^\beta))) \to U$ is the projection onto U and the operator $F^* \in \mathcal{L}(D(A^\beta))$ is the adjoint of $F \in \mathcal{L}(D(A^\beta))$.

Proof. Using inequality (16) and Theorem 51, the optimal control $\overline{u}_m \in U$ satisfies for every $u \in U$

$$\mathbb{E} \int_0^{\tau_m^{\overline{u}_m}} \langle z_m^*(t; \overline{u}_m), F(u(t) - \overline{u}_m(t)) \rangle_H \, dt$$

$$+ \mathbb{E} \int_0^T \left\langle A^\beta \overline{u}_m(t), A^\beta(u(t) - \overline{u}_m(t)) \right\rangle_H \, dt \geq 0.$$

By equation (19), we have $\mathbb{1}_{[0, \tau_m^{\overline{u}_m})}(t) z_m^*(t; \overline{u}_m) = z_m^*(t; \overline{u}_m)$ for all $t \in [0, T]$ and $\mathbb{P}$-almost surely. Using Lemma 8, we obtain for every $u \in U$

$$\mathbb{E} \int_0^{\tau_m^{\overline{u}_m}} \langle z_m^*(t; \overline{u}_m), F(u(t) - \overline{u}_m(t)) \rangle_H \, dt$$

$$= \mathbb{E} \int_0^T \left\langle A^\beta F^* A^{-2\beta} \mathbb{E}[z_m^*(t; \overline{u}_m) | \mathcal{F}_t], A^\beta(u(t) - \overline{u}_m(t)) \right\rangle_H \, dt.$$

Hence, we find for every $u \in U$

$$\mathbb{E} \int_0^T \left\langle -F^* A^{-2\beta} \mathbb{E}[z_m^*(t; \overline{u}_m) | \mathcal{F}_t] - \overline{u}_m(t), u(t) - \overline{u}_m(t) \right\rangle_{D(A^\beta)} \, dt \leq 0.$$

We obtain inequality (34) and thus, equation (35) holds. $\qquad \square$

Remark 28. Let us denote by $(\overline{y}(t))_{t \in [0, \overline{\tau})}$ and $(\overline{z}_m^*(t))_{t \in [0, T]}$ the local mild solutions of system (2) and the mild solution of system (17), respectively, corresponding to the optimal control $\overline{u}_m \in U$. As a consequence of the previous theorem, the optimal velocity field $(\overline{y}(t))_{t \in [0, \overline{\tau})}$ can be computed by solving the following system of coupled forward-backward SPDEs:

$$\begin{cases} d\overline{y}(t) = -[A\overline{y}(t) + B(\overline{y}(t)) + F P_U \left(F^* A^{-2\beta} \overline{z}_m^*(t) \right)] \, dt + G(t) \, dW(t), \\ d\overline{z}_m^*(t) = -\mathbb{1}_{[0, \tau_m)}(t)[-A\overline{z}_m^*(t) - A^{2\alpha} B_\delta^* \left(\overline{y}(t), A^\delta \overline{z}_m^*(t) \right)] \, dt \\ \qquad\qquad - \mathbb{1}_{[0, \tau_m)}(t) A^{2\gamma} \left(\overline{y}(t) - y_d(t) \right) \, dt \\ \overline{y}(0) = \xi, \quad \overline{z}_m^*(T) = 0. \end{cases}$$

Here, we derived an explicit formula for the optimal control. Next, we will show that this optimal control satisfies also a sufficient optimality condition.

2.4.4. *Sufficient Optimality Condition*

To show that the optimal control $\overline{u}_m \in U$ given by equation (35) satisfies a sufficient optimality condition, we apply the following criteria.

Proposition 2 (cf. Theorem 4.23, [47]). *Let X be an arbitrary Banach space and let $K \subset X$ be convex. Moreover, let the functional $f \colon X \to \mathbb{R}$ be twice continuous Fréchet differentiable in a neighborhood of $\overline{x} \in K$. If $\overline{x} \in K$ satisfies $d^F f(\overline{x})[x - \overline{x}] \geq 0$ for every $x \in K$ and there exists a constant $\delta > 0$ such that $d^F (f(\overline{x}))^2[h, h] \geq \delta \|h\|_X^2$ for every $h \in X$, then there exist constants $\varepsilon_1, \varepsilon_2 > 0$ such that*

$$f(x) \geq f(\overline{x}) + \varepsilon_1 \|x - \overline{x}\|_X^2$$

for every $x \in K$ with $\|x - \overline{x}\|_X \leq \varepsilon_2$.

Note that the set of admissible controls U is a convex subset of the Hilbert space $L^2_{\mathcal{F}}(\Omega; L^2([0,T]; D(A^\beta)))$. According to Section 2.3, the cost functional J_m given by equation (10) is twice continuous Fréchet differentiable in a neighborhood of the optimal control $\overline{u}_m \in U$. Recall that $\overline{u}_m \in U$ satisfies the necessary optimality condition (15), which is also valid for the Fréchet derivative due to Corollary 2. Moreover, we have for every $v \in L^2(\Omega; L^2([0,T]; D(A^\beta)))$

$$d^F(J_m(\overline{u}_m))^2[v, v] \geq \mathbb{E} \int_0^T \|v(t)\|_{D(A^\beta)}^2 \, dt.$$

Hence, the assumptions of Proposition 2 are fulfilled and the optimal control $\overline{u}_m \in U$ given by equation (35) is a local minimum of the cost functional J_m. Due to Theorem 48, we can conclude that the minimum is also global.

We derived an explicit formula for the optimal control of a control problem constrained by the stochastic Navier-Stokes equations driven by additive Wiener noise. Here, this formula is based on the corresponding adjoint equation, which is characterized by a deterministic backward equation. Moreover, we showed that the optimal control also satisfies a sufficient optimality condition. In the remaining part of this chapter, we will discuss the case of multiplicative noise and clarify the main differences. The results are mainly based on [2, 3]. In [3], the existence of a unique local mild solution of the stochastic Navier-Stokes equations driven by multiplicative Lévy noise as well as the existence of a unique optimal control of the control problem are provided. In [2], a stochastic maximum principle was applied to this control problem with the restriction to the stochastic Navier-Stokes equations driven by multiplicative Wiener noise. In the following section, we demonstrate this approach in abridged form taking into account the additional restriction.

3. Multiplicative Noise

3.1. *The State Equation*

In this section, we consider the stochastic Navier-Stokes equations with distributed control driven by multiplicative noise. The adjoint equation of the control problem introduced in the following section becomes a backward SPDE, which characterizes the main difference to the case of additive noise. Hence, the basic assertions of the state equation are slightly extensions of the results from Section 2.1 and can be formulated analogously.

We assume that the external random force f in system (1) can be decomposed into a sum of a control term and a noise term dependent on the velocity field. Hence, we obtain the stochastic Navier-Stokes equations with multiplicative noise in $D(A^\alpha)$:

$$\begin{cases} dy(t) = -[Ay(t) + B(y(t)) - Fu(t)]\, dt + G(y(t))\, dW(t), \\ y(0) = \xi, \end{cases} \tag{36}$$

where the operators A, B, F are defined in Section 2.1. We assume that $u \in L^2_{\mathcal{F}}(\Omega; L^2([0,T]; D(A^\beta)))$ and the process $(W(t))_{t\geq 0}$ is a Q-Wiener process with values in H and covariance operator $Q \in \mathcal{L}(H)$. The operator $G\colon H \to \mathcal{L}_{(HS)}(Q^{1/2}(H); D(A^\alpha))$ is linear and bounded. According to Definition 28, we introduce a solution to system (36) in a local mild sense with stopping times denoted by $(\tau_m)_{m\in\mathbb{N}}$. The idea for the proof of the existence and uniqueness to this local mild solution can be adopted from Theorem 47. Especially, the stopping times $(\tau_m)_{m\in\mathbb{N}}$ are given by

$$\tau_m = \inf\{t \in (0,T) : \|y_m(t)\|_{D(A^\alpha)} > m\} \wedge T \tag{37}$$

$\mathbb{P}$-a.s. with the usual convention that $\inf\{\emptyset\} = +\infty$. Here, the process $(y_m(t))_{t\in[0,T]}$ is the mild solution of the following modified system in $D(A^\alpha)$:

$$\begin{cases} dy_m(t) = -[Ay_m(t) + B(\pi_m(y_m(t))) - Fu(t)]\, dt \\ \qquad\qquad + G(y_m(t))\, dW(t), \\ y_m(0) = \xi, \end{cases} \tag{38}$$

where π_m is given by equation (5). The solution of system (38) is given in a mild sense according to Definition 29. We obtain $y(t) = y_m(t)$ for each $m \in \mathbb{N}$, all $t \in [0, \tau_m)$, and $\mathbb{P}$-almost surely. To illustrate the dependence on the control $u \in L^2_{\mathcal{F}}(\Omega; L^2([0,T]; D(A^\beta)))$, we denote denote by $(y_m(t; u))_{t \in [0,T]}$ and $(y(t; u))_{t \in [0, \tau^u)}$ the mild solution of system (38) and the local mild solution of system (36), respectively. Recall that the stopping times $(\tau^u_m)_{m \in \mathbb{N}}$ depend on the control as well. For fixed control, we use the notation introduced above. Furthermore, the continuity properties stated in Lemma 11 – 13 hold here as well.

Here, we introduced the stochastic controlled Navier-Stokes equations driven by multiplicative Wiener noise on bounded and connected domains. In the following section, we introduce the control problem and we state the derivatives of the corresponding cost function such that we obtain necessary and sufficient optimality conditions the optimal control has to satisfy.

3.2. The Control Problem

Following Section 2.2 and Section 2.3, we introduce the control problem and state a necessary optimality condition.

We define the cost functional $J_m \colon L^2_{\mathcal{F}}(\Omega; L^2([0,T]; D(A^\beta))) \to \mathbb{R}$ by

$$J_m(u) = \frac{1}{2} \mathbb{E} \int_0^{\tau^u_m} \|A^\gamma(y(t; u) - y_d(t))\|^2_H \, dt + \frac{1}{2} \mathbb{E} \int_0^T \|A^\beta u(t)\|^2_H \, dt, \quad (39)$$

where $m \in \mathbb{N}$ is fixed and $\gamma \in [0, \alpha]$. Moreover, the process $(y(t; u))_{t \in [0, \tau^u)}$ is the local mild solution of system (36) corresponding to the control $u \in L^2_{\mathcal{F}}(\Omega; L^2([0,T]; D(A^\beta)))$ and $y_d \in L^2([0,T]; D(A^\gamma))$ is a given desired velocity field. The set of admissible controls U is a nonempty, closed, bounded, and convex subset of the Hilbert space $L^2_{\mathcal{F}}(\Omega; L^2([0,T]; D(A^\beta)))$ such that $0 \in U$. The task is to find an optimal control $\overline{u}_m \in U$, i.e. $\overline{u}_m \in U$ satisfies

$$J_m(\overline{u}_m) = \inf_{u \in U} J_m(u).$$

A motivation of the control problem can be found in Section 2.2. Moreover, we can easily adopt the proof of Theorem 48 to obtain the existence of a unique optimal control.

In order to state necessary and sufficient optimality, we first introduce the following linearized system in $D(A^\alpha)$:

$$\begin{cases} dz(t) = -[Az(t) + B(z(t), y(t)) + B(y(t), z(t)) - Fv(t)] \, dt \\ \qquad\quad + G(z(t)) \, dW(t), \\ z(0) = 0, \end{cases} \tag{40}$$

where $v \in L^2_{\mathcal{F}}(\Omega; L^2([0, T]; D(A^\beta)))$, the process $(y(t))_{t \in [0,\tau)}$ is the local mild solution of system (36) and the process $(W(t))_{t \geq 0}$ is a Q-Wiener process with values in H and covariance operator $Q \in \mathcal{L}(H)$. The operators A, B, F, G are introduced in Section 2.1 and Section 3.1, respectively. Again, we define the solution of system (40) in a local mild sense according to Definition 28 with stopping times $(\tau_m)_{m \in \mathbb{N}}$ given by equation (37). In this context, we also introduce the following modified system in $D(A^\alpha)$:

$$\begin{cases} dz_m(t) = -[Az_m(t) + B(z_m(t), \pi_m(y_m(t)))] \, dt \\ \qquad\quad - [B(\pi_m(y_m(t)), z_m(t)) - Fv(t)] \, dt + G(z_m(t)) \, dW(t), \\ z_m(0) = 0, \end{cases} \tag{41}$$

where the process $(y_m(t))_{t \in [0,T]}$ is the mild solution of system (38) and π_m is given by (5). A solution of system (41) is given in a mild sense according to Definition 29 and we have $z(t) = z_m(t)$ for each $m \in \mathbb{N}$, all $t \in [0, \tau_m)$, and $\mathbb{P}$-almost surely. Similarly to Section 2.3.1, the processes denoted by $(z(t; u, v))_{t \in [0,\tau)}$ and $(z_m(t; u, v))_{t \in [0,T]}$ are the local mild solution of system (40) and the mild solution of system (41), respectively, corresponding to the controls $u, v \in L^2_{\mathcal{F}}(\Omega; L^2([0, T]; D(A^\beta)))$. For fixed controls, we use the notation introduced above.

Furthermore, the Gâteaux derivative of the local mild solution $(y(t; u))_{t \in [0,\tau^u)}$ to system (36) at $u \in L^2_{\mathcal{F}}(\Omega; L^2([0, T]; D(A^\beta)))$ in direction $v \in L^2_{\mathcal{F}}(\Omega; L^2([0, T]; D(A^\beta)))$ satisfies for all $t \in [0, \tau^u_m)$ and $\mathbb{P}$-a.s.

$$d^G_u y(t; u)[v] = z(t; u, v),$$

which can be obtained similarly to Theorem 49. As an immediate consequence of Theorem 50, the Gâteaux derivative of the cost functional J_m defined by equation (39) at $u \in L^2_{\mathcal{F}}(\Omega; L^2([0, T]; D(A^\beta)))$ in direction $v \in L^2_{\mathcal{F}}(\Omega; L^2([0, T]; D(A^\beta)))$ satisfies

$$d^G J_m(u)[v] = \mathbb{E} \int_0^{\tau^u_m} \langle A^\gamma(y(t; u) - y_d(t)), A^\gamma z(t; u, v) \rangle_H \, dt$$

$$+ \mathbb{E} \int_0^T \langle A^\beta u(t), A^\beta v(t) \rangle_H \, dt$$

and the Gâteaux derivative of order two at $u \in L^2_{\mathcal{F}}(\Omega; L^2([0,T]; D(A^\beta)))$ in directions $v_1, v_2 \in L^2_{\mathcal{F}}(\Omega; L^2([0,T]; D(A^\beta)))$ satisfies

$$d^G(J_m(u))^2[v_1, v_2] = \mathbb{E} \int_0^{\tau_m^u} \left\langle A^\gamma z(t; u, v_1), A^\gamma z(t; u, v_2) \right\rangle_H dt$$

$$+ \mathbb{E} \int_0^T \left\langle A^\beta v_1(t), A^\beta v_2(t) \right\rangle_H dt.$$

According to Section 2.3, these Gâteaux derivatives coincide with its Fréchet derivatives and are continuous with respect to u.

Since the set of admissible controls U is a closed, bounded, and convex subset of the Hilbert space $L^2_{\mathcal{F}}(\Omega; L^2([0,T]; D(A^\beta)))$ such that $0 \in U$, the optimal control $\overline{u}_m \in U$ satisfies the necessary optimality condition

$$d^G J_m(\overline{u}_m)[u - \overline{u}_m] \geq 0$$

for fixed $m \in \mathbb{N}$ and every $u \in U$. Thus, we get for fixed $m \in \mathbb{N}$ and every $u \in U$:

$$\mathbb{E} \int_0^{\tau_m^{\overline{u}_m}} \left\langle A^\gamma(y(t; \overline{u}_m) - y_d(t)), A^\gamma z(t; \overline{u}_m, u - \overline{u}_m) \right\rangle_H dt$$

$$+ \mathbb{E} \int_0^T \left\langle A^\beta \overline{u}_m(t), A^\beta(u(t) - \overline{u}_m(t)) \right\rangle_H dt \geq 0. \tag{42}$$

In this section, we stated the derivatives of the cost function in order to obtain necessary and sufficient optimality conditions the optimal control has to satisfy. We will utilize this necessary optimality condition to derive an explicit formula of the optimal control such that the sufficient optimality condition holds.

3.3. The Optimal Control

Following Section 2.4, we derive a duality principle to obtain an explicit formula of the optimal control based on a suitable adjoint equation. Since the control problem is constrained by a SPDE with linear multiplicative noise, the adjoint equation is specified by a backward SPDE. We note that this is the main difference to the control problem considered in Section 2. For mild solutions of backward SPDE's, the existence and uniqueness result is based on a martingale representation theorem, see [34].

3.3.1. *The Adjoint Equation*

We introduce the following backward SPDE in $D(A^\delta)$:

$$
\begin{cases}
dz_m^*(t) = -\mathbb{1}_{[0,\tau_m)}(t)[-Az_m^*(t) - A^{2\alpha}B_\delta^*\left(y(t), A^\delta z_m^*(t)\right)]\,dt \\
\qquad - \mathbb{1}_{[0,\tau_m)}(t)[G^*(A^{-2\alpha}\Phi_m(t)) + A^{2\gamma}\left(y(t) - y_d(t)\right)]\,dt \\
\qquad + \Phi_m(t)\,dW(t), \\
z_m^*(T) = 0,
\end{cases}
\tag{43}
$$

where $m \in \mathbb{N}$ and $(y(t))_{t \in [0,\tau)}$ is the local mild solution of system (36). The stopping times $(\tau_m)_{m \in \mathbb{N}}$ are defined by (37) and $y_d \in L^2([0,T]; D(A^\gamma))$ is the given desired velocity field. The operators A, B_δ^* are introduced in Section 2.1 and Section 3.3.1, respectively. The process $(W(t))_{t \geq 0}$ is a Q-Wiener process with values in H and covariance operator $Q \in \mathcal{L}(H)$. Moreover, the operator $G^*: \mathcal{L}_{(HS)}(Q^{1/2}(H); D(A^\alpha)) \to H$ is linear and bounded. A precise meaning for the operator G^* is given in the following remark.

Remark 29. Since the operator $G: H \to \mathcal{L}_{(HS)}(Q^{1/2}(H); D(A^\alpha))$ is linear and bounded, there exists a linear and bounded operator $G^*: \mathcal{L}_{(HS)}(Q^{1/2}(H); D(A^\alpha)) \to H$ satisfying for every $h \in H$ and every $\Phi \in \mathcal{L}_{(HS)}(Q^{1/2}(H); D(A^\alpha))$

$$
\langle G(h), \Phi \rangle_{\mathcal{L}_{(HS)}(Q^{1/2}(H); D(A^\alpha))} = \langle h, G^*(\Phi) \rangle_H.
$$

We can rewrite this equivalently as

$$
\langle A^\alpha G(h), A^\alpha \Phi \rangle_{\mathcal{L}_{(HS)}(Q^{1/2}(H); H)} = \langle h, G^*(\Phi) \rangle_H
$$

for every $h \in H$ and every $\Phi \in \mathcal{L}_{(HS)}(Q^{1/2}(H); D(A^\alpha))$.

The solution of system (43) is given in a mild sense according to the following definition.

Definition 31. A pair of predictable processes $(z_m^*(t), \Phi_m(t))_{t \in [0,T]}$ with values in $D(A^\delta) \times \mathcal{L}_{(HS)}(Q^{1/2}(H); H)$ is called a *mild solution of system (43)* if

$$
\mathbb{E} \sup_{t \in [0,T]} \|z_m^*(t)\|_{D(A^\delta)}^2 < \infty,
$$

$$
\mathbb{E} \int_0^T \|\Phi_m(t)\|_{\mathcal{L}_{(HS)}(Q^{1/2}(H); H)}^2\,dt < \infty,
$$

and we have for all $t \in [0,T]$ and $\mathbb{P}$-a.s.

$$
z_m^*(t) = - \int_t^T \mathbb{1}_{[0,\tau_m)}(s) A^\alpha e^{-A(s-t)} A^\alpha B_\delta^* \left(y(s \wedge \tau_m), A^\delta z_m^*(s) \right) ds
$$

$$
+ \int_t^T \mathbb{1}_{[0,\tau_m)}(s) e^{-A(s-t)} G^* (A^{-2\alpha} \Phi_m(s)) ds
$$

$$
+ \int_t^T \mathbb{1}_{[0,\tau_m)}(s) A^\gamma e^{-A(s-t)} A^\gamma \left(y(s \wedge \tau_m) - y_d(s) \right) ds
$$

$$
- \int_t^T e^{-A(s-t)} \Phi_m(s) \, dW(s). \tag{44}
$$

To prove the existence and uniqueness of the mild solution to system (43), we need the following auxiliary results.

Lemma 18. *Let* $\delta, \varepsilon \in [0, \frac{1}{2})$ *such that* $\delta + \varepsilon < \frac{1}{2}$. *Moreover, assume that* $z \in L^2(\Omega; D(A^\delta))$ *is* $\mathcal{F}_T$-*measurable and* $(f(t))_{t \in [0,T]}$ *is a predictable process with values in* H *such that* $\mathbb{E} \int_0^T \|f(t)\|_H^2 \, dt < \infty$. *Then there exists a unique pair of predictable processes* $(\varphi(t), \Phi(t))_{t \in [0,T]}$ *with values in the space* $D(A^\delta) \times \mathcal{L}_{(HS)}(Q^{1/2}(H); D(A^\varepsilon))$ *such that for all* $t \in [0,T]$ *and* $\mathbb{P}$-a.s.

$$
\varphi(t) = e^{-A(T-t)} z + \int_t^T A^\varepsilon e^{-A(s-t)} f(s) \, ds - \int_t^T e^{-A(s-t)} A^\varepsilon \Phi(s) \, dW(s).
$$

Furthermore, there exists a constant $c^* > 0$ *such that for all* $t \in [0,T]$

$$
\mathbb{E} \sup_{s \in [t,T]} \|\varphi(s)\|_{D(A^\delta)}^2
$$

$$
\leq c^* \left[\mathbb{E} \|z\|_{D(A^\varepsilon)}^2 + (T-t)^{1-2\delta-2\varepsilon} \, \mathbb{E} \int_t^T \|f(s)\|_H^2 \, ds \right], \tag{45}
$$

$$
\mathbb{E} \int_t^T \|\Phi(s)\|_{\mathcal{L}_{(HS)}(Q^{1/2}(H); D(A^\varepsilon))}^2 \, ds
$$

$$
\leq c^* \left[\mathbb{E} \|z\|_{D(A^\varepsilon)}^2 + (T-t)^{1-2\varepsilon} \, \mathbb{E} \int_t^T \|f(s)\|_H^2 \, ds \right]. \tag{46}
$$

Proof. For $\delta = \varepsilon = 0$, a proof can be found in [34, Lemma 2.1]. For arbitrary $\varepsilon \in [0, \frac{1}{2})$ and $\delta \in [0, \frac{1}{2} - \varepsilon)$, one can show the result similarly using the properties of fractional powers of the operator A provided by Lemma 8. □

Theorem 53. *Let* $\alpha \in (0, \frac{1}{2})$ *and* $\delta \in [0, \frac{1}{2})$ *satisfy* $1 > \delta + \alpha > \frac{1}{2}$ *and* $\delta + 2\alpha \geq \frac{n}{4} + \frac{1}{2}$, *and let* $\gamma \in [0, \alpha]$ *such that* $\gamma + \delta < \frac{1}{2}$. *Then for fixed* $m \in \mathbb{N}$ *and fixed* $u \in L_{\mathcal{F}}^2(\Omega; L^2([0, T]; D(A^\beta)))$, *there exists a unique mild solution* $(z_m^*(t), \Phi_m(t))_{t \in [0,T]}$ *of system (43).*

Proof. Let $m \in \mathbb{N}$ be fixed. Moreover, let $\mathcal{Z}_T^1$ denote the space of all predictable processes $(z(t))_{t \in [0,T]}$ with values in $D(A^\delta)$ such that $\mathbb{E} \sup_{t \in [0,T]} \|z(t)\|_{D(A^\delta)}^2 < \infty$. The space $\mathcal{Z}_T^1$ equipped with the norm

$$\|z\|_{\mathcal{Z}_T^1}^2 = \mathbb{E} \sup_{t \in [0,T]} \|z(t)\|_{D(A^\delta)}^2$$

for every $z \in \mathcal{Z}_T^1$ becomes a Banach space. Similarly, let $\mathcal{Z}_T^2$ denote the space of all predictable processes $(\Phi(t))_{t \in [0,T]}$ with values in $\mathcal{L}_{(HS)}(Q^{1/2}(H); H)$ such that $\mathbb{E} \int_0^T \|\Phi(t)\|_{\mathcal{L}_{(HS)}(Q^{1/2}(H);H)}^2 dt < \infty$. The space $\mathcal{Z}_T^2$ equipped with the inner product

$$\langle \Phi_1, \Phi_2 \rangle_{\mathcal{Z}_T^2}^2 = \mathbb{E} \int_0^T \langle \Phi_1(t), \Phi_2(t) \rangle_{\mathcal{L}_{(HS)}(Q^{1/2}(H);H)}^2 dt$$

for every $\Phi_1, \Phi_2 \in \mathcal{Z}_T^2$ becomes a Hilbert space.

Next, we define a sequence $(z_m^k, \Phi_m^k)_{k \in \mathbb{N}} \subset \mathcal{Z}_T^1 \times \mathcal{Z}_T^2$ satisfying for each $k \in \mathbb{N}$, all $t \in [0, T]$, and $\mathbb{P}$-a.s.

$$z_m^k(t) = -\int_t^T \mathbb{1}_{[0,\tau_m)}(s) A^\alpha e^{-A(s-t)} A^\alpha B_\delta^* \left(y(s \wedge \tau_m), A^\delta z_m^{k-1}(s) \right) ds$$

$$+ \int_t^T \mathbb{1}_{[0,\tau_m)}(s) e^{-A(s-t)} G^* (A^{-2\alpha} \Phi_m^{k-1}(s)) ds$$

$$+ \int_t^T \mathbb{1}_{[0,\tau_m)}(s) A^\gamma e^{-A(s-t)} A^\gamma \left(y(s \wedge \tau_m) - y_d(s) \right) ds$$

$$- \int_t^T e^{-A(s-t)} \Phi_m^k(s) \, dW(s), \tag{47}$$

where $z_m^0(t) = 0$ and $\Phi_m^0(t) = 0$ for all $t \in [0, T]$. Indeed, one can easily verify that $(z_m^k, \Phi_m^k)_{k \in \mathbb{N}} \subset \mathcal{Z}_T^1 \times \mathcal{Z}_T^2$ using Lemma 18. Recall that the operators $A^\alpha B_\delta^*(y(t), \cdot) \colon H \to H$ and $G^* \colon \mathcal{L}_{(HS)}(Q^{1/2}(H); D(A^\alpha)) \to H$ are linear and bounded. We obtain for each $k \in \mathbb{N}$, all $t \in [0, T]$, and $\mathbb{P}$-a.s.

$$z_m^{k+1}(t) - z_m^k(t)$$

$$= -\int_t^T \mathbb{1}_{[0, \tau_m)}(s) A^\alpha e^{-A(s-t)} A^\alpha B_\delta^* \left(y(s \wedge \tau_m), A^\delta \left[z_m^k(s) - z_m^{k-1}(s) \right] \right) ds$$

$$+ \int_t^T \mathbb{1}_{[0, \tau_m)}(s) e^{-A(s-t)} G^* \left(A^{-2\alpha} \left[\Phi_m^k(s) - \Phi_m^{k-1}(s) \right] \right) ds$$

$$- \int_t^T e^{-A(s-t)} \left(\Phi_m^{k+1}(s) - \Phi_m^k(s) \right) dW(s), \tag{48}$$

which satisfies the assumptions of Lemma 18. Assume that $T_1 \in [0, T)$. Due to inequality (45) and inequality (46), we have for each $k \in \mathbb{N}$

$$\mathbb{E} \sup_{t \in [T_1, T]} \| z_m^{k+1}(t) - z_m^k(t) \|_{D(A^\delta)}^2 \le C_{T_1} \, \mathbb{E} \sup_{t \in [T_1, T]} \| z_m^k(t) - z_m^{k-1}(t) \|_{D(A^\delta)}^2,$$

$$\mathbb{E} \int_{T_1}^T \left\| \Phi_m^{k+1}(t) - \Phi_m^k(t) \right\|_{\mathcal{L}_{(HS)}(Q^{1/2}(H); H)}^2 dt$$

$$\le C_{T_1} \, \mathbb{E} \int_{T_1}^T \left\| \Phi_m^k(t) - \Phi_m^{k-1}(t) \right\|_{\mathcal{L}_{(HS)}(Q^{1/2}(H); H)}^2 dt,$$

where $C_{T_1} > 0$ is a constant dependent on T_1. Therefore, we find for each $k \in \mathbb{N}$

$$\mathbb{E} \sup_{t \in [T_1, T]} \| z_m^{k+1}(t) - z_m^k(t) \|_{D(A^\delta)}^2 \le C_{T_1}^k \, \mathbb{E} \sup_{t \in [T_1, T]} \| z_m^1(t) \|_{D(A^\delta)}^2,$$

$$\mathbb{E} \int_{T_1}^T \left\| \Phi_m^{k+1}(t) - \Phi_m^k(t) \right\|_{\mathcal{L}_{(HS)}(Q^{1/2}(H); H)}^2 dt$$

$$\le C_{T_1}^k \, \mathbb{E} \int_{T_1}^T \left\| \Phi_m^1(t) \right\|_{\mathcal{L}_{(HS)}(Q^{1/2}(H); H)}^2 dt.$$

We choose $T_1 \in [0, T)$ such that $C_{T_1} < 1$. Thus, we can conclude that the sequence $(z_m^k, \Phi_m^k)_{k \in \mathbb{N}} \subset \mathcal{Z}_T^1 \times \mathcal{Z}_T^2$ is a Cauchy sequence on the interval

$[T_1, T]$. Using equation (48), we have for each $k \in \mathbb{N}$, all $t \in [0, T_1]$, and $\mathbb{P}$-a.s.

$$z_m^{k+1}(t) - z_m^k(t)$$
$$= e^{-A(T_1-t)}[z_m^{k+1}(T_1) - z_m^k(T_1)]$$
$$- \int_t^{T_1} \mathbb{1}_{[0,\tau_m)}(s) A^\alpha e^{-A(s-t)} A^\alpha B_\delta^* \left(y(s \wedge \tau_m), A^\delta \left[z_m^k(s) - z_m^{k-1}(s)\right]\right) ds$$
$$+ \int_t^{T_1} \mathbb{1}_{[0,\tau_m)}(s) e^{-A(s-t)} G^* \left(A^{-2\alpha} \left[\Phi_m^k(s) - \Phi_m^{k-1}(s)\right]\right) ds$$
$$- \int_t^{T_1} e^{-A(s-t)} \left(\Phi_m^{k+1}(s) - \Phi_m^k(s)\right) dW(s).$$

Again, we find $T_2 \in [0, T_1]$ such that the sequence $(z_m^k, \Phi_m^k)_{k \in \mathbb{N}} \subset \mathcal{Z}_T^1 \times \mathcal{Z}_T^2$ is a Cauchy sequence on the interval $[T_2, T_1]$. By continuing, we can conclude that the sequence $(z_m^k, \Phi_m^k)_{k \in \mathbb{N}} \subset \mathcal{Z}_T^1 \times \mathcal{Z}_T^2$ is a Cauchy sequence on the interval $[0, T]$. Hence, there exist $z_m^* \in \mathcal{Z}_T^1$ and $\Phi_m \in \mathcal{Z}_T^2$ such that

$$z_m^* = \lim_{k \to \infty} z_m^k, \quad \Phi_m = \lim_{k \to \infty} \Phi_m^k.$$

By equation (47), one can easily verify that the pair of processes $(z_m^*(t), \Phi_m(t))_{t \in [0,T]}$ fulfills equation (44). $\qquad \square$

Remark 30. If $y_d \in L^\infty([0, T]; D(A^\gamma))$, then the restriction $\gamma + \delta < \frac{1}{2}$ vanishes in the previous theorem. Moreover, note that we have the additional restrictions $\alpha, \delta < \frac{1}{2}$.

Corollary 3. *Let* $(z_m^*(t), \Phi_m(t))_{t \in [0,T]}$ *be the mild solution of system (43). Then we have for fixed* $m \in \mathbb{N}$

$$\mathbb{E} \sup_{t \in [\tau_m, T]} \|z_m^*(t)\|_{D(A^\delta)}^2 = 0 \quad and \quad \mathbb{E} \int_{\tau_m}^T \|\Phi_m(t)\|_{\mathcal{L}_{(HS)}(Q^{1/2}(H);H)}^2 dt = 0.$$

Here, we proved the existence and uniqueness of a mild solution to the adjoint equation. Next, we will approximate the local mild solution of the linearized stochastic Navier-Stokes equations as well as the mild solution of the adjoint equation by strong formulations.

3.3.2. Approximation by a Strong Formulation

Here, we omit the dependence on the controls for the sake of simplicity. First, we give an approximation of the mild solution of system (41). We introduce the following system in $D(A^{1+\alpha})$:

$$
\begin{cases}
dz_m(t,\lambda) = -[Az_m(t,\lambda) + R(\lambda)B(R(\lambda)z_m(t,\lambda), \pi_m(y_m(t)))]\, dt \\
\quad - [R(\lambda)B(\pi_m(y_m(t)), R(\lambda)z_m(t,\lambda)) - R(\lambda)Fv(t)]\, dt \\
\quad + R(\lambda)G(R(\lambda)z_m(t,\lambda))\, dW(t), \\
z_m(0,\lambda) = 0,
\end{cases}
\tag{49}
$$

where $m \in \mathbb{N}$, $\lambda > 0$, and $v \in L^2_{\mathcal{F}}(\Omega; L^2([0,T]; D(A^\beta)))$. The operators $A, B, R(\lambda), F, G$ are introduced in Sections 2.1, 2.4.2, and 3.1, respectively. The mapping π_m is given by (5) and the process $(y_m(t))_{t \in [0,T]}$ is the mild solution of system (38). The process $(W(t))_{t \geq 0}$ is a Q-Wiener process with values in H and covariance operator $Q \in \mathcal{L}(H)$. We consider the solution of system (49) in a mild sense according to Definition 29. Since the operators $R(\lambda)$ and $AR(\lambda)$ are linear and bounded on H, an existence and uniqueness result with fixed $m \in \mathbb{N}$ and fixed $\lambda > 0$ can be obtained similarly to Theorem 47. Analog to Section 2.4.2, we have for fixed $m \in \mathbb{N}$, fixed $\lambda > 0$, all $t \in [0,T]$, and $\mathbb{P}$-a.s.

$$
z_m(t,\lambda) = - \int_0^t Az_m(s,\lambda) + A^\delta R(\lambda)A^{-\delta}B(R(\lambda)z_m(s,\lambda), \pi_m(y_m(s)))\, ds
$$

$$
- \int_0^t A^\delta R(\lambda)A^{-\delta}B(\pi_m(y_m(s)), R(\lambda)z_m(s,\lambda)) - R(\lambda)Fv(s)\, ds
$$

$$
+ \int_0^t R(\lambda)G(R(\lambda)z_m(s,\lambda))\, dW(s)
\tag{50}
$$

and

$$
\lim_{\lambda \to \infty} \mathbb{E} \sup_{t \in [0,T]} \|z_m(t) - z_m(t,\lambda)\|^2_{D(A^\alpha)} = 0,
\tag{51}
$$

where $(z_m(t))_{t \in [0,T]}$ is the mild solution of system (41).

Next, we give an approximation of the mild solution to system (43). We introduce the following backward SPDE in $D(A^{1+\delta})$:

$$
\begin{cases}
dz_m^*(t,\lambda) = \mathbb{1}_{[0,\tau_m)}(t)[Az_m^*(t,\lambda) \\
\qquad + A^\alpha R(\lambda)A^\alpha B_\delta^* \left(y(t), R(\lambda)A^\delta z_m^*(t,\lambda)\right) \\
\qquad - R(\lambda)G^*(A^{-2\alpha}R(\lambda)\Phi_m(t,\lambda)) \\
\qquad - A^\gamma R(\lambda)A^\gamma \left(y(t) - y_d(t)\right)]\,dt + \Phi_m(t,\lambda)\,dW(t), \\
z_m^*(T,\lambda) = 0,
\end{cases}
\tag{52}
$$

where $m \in \mathbb{N}$ and $\lambda > 0$. The operators $A, R(\lambda), B_\delta^*, G^*$ are introduced in Sections 2.1, 2.4.1, 2.4.2, and 3.3.1, respectively. The process $(y(t))_{t\in[0,\tau)}$ is the local mild solution of system (36) with stopping times $(\tau_m)_{m\in\mathbb{N}}$ defined by (37) and $y_d \in L^2([0,T]; D(A^\gamma))$ is the given desired velocity field. The process $(W(t))_{t\geq 0}$ is a Q-Wiener process with values in H and covariance operator $Q \in \mathcal{L}(H)$. The solution of system (52) is defined in a mild sense according to Definition 31. Recall that the operators $R(\lambda)$ and $AR(\lambda)$ are linear and bounded in H. Hence, an existence and uniqueness result of a mild solution $(z_m^*(t,\lambda), \Phi_m(t,\lambda))_{t\in[0,T]}$ to system (52) can be obtained similarly to Theorem 53 for fixed $m \in \mathbb{N}$ and fixed $\lambda > 0$. Moreover, we have for fixed $m \in \mathbb{N}$, fixed $\lambda > 0$, all $t \in [0,T]$, and $\mathbb{P}$-a.s.

$$
z_m^*(t,\lambda) = -\int_t^T \mathbb{1}_{[0,\tau_m)}(s)Az_m^*(s,\lambda)\,ds
$$

$$
-\int_t^T \mathbb{1}_{[0,\tau_m)}(s)A^\alpha R(\lambda)A^\alpha B_\delta^* \left(y(s \wedge \tau_m), R(\lambda)A^\delta z_m^*(s,\lambda)\right)\,ds
$$

$$
+\int_t^T \mathbb{1}_{[0,\tau_m)}(s)R(\lambda)G^*(A^{-2\alpha}R(\lambda)\Phi_m(s,\lambda))\,ds
$$

$$
+\int_t^T \mathbb{1}_{[0,\tau_m)}(s)A^\gamma R(\lambda)A^\gamma \left(y(s \wedge \tau_m) - y_d(s)\right)\,ds
$$

$$
-\int_t^T \Phi_m(s,\lambda)\,dW(s),
\tag{53}
$$

which is an immediate consequence of [1, Theorem 3.4 and Theorem 4.1].

Lemma 19. Let $(z_m^*(t), \Phi_m(t))_{t\in[0,T]}$ and $(z_m^*(t,\lambda), \Phi_m(t,\lambda))_{t\in[0,T]}$ be the mild solutions of (43) and (52), respectively. We have for fixed $m \in \mathbb{N}$

$$
\lim_{\lambda\to\infty} \mathbb{E}\sup_{t\in[0,T]} \|z_m^*(t) - z_m^*(t,\lambda)\|_{D(A^\delta)}^2 = 0,
$$

$$\lim_{\lambda \to \infty} \mathbb{E} \int_0^T \|\Phi_m(t) - \Phi_m(t, \lambda)\|^2_{\mathcal{L}_{(HS)}(Q^{1/2}(H);H)} \, dt = 0.$$

Proof. We introduce the stochastic process $(\tilde{z}_m^*(t, \lambda))_{t \in [0,T]}$ defined by

$$\tilde{z}_m^*(t, \lambda) = z_m^*(t) - z_m^*(t, \lambda)$$

for all $\lambda > 0$, all $t \in [0, T]$, and $\mathbb{P}$-almost surely. By definition, we have for all $\lambda > 0$, all $t \in [0, T]$, and $\mathbb{P}$-a.s.

$\tilde{z}_m^*(t, \lambda)$

$$= - \int_t^T \mathbb{1}_{[0,\tau_m)}(s) A^\alpha e^{-A(s-t)} A^\alpha B_\delta^* \left(y(s \wedge \tau_m), [I - R(\lambda)] A^\delta z_m^*(s) \right) ds$$

$$- \int_t^T \mathbb{1}_{[0,\tau_m)}(s) A^\alpha e^{-A(s-t)} [I - R(\lambda)] A^\alpha B_\delta^* \left(y(s \wedge \tau_m), R(\lambda) A^\delta z_m^*(s) \right) ds$$

$$- \int_t^T \mathbb{1}_{[0,\tau_m)}(s) A^\alpha e^{-A(s-t)} R(\lambda) A^\alpha B_\delta^* \left(y(s \wedge \tau_m), R(\lambda) A^\delta \tilde{z}_m^*(s, \lambda) \right) ds$$

$$+ \int_t^T \mathbb{1}_{[0,\tau_m)}(s) e^{-A(s-t)} G^* (A^{-2\alpha}[I - R(\lambda)] \Phi_m(s)) \, ds$$

$$+ \int_t^T \mathbb{1}_{[0,\tau_m)}(s) e^{-A(s-t)} [I - R(\lambda)] G^* (A^{-2\alpha} R(\lambda) \Phi_m(s)) \, ds$$

$$+ \int_t^T \mathbb{1}_{[0,\tau_m)}(s) e^{-A(s-t)} R(\lambda) G^* (A^{-2\alpha} R(\lambda) [\Phi_m(s) - \Phi_m(s, \lambda)]) \, ds$$

$$+ \int_t^T \mathbb{1}_{[0,\tau_m)}(s) A^\gamma e^{-A(s-t)} [I - R(\lambda)] A^\gamma \left(y(s \wedge \tau_m) - y_d(s) \right) ds$$

$$- \int_t^T e^{-A(s-t)} [\Phi_m(s) - \Phi_m(s, \lambda)] \, dW(s),$$

where I is the identity operator on H. Note that the assumptions of Lemma 18 are fulfilled. Let $T_1 \in [0, T)$ be a point of time, which we specify below. Using inequalities (21), (45), and (46), there exist a constant $C_{T_1} > 0$

dependent on T_1 such that for all $\lambda > 0$

$$\mathbb{E} \sup_{t \in [T_1, T]} \|z_m^*(t) - z_m^*(t, \lambda)\|_{D(A^\delta)}^2$$

$$+ \mathbb{E} \int_{T_1}^{T} \|\Phi_m(t) - \Phi_m(t, \lambda)\|_{\mathcal{L}_{(HS)}(Q^{1/2}(H); H)}^2 \, dt$$

$$\leq C_{T_1} \mathbb{E} \sup_{t \in [T_1, T]} \|z_m^*(t) - z_m^*(t, \lambda)\|_{D(A^\delta)}^2$$

$$+ C_{T_1} \mathbb{E} \int_{T_1}^{T} \|\Phi_m(t) - \Phi_m(t, \lambda)\|_{\mathcal{L}_{(HS)}(Q^{1/2}(H); H)}^2 \, dt + \mathcal{I}_2(\lambda),$$

where $\mathcal{I}_2(\lambda)$ satisfies $\mathcal{I}_2(\lambda) = 0$ as $\lambda \to \infty$. We chose the point of time $T_1 \in [0, T)$ such that $C_{T_1} < 1$. Thus, we can conclude

$$\lim_{\lambda \to \infty} \mathbb{E} \sup_{t \in [T_1, T]} \|z_m^*(t) - z_m^*(t, \lambda)\|_{D(A^\delta)}^2 = 0,$$

$$\lim_{\lambda \to \infty} \mathbb{E} \int_{T_1}^{T} \|\Phi_m(t) - \Phi_m(t, \lambda)\|_{\mathcal{L}_{(HS)}(Q^{1/2}(H); H)}^2 \, dt = 0.$$

Similarly to Theorem 53, we can conclude that the convergence results hold on the whole interval $[0, T]$. $\square$

Here, we provided approximations of the local mild solution of the linearized stochastic Navier-Stokes equations and the mild solution of the adjoint equation by strong formulations. Next, we will show the duality principle and derive an explicit formula for the optimal control, which also satisfies a sufficient optimality condition.

3.3.3. *The Duality Principle and Derivation of the Optimal Control*

Based on the results provided in the previous sections, we are able to state a duality principle, which gives us a relation between the local mild solution of system (40) and the mild solution of system (43). Recall that the local mild solution of system (38) depends on the control $u \in L_{\mathcal{F}}^2(\Omega; L^2([0, T]; D(A^\beta)))$. Hence, the mild solution of system (43) depends on the control $u \in L_{\mathcal{F}}^2(\Omega; L^2([0, T]; D(A^\beta)))$ as well. Let us denote this solution by $(z_m^*(t; u), \Phi_m(t; u))_{t \in [0, T]}$.

Theorem 54. *Let* $(y(t; u))_{t \in [0, \tau^u)}$ *and* $(z(t; u, v))_{t \in [0, \tau^u)}$ *be the local mild solutions of system (36) and system (40), respectively, corresponding to the controls* $u, v \in L_{\mathcal{F}}^2(\Omega; L^2([0, T]; D(A^\beta)))$. *Moreover, let*

$(z_m^*(t; u), \Phi_m(t; u))_{t \in [0,T]}$ *be the mild solution of system (43) correspond-ing to the control* $u \in L_{\mathcal{F}}^2(\Omega; L^2([0, T]; D(A^\beta)))$. *Then we have for fixed* $m \in \mathbb{N}$

$$\mathbb{E} \int_0^{\tau_m^u} \langle A^\gamma(y(t; u) - y_d(t)), A^\gamma z(t; u, v) \rangle_H \, dt = \mathbb{E} \int_0^{\tau_m^u} \langle z_m^*(t; u), Fv(t) \rangle_H \, dt.$$

Proof. The idea of the proof is given in Theorem 51. Indeed, one can show that the result holds for the mild solutions by system (49) and system (52) by applying of the product formula stated in Proposition 17 to the strong formulation given by equation (50) and equation (53). The convergence results from equation (51) and Lemma 19, which completes the proof. □

Based on the necessary optimality condition formulated as the varia-tional inequality (42) and the duality principle given by the previous the-orem, we are able to deduce a formula the optimal control has to satisfy. The proof can be adopted from Theorem 52.

Theorem 55. *Let* $(z_m^*(t; u), \Phi_m(t; u))_{t \in [0,T]}$ *be the mild solution of system (43) corresponding to the control* $u \in L_{\mathcal{F}}^2(\Omega; L^2([0, T]; D(A^\beta)))$. *Then for fixed* $m \in \mathbb{N}_0$, *the optimal control* $\bar{u}_m \in U$ *satisfies for almost all* $t \in [0, T]$ *and* $\mathbb{P}$*-a.s.*

$$\bar{u}_m(t) = -P_U \left(F^* A^{-2\beta} z_m^*(t; \bar{u}_m) \right), \tag{54}$$

where $P_U: L_{\mathcal{F}}^2(\Omega; L^2([0, T]; D(A^\beta))) \to U$ *is the projection onto* U *intro-duced in Section 2.4.3 and* $F^* \in \mathcal{L}(D(A^\beta))$ *is the adjoint operator of* $F \in \mathcal{L}(D(A^\beta))$.

According to Section 2.4.4, the optimal control $\bar{u}_m \in U$ given by equa-tion (54) satisfies the sufficient optimality condition stated in Proposition 2. Furthermore, let us denote by $(\bar{y}(t))_{t \in [0,\bar{\tau})}$ and $(\bar{z}_m^*(t), \bar{\Phi}_m(t))_{t \in [0,T]}$ the local mild solutions of system (36) and the mild solution of system (43), respectively, corresponding to the optimal control $\bar{u}_m \in U$. As a conse-quence of the previous theorem, the optimal velocity field $(\bar{y}(t))_{t \in [0,\bar{\tau})}$ can be computed by solving the following system of coupled forward-backward SPDEs:

$$\begin{cases} d\bar{y}(t) = -[A\bar{y}(t) + B(\bar{y}(t)) + FP_U \left(F^* A^{-2\beta} \bar{z}_m^*(t) \right)] \, dt + G(\bar{y}(t)) \, dW(t), \\ d\bar{z}_m^*(t) = -\mathbb{1}_{[0,\tau_m)}(t)[-A\bar{z}_m^*(t) - A^{2\alpha} B_\delta^* \left(\bar{y}(t), A^\delta \bar{z}_m^*(t) \right)] \, dt \\ \qquad - \mathbb{1}_{[0,\tau_m)}(t)[G^*(A^{-2\alpha}\bar{\Phi}_m(t)) + A^{2\gamma} \left(\bar{y}(t) - y_d(t) \right)] \, dt \\ \qquad + \bar{\Phi}_m(t) \, dW(t), \\ \bar{y}(0) = \xi, \quad \bar{z}_m^*(T) = 0. \end{cases}$$

4. Conclusion

In this chapter, we considered the stochastic Navier-Stokes equations with distributed control driven by additive as well as linear multiplicative Wiener noise. The existence and uniqueness of a local mild solution is shown on bounded and connected multidimensional domains. We discussed a control problem constrained by the stochastic Navier-Stokes equations formulated as an non-convex optimization problem. The existence of a unique optimal control of this control problem was provided. We stated a necessary optimality condition as a variational inequality using the Gâteaux derivative of the cost functional related to the control problem. By a suitable duality principle, we derived an explicit formula for the optimal control based on the corresponding adjoint equation. For the stochastic Navier-Stokes equations driven by additive Wiener noise, the adjoint equation is given by a deterministic backward equation, while for the stochastic Navier-Stokes equations driven by multiplicative Wiener noise, the adjoint equation is given by a backward SPDE. This characterizes the main difference of additive and multiplicative noise. For both cases, one can obtain that the optimal velocity field can be computed by solving a system of coupled forward and backward SPDEs. Moreover, we showed that the optimal control satisfies a sufficient optimality condition using the second-order Fréchet derivative of the cost functional.

References

[1] A. Al-Hussein, Strong, mild and weak solutions of backward stochastic evolution equations, *Random Oper. Stoch. Equ.* **13**, 2, 129–138 (2005).

[2] P. Benner and C. Trautwein, A stochastic maximum principle for control problems constrained by the stochastic Navier-Stokes equations, *arXiv:1810.12119* (2018).

[3] P. Benner and C. Trautwein, Optimal control problems constrained by the stochastic Navier-Stokes equations with multiplicative Lévy noise, *Math. Nachr.* **292**, 7, 1444–1461 (2019).

[4] P. Benner and C. Trautwein. Optimal distributed and tangential boundary control for the unsteady stochastic Stokes equations. Accepted for publication in ESAIM: COCV (2019).

[5] A. Bensoussan and J. Frehse, Local solutions for stochastic Navier Stokes equations, *M2AN Math. Model. Numer. Anal.* **34**, 2, 241–273 (2000).

[6] M. F. Bidaut. *Théorèmes d'existence et d'existence "en général" d'un contrôle optimal pour des systèmes régis par des équations aux dérivées partielles non linéaires.* PhD Thesis, Université Paris VI (1973).

[7] P. Billingsley, *Probability and Measure.* John Wiley & Sons (1995).

[8] H. Breckner, Galerkin approximation and the strong solution of the Navier-Stokes equation, J. Appl. Math. Stochastic Anal. **13**, 239–259 (2000).

[9] Z. Brzeźniak, E. Hausenblas, and J. Zhu, 2D stochastic Navier-Stokes equations driven by jump noise, *Nonlinear Anal.* **79**, 122–139 (2013).

[10] Z. Brzeźniak, B. Maslowski, and J. Seidler, Stochastic nonlinear beam equations, *Probab. Theory Relat. Fields.* **132**, 119–149 (2005).

[11] M. Capinski and D. Gaterek, Stochastic equations in Hilbert space with application to Navier-Stokes equations in any dimension, *J. Funct. Anal.* **126**, 26–35 (1994).

[12] G. Da Prato and A. Debussche, Dynamic programming for the stochastic Navier-Stokes equations, *M2AN Math. Model. Numer. Anal.* **34**, 2, 459–475 (2000).

[13] G. Da Prato and J. Zabczyk, *Ergodicity for Infinite Dimensional Systems.* Cambridge University Press (1996).

[14] G. Da Prato and J. Zabczyk, *Stochastic Equations in Infinite Dimensions.* Cambridge University Press (2014).

[15] L. Debbi, Well-posedness of the multidimensional fractional stochastic Navier-Stokes equations on the torus and on bounded domains, *J. Math. Fluid Mech.* **18**, 25–69 (2016).

[16] G. Fabbri, F. Gozzi, and A. Swiech, *Stochastic Optimal Control in Infinite Dimension: Dynamic Programming and HJB Equations.* Springer, Berlin (2017).

[17] B. P. W. Fernando, B. Rüdiger, and S. S. Sritharan, Mild solutions of stochastic Navier-Stokes equation with jump noise in $\mathbb{L}^p$-spaces, *Math. Nachr.* **288**, 1615 – 1621 (2015).

[18] B. P. W. Fernando and S. S. Sritharan, Nonlinear filtering of stochastic Navier-Stokes equation with Itô-Lévy noise, *Stoch. Anal. Appl.* **31**, 3, 381–426 (2013).

[19] F. Flandoli, *An Introduction to 3D Stochastic Fluid Dynamics. In: Proceedings of the CIME Course on SPDE in Hydrodynamics: Recent Progress and Prospects.* Lecture Notes in Math., vol. 1942, pp. 51–150. Springer, Berlin (2008).

[20] F. Flandoli and D. Gatarek, Martingale and stationary solutions for stochastic Navier-Stokes equations, *Probab. Theory Relat. Fields.* **102**, 367–391 (1995).

[21] M. Fuhrman and C. Orrieri, Stochastic maximum principle for optimal control of a class of nonlinear SPDEs with dissipative drift, *SIAM J. Control Optim.* **54**, 1, 341–371 (2016).

[22] H. Fujita and H. Morimoto, On fractional powers of the Stokes operator, Proc. Japan Acad. **46**, 1141–1143 (1970).

[23] D. Fujiwara and H. Morimoto, An L_r-theorem of the Helmholtz decomposition of vector fields, *J. Fac. Sci. Univ. Tokyo Sec. 1 A.* **24**, 3, 685–700 (1977).

[24] L. Gawarecki and V. Mandrekar, *Stochastic Differential Equations in Infinite Dimensions: with Applications to Stochastic Partial Dierential Equations.* Springer, Berlin (2011).

[25] Y. Giga, Analyticity of the semigroup generated by the Stokes operator in L_r spaces, *Math. Z.* **178**, 297–329 (1981).

[26] Y. Giga and T. Miyakawa, Solutions in L_r of the Navier-Stokes initial value problem, *Arch. Ration. Mech. Anal.* **89**, 3, 267–281 (1985).

[27] N. Glatt-Holtz and M. Ziane, Strong pathwise solutions of the stochastic Navier-Stokes system, *Adv. Differential Equations.* **14**, 5–6, 567–600 (2009).

[28] M. Goebel, On existence of optimal control, *Math. Nachr.* **93**, 67–73 (1979).

[29] T. E. Govindan, *Yosida Approximations of Stochastic Differential Equations in Infinite Dimensions and Applications.* Springer, New York (2016).

[30] F. Gozzi, S. S. Sritharan, and A. Swiech, Bellman equations asociated to the optimal feedback control of stochastic Navier-Stokes equations, *Comm. Pure Appl. Math.* **58**, 671–700 (2005).

[31] E. Hausenblas and J. Seidler, Stochastic convolutions driven by martingales: Maximal inequalities and exponential integrability, *Stoch. Anal. Appl.* **26**, 98–119 (2008).

[32] M. Hintermüller and M. Hinze, A SQP-semismooth Newton-type algorithm applied to control of the instationary Navier-Stokes system subject to control constraints, *SIAM J. Optim.* **16**, 4, 1177–1200 (2006).

[33] M. Hinze, R. Pinnau, M. Ulbrich, and S. Ulbrich, *Optimization with PDE Constraints.* Springer, New York (2009).

[34] Y. Hu and S. Peng, Adapted solution of a backward semilinear stochastic evolution equation, *Stoch. Anal. Appl.* **9**,4 , 445–459 (1991).

[35] A. Ichikawa, Stability of semilinear stochastic evolution equations, *J. Math. Anal. Appl.* **90**, 12–44 (1982).

[36] A. J. Kurdila and M. Zabarankin, *Convex Functional Analysis.* Birkhäuser (2005).

[37] H. Liang, L. Hou, and J. Ming, The velocity tracking problem for Wick-stochastic Navier-Stokes flows using Wiener chaos expansion, *J. Comput. Appl. Math.* **307**, 25–36 (2016).

[38] J. L. Menaldi and S. S. Sritharan, Stochastic 2-D Navier-Stokes equation, *Appl. Math. Optim.* **46**, 31–53 (2002).

[39] M. T. Mohan and S. S. Sritharan, $\mathbb{L}^p$-solutions of the stochastic Navier-Stokes equations subject to Lévy noise with $\mathbb{L}^m(\mathbb{R}^m)$ initial data, *Evol. Equ. Control Theory.* **6**, 3, 409–425 (2017).

[40] E. Motyl, Stochastic Navier-Stokes equations driven by Lévy noise in unbounded 3D domains, *Potential Anal.* **38**, 863–912 (2013).

[41] Y. R. Ou, *Design of Feedback Compensators for Viscous Flow. In: Optimal Control of Viscous Flow*, pp. 151–180. SIAM, Philadelphia (1998).

[42] A. Pazy, *Semigroups of Linear Operators and Applications to Partial Differential Equations.* Springer, New York (1983).

[43] S. S. Sritharan, *An Introduction to Deterministic and Stochastic Control of Viscous Flow. In: Optimal Control of Viscous Flow*, pp. 1–42. SIAM, Philadelphia (1998).

[44] S. S. Sritharan and P. Sundar, Large deviations for the two-dimensional Navier-Stokes equations with multiplicative noise, *Stochastic Process. Appl.* **116**, 1636–1659 (2006).

[45] G. Tessitore, Existence, uniqueness and space regularity of the adapted solutions of a backward SPDE, *Stochastic Anal. Appl.* **14**, 4, 461–486 (1996).

[46] C. Trautwein. *Optimal Control Problems Constrained by Stochastic Partial Differential Equations.* PhD Thesis, Otto von Guericke University, Magdeburg (2018).

[47] F. Tröltzsch, *Optimal Control of Partial Differential Equations: Theory, Methods and Applications.* American Mathematical Society (2010).

[48] M. Ulbrich, Constrained optimal control of Navier-Stokes flow by semismooth Newton methods, *Systems Control Lett.* **48**, 297–311 (2003).

[49] W. von Wahl, *The Equations of Navier-Stokes and Abstract Parabolic Equations.* Vieweg + Teubner Verlag (1985).

[50] I. Vrabie, *C_0-Semigroups and Applications.* Elsevier, Amsterdam (2003).

[51] D. Wachsmuth, Regularity and stability of optimal controls of nonstationary Navier-Stokes equations, *Control and Cybernetics.* **34**, 2, 387–409 (2005).

[52] E. Zeidler, *Nonlinear Functional Analysis and its Applications III: Variational Methods and Optimization.* Springer, New York (1985).

© 2020 World Scientific Publishing Company
https://doi.org/10.1142/9789811209796_0005

Chapter 5

Quantum Hamilton Equations from Stochastic Optimal Control Theory

Jeanette Köppe[1,2*], Markus Patzold[2†], Michael Beyer[2‡],
Wilfried Grecksch[3§] and Wolfgang Paul[2¶]

[1] *Institute of Biostatistics and Clinical Research, Westfälische
Wilhelms-Universität Münster, 48149 Münster, Germany*
[2] *Institute of Physics, Martin-Luther-Universität Halle-Wittenberg,
06099 Halle (Saale), Germany*
[3] *Institute of Mathematics, Martin-Luther-Universität Halle-Wittenberg,
06099 Halle (Saale), Germany*

In 1966, E. Nelson laid the foundations for an understanding of quantum mechanics in terms of stochastic processes. He showed that a description of the motion of quantum particles based on time-reversible Brownian motion leads to a set of hydrodynamic equations derived by Madelung, which in turn can be transformed into the Schrödinger equation. We recently established that the generalization of Hamilton's equations of motion to the quantum world in terms of coupled forward-backward stochastic differential equations can be derived, when the quantum Hamilton principle, which is the quantum analogue of the Hamilton principle of classical mechanics, is interpreted as a stochastic optimal control problem for the two velocity fields in Nelson's equations.

The quantum Hamilton equations typically have to be solved numerically. Additionally, we will establish links to the theory of supersymmetric Hamiltonians and explain how this allows to obtain the complete bound spectrum of a quantum problem from the quantum Hamilton equations. All mathematical results will be exemplified by applications to suitable quantum mechanical problems from physics.

*E-mail (corresponding author): jeanette.koeppe@ukmuenster.de
†E-mail: markus.patzold@student.uni-halle.de
‡E-mail: michael.beyer@physik.uni-halle.de
§E-mail: wilfried.grecksch@mathematik.uni-halle.de
¶E-mail: wolfgang.paul@physik.uni-halle.de

Contents

1. Introduction

Classical analytical mechanics is a conceptually and mathematically highly developed part of theoretical physics concerned with the motion of (in an idealized form) point particles under the influence of mutual interactions. It's aim is to predict the paths $x(t)$ of these particles, i.e., their positions $x \in \mathbb{R}^3$ as a function of time $t \in \mathbb{R}$ from the knowledge of the relevant interactions. Isaac Newton (1643-1727) was the first to formulate a physical law allowing for this prediction. It takes the form of a second order ordinary differential equation (see Fig. 1), where the force, F, is given as the negative gradient of the interaction potential, V, which is typical for the fundamental interactions in physics. Newton's equation in this case gives rise to paths $x(t) \in C^2(\mathbb{R})$ which additionally conserve at all times the total energy $H = \frac{1}{2}m\dot{x}^2(t) + V(x(t))$ of the particle. William Rowan Hamilton (1805-1865) recognized that one can obtain Newton's equation as the law fulfilled by the paths extremizing the action functional $S[x]$ (see Fig. 1). Together with Carl Gustav Jacob Jacobi (1804-1851) he derived the nonlinear partial differential equation for the cost function of this variational principle (see Fig. 1). The Newton equation is, of course, equivalent to a set of two coupled first order differential equations, $\dot{x}(t) = v(t)$, $m\dot{v}(t) = F$, with $v(t)$ being the velocity of the particle. Written in the form $\dot{x} = \partial H/\partial p$ and $\dot{p} = -\partial H/\partial x$, with $p = mv$ in the most simple case, these equations are also called Hamilton's equations of motion. Interestingly, it took a hundred years until it was recognized that these equations also result when one understands Hamilton's principle as a deterministic optimal control problem, i.e., searching not for the optimal path $x(t)$ but for the optimal control $v(t)$ in Newton's equation. In this form the principle is called Pontryagin principle [6] and its Hamilton-Jacobi-Bellman equation is again

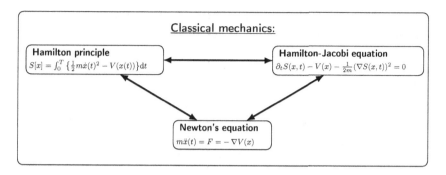

Fig. 1. Scheme of the structure of classical analytical mechanics.

the Hamilton-Jacobi equation given in Fig. 1. All these equations describing analytical mechanics are mutually equivalent and their interrelations constitute powerful tools to describe and to interpret physical phenomena described by classical mechanics.

This theoretical approach and its concomitant picture of point particles moving in space, however, completely failed when physicists at the beginning of the 20th century began to study quantum phenomena involving atomic properties. If one imagines electrons circling atomic nuclei on something resembling the Keplerian orbits of planets around the sun, electromagnetism tells us that they should emit radiation, thereby loosing energy and dropping into the nucleus. Also, in some experiments, beams of particles behaved as if they were waves, creating an interference pattern. All this could be captured by the Schrödinger equation

$$- i\hbar \partial_t \psi(x,t) = \left[-\frac{\hbar^2}{2m} \Delta + V(x) \right] \psi(x,t). \qquad (1)$$

Schrödinger arrived at this equation by a complex line of arguments trying to generalize the Hamilton-Jacobi equation of classical mechanics to the quantum world. $\psi(x,t)$ is a complex-valued function called the wave function, and the operator on the right hand side is the representation of the total energy in quantum mechanics $H = -\frac{\hbar^2}{2m} \Delta + V(x)$. However, the physical meaning of the wave function only emerged with Born's postulate, that $\|\psi(x,t)\|^2 dx$ is the probability to find a particle in the volume element dx around point x at time t. Hence, a particle is somehow represented by a wave function which obeys a normalization condition $\int \|\psi(x,t)\|^2 dx = 1$, i.e., the wave function has to be an element of the separable Hilbert space of square integrable complex-valued functions. Physical observables like energy, momentum and any other then have to be represented by (essentially) self-adjoined operators on this Hilbert space to possess a spectrum of real eigenvalues (as measured). In the quantum world, it appears that mechanics has to be formulated in the language of functional analysis, first beautifully subsumized in the book by von Neumann [27]. While this was mathematically satisfying, quantum physics was confined to the equivalent of the upper right corner of Fig. 1, and that brought about a slew of interpretation problems. While electrons, for example, in many experiments behave as well-defined negatively charged particles, one has to model them with a wave function if one wants to describe, e.g., their scattering in a double slit experiment. So what are they: particles or waves? Is the electron the wave function (in an ontological meaning) or does the wave function only describe what we know about the electron? And how subjective would such a knowledge be? Niels Bohr (1885-1962), one of the fathers of quantum theory, was very philosophically inclined and tried to solve this conundrum by postulating that in quantum mechanics we need to use complementary pictures for different questions, promoting this even to an underlying principle of the quantum world. Lack of a more diverse and complete mathematical structure such as the one existing for classical mechanics led to the development of many conflicting ways to interpret quantum mechanics and to a discussion about the correct physical understanding of quantum mechanics which has been going on unabatedly for almost a hundred years now.

However, in the early 1960s, E. Nelson, a mathematician in Princeton, was working on theories of Brownian motion and found that, contrary to the believe (at least among physicists) that Brownian motion describes dissipative processes, one can formulate conservative Brownian motion processes, i.e., processes in which the energy is conserved on average. These, he recognized, could form a tool for a description of quantum phenomena based on continuous, but not differentiable paths of particles. In 1966, he published a paper [21] entitled *Derivation of the Schrödinger equation from Newtonian mechanics* providing quantum mechanics with the lower corner of the triangle in Fig. 1. Particle paths were back in existence, physical observables could be described by stochastic variables and the mathematical foundation of quantum mechanics was augmented by stochastic theory.

In the next section, we will present an introduction to Nelson's stochastic mechanics and Section 3 is dedicated to the derivation of the quantum

Hamilton equations of motion. Section 4 will present some applications followed by an outlook in Section 5. In an appendix, we will discuss some more technical aspects on the numerical solution of the quantum Hamilton equations for higher dimensional systems.

2. Nelson Mechanics

Energy conservation, exact or in the mean, is connected with time reversal symmetry. A conservative Brownian motion therefore can be described by two coupled forward-backward stochastic differential equations (FBSDEs)

$$
\begin{aligned}
dX(t) &= C_f(X(t), t)dt + \sigma dW_f(t) \\
dX(t) &= C_b(X(t), t)dt + \sigma dW_b(t)
\end{aligned}
\tag{2}
$$

where C_f and C_b are forward and backward drift velocities. $(W_f(t))_{t \geq 0}$ is a forward in time Wiener process and $W_f(t)$ is independent of all $X(s)$ with $s \leq t$. $(W_b(t))_{t \geq 0}$ is a backward in time Wiener process and $W_b(t)$ is independent of all $X(s)$ with $s \geq t$. Finally, σ^2 is the diffusion coefficient to be identified shortly. Adding the corresponding Fokker-Planck equations, one arrives at the continuity equation for the probability density $p(x, t)$ of the process $(X(t))_{t \geq 0}$, i.e.

$$
\partial_t p(x, t) + \nabla \cdot [v(x, t)p(x, t)] = 0,
\tag{3}
$$

with $v(x, t) = \frac{1}{2}(C_f(x, t) + C_b(x, t))$ as the velocity of the probability current. The osmotic velocity $u(x, t) = \frac{1}{2}(C_f(x, t) - C_b(x, t))$ is determined by the probability density of the process

$$
u(x, t) = \frac{\sigma^2}{2} \nabla \ln[p(x, t)] =: \sigma^2 \nabla R(x, t).
\tag{4}
$$

Classical mechanics needs to be the deterministic limit of this formulation of quantum theory, obtained as $\sigma^2 \to 0$. It is valid for heavy, macroscopic objects of our everyday experience. Assuming $\sigma^2 \propto 1/m$, with the mass m of the particle, the proportionality constant has to have units of an action. Setting $\sigma^2 = \hbar/m$ with $\hbar$ being Plancks constant devided by 2π, one obtains the correct constants in the Schrödinger equation. In the classical limit one has $v(x, t) = 1/m \nabla S(x, t)$ with the action $S(x, t)$, which can be shown to hold here also, with a modified action function, so both velocities are curl free and given as gradients of scalar functions.

   Partial differential equations for the velocities $v(x, t)$ and $u(x, t)$ or, equivalently, the action $S(x, t)$ and the logarithm of the probability density

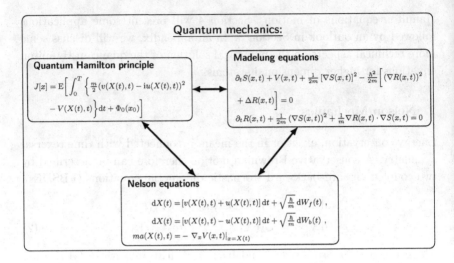

Fig. 2.   Scheme of the structure of quantum analytical mechanics.

$R(x,t)$ can be obtained using Itô calculus. One important ingredient for this is the extension of the standard time derivative of the path $x(t)$ used in classical mechanics to the Brownian paths - continuous, but not differentiable - of quantum mechanics. Nelson [21] defined mean forward and backward derivatives

$$D_f X(t) := \lim_{\Delta t \to 0} \mathbb{E}\left[\left.\frac{\mathbf{X}(t+\Delta t) - \mathbf{X}(t)}{\Delta t}\right| \mathbf{X}(t)\right],$$

$$D_b X(t) := \lim_{\Delta t \to 0} \mathbb{E}\left[\left.\frac{\mathbf{X}(t) - \mathbf{X}(t-\Delta t)}{\Delta t}\right| \mathbf{X}(t)\right].$$

The symmetric second order derivative $\frac{1}{2}(D_b D_f + D_f D_b)X(t) =: a(t)$ defines the mean acceleration of the quantum particle. As a final physical input Nelson assumed that Newton's second law relating this acceleration to the force on the particle also holds for quantum systems (third of Nelson's equations in Fig. 2). Using this physical assumption and Itô calculus one arrives at the Madelung equations for $S(x,t)$ and $R(x,t)$ as the quantum analogue of the Hamilton-Jacobi equation (upper right corner of Fig. 2). For node free wave functions $\psi(x,t) = \exp\{R(x,t) + \frac{i}{\hbar}S(x,t)\}$ - and only for those - the solutions to the Madelung equations define a function which solves the Schrödinger equation Eq. (1). This construction of the wave function $\psi(x,t)$ automatically leads to $p(x,t) = \|\psi(x,t)\|^2$ and Born's postulate is no longer needed as an independent ingredient of quantum theory.

Soon after Nelson's original publication, people established the existence of a variation principle equivalent to Hamilton's principle [12, 24, 29], which has the Madelung equations as its Hamilton-Jacobi-Bellman equations. The formulation we present in Fig. 2 follows the derivation presented by M. Pavon [24]. He showed that one actually needs two extremal principles, a stationary action principle which leads to the Madelung equation for $S(x,t)$ and a stationary entropy production principle, which leads to the equation for $R(x,t)$. A beautiful way to formulate this is to combine the two into a complex valued variational principle which turns out to be a direct generalization of Hamilton's principle of classical mechanics to a complex valued velocity, $v(x,t) - iu(x,t)$, which reduces to the classical one in the limit $\frac{\hbar}{m} \to 0$.

Interpreting the functional $J[x]$ in Fig. 2 as a functional of the velocities instead, $J[v,u]$, we were able to establish a stochastic optimal control problem as the equivalent to the Pontryagin principle in classical mechanics leading to quantum Hamilton equations of motion [18]. This will be presented in the next section. With that, we now have arrived at what one could call quantum analytical mechanics.

3. Derivation of Quantum Hamilton Equations of Motion

When we generalize the discussion of the last section to many particle systems, non-stationary systems of N particles are described by stochastic processes $(X(t))_{t \in [0,T]} \in \mathbb{R}^{3N}$ that are characterized by the following forward as well as backward SDE:

$$\begin{cases} dX(t) = [v(X(t),t) + u(X(t),t)]\,dt + \underline{\sigma}\,dW_f(t)\,, \quad t \in [0,T] \\ X(t=0) = x_0 \in \mathbb{R}^{3N} \end{cases}, \quad (5)$$

$$\begin{cases} dX(t) = [v(X(t),t) - u(X(t),t)]\,dt + \underline{\sigma}\,dW_b(t)\,, \quad t \in [0,T] \\ X(t=T) = x_T(\omega) \end{cases} \quad (6)$$

The starting point of the forward motion is arbitrary and the starting point of the backward motion has to be the end point of the forward process and is affected by random events. The Wiener processes $(W_f(t))_{t \in [0,T]}$ and $(W_b(t))_{t \in [0,T]}$ were explained above. The diffusion coefficient is given by the diagonal matrix $\underline{\sigma} := \hbar/m\,\mathbb{I}_{3N}$. The current and the osmotic velocity can be obtained from the complex variational principle introduced by Pavon [24]:

$$J[u,v] = \mathrm{E}\left[\int_0^T \left\{\frac{m}{2}(v(X(t),t) - iu(X(t),t))^2 - V(X(t),t)\right\}\mathrm{d}t + \Phi_0(x_0)\right].$$

$$(7)$$

The function $\Phi_0(x)$ is given by the initial condition of the wave function $\Psi(x,0) = \exp\left[\frac{i}{\hbar}\Phi_0(x)\right]$. Separating real and imaginary part of this variational principle leads to two variational problems, where the current and osmotic velocity are given as the saddle-points of both.

$$J_\mathrm{R}[\hat{v},\hat{u}] = \min_v \max_u \mathrm{E}\left[\int_0^T \left\{\frac{m}{2}\left(v^2(X(t),t) - u^2(X(t),t)\right) - V(X(t),t)\right\}\mathrm{d}t\right.$$

$$\left. + S(x_0)\right],$$

$$(8)$$

$$J_\mathrm{I}[\hat{v},\hat{u}] = \max_v \min_u \mathrm{E}\left[\int_0^T m\, v(X(t),t) \cdot u(X(t),t)\mathrm{d}t + \hbar R(x_0)\right].$$

$$(9)$$

There are two methods available for finding the two velocities: Firstly, both saddle-point problems can be solved simultaneously by finding the Nash equilibrium and, secondly, the complex-valued quantum Hamilton problem can be analyzed [17, 18]. Before explaining the first method in full generality (the second one can be found in [17]), the derivation is explained for stationary systems for the sake of clarity.

3.1. Stationary Problems

For stationary systems, the wave function $\Psi(x)$ is given by the solution of the stationary Schrödinger equation,

$$\left[-\frac{\hbar^2}{2\,m}\Delta + V(x)\right]\Psi(x) = E\,\Psi(x),$$

$$(10)$$

and the current velocity equals zero. Hence, the motion process $(X(t))_{t\in[0,T]}$ is described by the following forward and backward SDEs,

$$\begin{cases} \mathrm{d}X(t) &= u(X(t))\mathrm{d}t + \underline{\sigma}\,\mathrm{d}W_f(t)\,, \quad t \in [0,T] \\ X(t=0) &= x_0 \in \mathbb{R}^{3N} \end{cases},$$

$$(11)$$

$$\begin{cases} \mathrm{d}X(t) &= -u(X(t))\mathrm{d}t + \underline{\sigma}\,\mathrm{d}W_b(t)\,, \quad t \in [0,T] \\ X(t=T) &= x_T(\omega) \end{cases}$$

$$(12)$$

Since the osmotic velocity is only a function of the motion processes in the stationary case, no explicit time-dependence arises in the drift terms of the two SDEs.

The current velocity is zero and the osmotic velocity is thus given by the maximum of the following action functional, i.e., searching for $\hat{u}(X(t))$ with

$$J[\hat{u}] = \max_u \mathbb{E}\left[\int_0^T \left(-\frac{m}{2}u^2(X(t)) - V(X(t))\right)dt + S_0(x_0)\right]. \quad (13)$$

The occurring time horizon is finite ($T \in \mathbb{R}$ is fixed) and is considered to be sufficiently large. The initial costs $S_0(x_0)$ are related to the initial condition of the Schrödinger equation, i.e., $\Psi(x_0, t = 0) = \exp\left\{R(x_0, 0) + \frac{i}{\hbar}S(x_0, 0)\right\}$. Since in the stationary case the function $S(x, t) = S(t)$ depends only on time, $S_0(x) := S(x, t = 0) = S_0(x_0)$ is constant for all $x \in \mathbb{R}^{3N}$ and one can choose $S_0(x_0) \equiv 0$.

The osmotic velocity is the solution of a so-called optimal feedback control problem, which is, in the stationary case, the quantum-mechanical counterpart of the least-action principle in classical mechanics. In contrast to the treatment by Pavon [24], who includes the SDEs for the motion processes as constraints for the search of the optimal paths in order to extremize the action functional, Eq. (13) is interpreted as an optimal control problem. Hence, the optimal control $u(X(t))$ can be determined using the maximum principle derived in [23]. The related Hamilton function reads

$$H(x, u, \lambda, p, q) = -\frac{m}{2}u^2 - V(x) + \langle\lambda + p, u\rangle + \text{Tr}(\sigma^T q), \quad (14)$$

where $\langle a, b\rangle$ denotes the scalar product of two vectors a, b. The adjoint processes are defined by the following forward as well as backward SDEs,

$$d\lambda(t) = 0, \quad \lambda(0) = 0, \quad (15)$$

$$dp(t) = \nabla_x V(x)|_{x=X(t)}\, dt + q(t)\, dW_b(t), \quad p(T) = \lambda(T). \quad (16)$$

Using the maximum principle [23], the osmotic velocity reads

$$m\, u(X(t)) = p(t). \quad (17)$$

Inserting Eq. (17) into Eq. (16), one ends up with

$$du(t) = \frac{1}{m}\nabla_x V(x)|_{x=X(t)}\, dt + \frac{q(t)}{m}\, dW_b(t). \quad (18)$$

It is not necessary to consider the backward SDE for the motion process, since the adjoint process $\lambda(t)$ vanishes due to the absence of any initial or final cost and thus the backward SDE of the motion process provides no further information to the optimal control problem. Eq. (11) and Eq. (18) build a system of coupled forward-backward stochastic differential equations, which constitute the stochastic Hamilton equations of motion in the stationary case:

$$dX(t) = u(X(t))dt + \sqrt{\frac{\hbar}{m}}\, dW_f(t)\,, \quad X(0) = x_0 \in \mathbb{R}\,, \tag{19}$$

$$du(t) = \frac{1}{m}\, \nabla_x V(x)|_{x=X(t)}\, dt + \frac{q(t)}{m}\, dW_b(t)\,, \quad u(X(T)) = 0 \tag{20}$$

As such, these coupled equations can now be solved independently of the wave function and describe the dynamics of quantum particles uniquely. Furthermore, since the osmotic velocity and the adjoint processes are functionals of the motion process, it is possible to determine PDEs for them using Itô's formula [20], i.e.

$$du(t) = \left[u_t(t,x) + u_x(x,t)\, u(t,x) + \frac{\hbar}{2m} u_{xx}(x,t)\right]_{x=X(t)} dt$$

$$+ \left[\sqrt{\frac{\hbar}{m}}\, u_x(x,t)\right]_{x=X(t)} dW_b(t)$$

$$= \frac{1}{m}\, \nabla_x V(x)|_{x=X(t)}\, dt + \frac{q(t)}{m}\, dW_b(t)$$

$$\Rightarrow 0 = u_t(t,x) + u_x(x,t)\, u(t,x) + \frac{\hbar}{2m} u_{xx}(x,t) - \frac{1}{m}\nabla_x V(x) \tag{21}$$

$$q(x,t) = \sqrt{\hbar m}\, u_x(x,t)\,, \tag{22}$$

with $u_x(t,x)$ being the Jacobian matrix and $u_{xx}(t,x)$ being the Hessian matrix of the osmotic velocity $u : [0,T] \times \mathbb{R}^{3N} \to \mathbb{R}^{3N}$. They can be simplified to first order ordinary differential equations, since $u(x)$ and $q(x)$ are only functions of the particles position without an explicit time-dependency:

$$0 = u(x)\, u_x(x)dx + \frac{\hbar}{2\,m} u_{xx}(x) - \frac{1}{m}\nabla_x V(x)\,, \tag{23}$$

$$q(x) = \sqrt{\hbar m}\, u_x(x)\,. \tag{24}$$

The ODE for the osmotic velocity equals the gradient of the time-independent Schrödinger equation and is in accordance with the equation derived in [21].

3.1.1. Excited States

As we noted, the diffusion process in Nelson's mechanics only generates the wave function if the latter is node-free, even though there are different approaches for diffusion processes with singular drifts [4, 7]. However, we can extend our theory to be able to iteratively generate the complete bound spectrum of a quantum problem based on the theory of supersymmetric Hamiltonians [26].

The concept of supersymmetry in one-dimensional quantum mechanics introduced in 1982 by Witten [28] is a well-known concept in mathematics. In 1912, Darboux [10] derived a method for Sturm-Liouville type differential equations to construct an infinite series of exactly solvable problems starting from a known base problem in one dimension. In 1955, Crum [9] presented a compact form of the expressions and investigated the Schrödinger equation. The work of Sabatier [25] generalizes this idea to multi-dimensional systems and a lot of work in multi-dimensional SUSY quantum mechanics followed [1].

In our case, the strategy is to transform the well-known formalism from wave functions to the velocity fields of Nelsons equation by starting with the stationary case, where the quantum mechanical Hamiltonian reads

$$\mathcal{H}_0 = -\frac{1}{2}\triangle + V_0(x) \ , \quad \triangle = \partial_1^2 + ... + \partial_d^2 \,, \tag{25}$$

$x = (x_1, ..., x_d)^T \in \mathbb{R}^d$, $d = 3N$. Note that atomic units are used for simplicity and that $\partial_i \equiv \partial_{x_i}$. To simplify the calculations, the potential is shifted by $V_0(x) \to V_0(x) + const.$ in such a way that the ground state energy becomes zero. In order to proceed, we will state the following: If $\mathcal{H}_0$ has a normalized ground state wave function, then, in any dimension, this function has zero nodes [8]. Thus, it's possible to find a well defined function $\chi(x)$ to write the ground state wave function in the following form

$$\psi_0^{(0)}(x) = e^{-\chi(x)} \,. \tag{26}$$

The superscript (n) in $\psi_k^{(n)}$ indicates the n-th excited state of the wave function belonging to the Hamiltonian $\mathcal{H}_k$. Note that n has to replaced by a vector in multiple dimensions for excited states.

To factorize the Hamiltonian $\mathcal{H}_0$ define the $2d$ operators

$$Q_{0,l}^{\pm} = \frac{1}{\sqrt{2}}(\mp\partial_l + \partial_l\chi) \tag{27}$$

fulfilling

$$\left[Q_{0,i}^-, Q_{0,k}^+\right] = \partial_i\partial_k\chi \,, \quad \left[Q_{0,i}^+, Q_{0,k}^+\right] = \left[Q_{0,i}^-, Q_{0,k}^-\right] = 0 \,, \tag{28}$$

$$V_0 = \frac{1}{2}\left[(\partial_l\chi)^2 - \partial_l^2\chi\right] \,, \quad \nabla\chi = -\nabla\ln\psi_0^{(0)}. \tag{29}$$

Consequently, the Hamiltonian can be factorized to (using Einstein's sum convention)

$$\mathcal{H}_0 = -\frac{1}{2}\triangle + V_0(x) = Q_{0,l}^+ Q_{0,l}^- \,. \tag{30}$$

If we apply the Darboux transformation to the Schrödinger equation with Hamilton operator $\mathcal{H}_0$, the partial differential equation can be written in terms of a partner Hamiltonian (all indices take d values)

$$\mathcal{H}_1^{i,k} = \mathcal{H}_0 \delta_{ik} + \left[Q_{0,i}^-, Q_{0,k}^+ \right] = -\frac{1}{2}\triangle \delta_{ik} + V_1^{i,k} \,, \tag{31}$$

$$V_1^{i,k} = V_0 \delta_{ik} + \partial_i \partial_k \chi \,. \tag{32}$$

Despite the fact that the result is a Hamiltonian in matrix form and it could be a difficult task to find eigenfunctions, there exist some practical intertwining relations. Start with

$$\mathcal{H}_1^{i,k} Q_{0,k}^- \psi_0^{(n)} = Q_{0,i}^- \mathcal{H}_0 \psi_0^{(n)} = Q_{0,i}^- E_0^n \psi_0^{(n)} = E_0^n Q_{0,i}^- \psi_0^{(n)} \,. \tag{33}$$

The diagonal elements $(i = k)$ fulfil

$$\mathcal{H}_1^{k,k}(Q_{0,k}^- \psi_0^{(n)}) = E_0^n (Q_{0,k}^- \psi_0^{(n)}) \,. \tag{34}$$

Following the procedure of [19], the relation

$$\psi_0^{(0_1, \ldots, n+1_k, \ldots, 0_d)} = c_{n,k} Q_{n,k}^+ \psi_1^{(n),k} \tag{35}$$

holds. The wave function $\psi_0^{(0_1, \ldots, n+1_k, \ldots, 0_d)}(x)$ is the $(n+1)$-th excited state wave function in the quantum number corresponding to the k-th coordinate and $\psi_1^{(n),k}$ is the n-th state for $\mathcal{H}_1^{k,k}$. The ground state of $\mathcal{H}_1^{k,k}$ is given by the QHEs with a modified potential $V_1^{k,k}(x)$

$$dX^k(t) = u_1^k(X^k(t))dt + \underline{\sigma}dW_f(t) \,, \tag{36}$$

$$du_1^k(t) = \partial_x V_1^{k,k}(x)\Big|_{x=X^k(t)} dt + \underline{q}_0(t)dW_b(t)$$

$$= \left(\partial_x V_0(x) - \partial_x \partial_k u_0^k(x) \right)_{x=X^k(t)} dt + \underline{q}_0(t)dW_b(t) \,. \tag{37}$$

Since $v(t) = 0$ for stationary systems, the ground-state solution of the adjoint process reads $\underline{q}_0(t) = \underline{\sigma}\, \partial_x u_0(x)|_{x=X^k(t)}$. The calculation of the excited states is reduced to the determination of ground states of the partner Hamiltonians. Due to the node theorem these states are node-free and therefore their corresponding velocities have no singularities, which in turn leads to diffusions covering the whole space. That allows to determine the

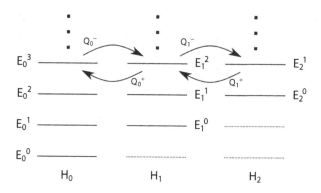

Fig. 3. A schematic view showing how to proceed to excited states using SUSY.

original excited wave functions with their nodes by the application of differential operators (see Fig. 3).

In the end, this leads to an iterative way for all bound states $\psi_0^{(a_1,\ldots,b_k,\ldots,c_d)}(x)$ for the Hamiltonian $\mathcal{H}_0$. For example, to calculate the state $\psi_0^{(1,1,0)}(x)$ in three dimensions, first determine the ground state of $\mathcal{H}_0$, $u_0^{(0),0}(x)$, then the one corresponding to $\mathcal{H}_1^{1,1}$, $u_1^{(0),1}(x)$. Then, we can determine a first excited state velocity w. r. t. the axis x_1, e.g.

$$\left(u_0^{(1),1}\right)_i = \sigma^2 \frac{\partial_1 \left(u_1^{(0),1}(x) + u_0^{(0),0}\right)_i}{\left(u_1^{(0),1} + u_0^{(0),0}\right)_i} + \left(u_1^{(0),1}(x)\right)_i. \tag{38}$$

Subsequently determine $u_2^{(0),2}(x)$ corresponding to $\mathcal{H}_2$. From there we can calculate first $u_1^{(1),2}(x)$ and finally, $u_0^{(2),(1,1,0)}(x)$ similar to Eq. (38). Thus, this iterative method allows us to calculate the excited states by simply adjusting the potential with the help of previously calculated velocities and, finally, by acting with operators on the newly determined osmotic velocities. The solutions of the QHEs in Eq. (37) can be found according to a numerical iteration similar to the presented algorithm in [18] but now in the case of a multi-dimensional problem.

3.2. Non-Stationary Problems

In the general case of non-stationary systems, where the current velocity is different from zero, both variational problems (Eq. (8) and Eq. (9)) have to be solved simultaneously. For the sake of clarity, the following derivation is only presented for one-dimensional systems, but can be directly generalized for higher dimensional systems (see also [3]). Further investigations are based on the maximum principle for forward-backward stochastic games, described in [23].

It can be shown, that the current and the osmotic velocity are given by the Nash equilibrium of the two saddle-point principles (Eq. (8) and Eq. (9)). The velocity fields $\hat{u}, \hat{v}$ are called Nash equilibrium, if

$$J_R[u, \hat{v}] \leq J_R[\hat{u}, \hat{v}] \quad \text{and} \quad J_I[\hat{u}, v] \leq J_I[\hat{u}, \hat{v}] \tag{39}$$

is fulfilled for all u, v.

In the case of quantum mechanical systems, the forward controlled process (the motion process) equals the backward one. However, in general, the two controlled processes are given by two different stochastic processes. Hence, the backward process is given by a second one $(Y(t))_{t \in [0,T]}$, with

$$\begin{cases} dY(t) & = [v(Y(t),t) - u(Y(t),t)] \, dt + \sqrt{\frac{\hbar}{m}} \, dW_b(t) \,, \quad t \in [0,T] \\ Y(t=T) & = x_T(\omega) \end{cases}$$
$$\tag{40}$$

The equivalence of the forward process $(X(t))_{t \in [0,T]}$ and the backward process $(Y(t))_{t \in [0,T]}$ in each time step is included as a constraint

$$Y(t) = X(t) \quad \forall t \in [0,T]. \tag{41}$$

Therefore, the two related Hamiltonians are given by

$$H_R(t,x,y,u,v) = \frac{m}{2}(v^2 - u^2) - V(x,t) + \lambda_R(u-v) + p_R(u+v)$$

$$+ \sqrt{\frac{\hbar}{m}} q_R + \alpha_R(x-y) \tag{42}$$

and

$$H_I(t,x,y,u,v) = m\,uv + \lambda_I(u-v) + p_R(u+v) + \sqrt{\frac{\hbar}{m}} q_I + \alpha_I(x-y), \tag{43}$$

where α_R and α_I are two Lagrange multipliers. Following [23], the adjoint processes are given by the forward SDEs,

$$\begin{cases} d\lambda_R(t) & = -\alpha_R \, dt \\ \lambda_R(t=0) & = S'(x_0) \end{cases}, \quad \begin{cases} d\lambda_I(t) & = -\alpha_I \, dt \\ \lambda_I(t=0) & = R'(x_0) \end{cases}, \tag{44}$$

and the backward SDEs,

$$\begin{cases} dp_R(t) & = \left[\left. \frac{\partial V(x,t)}{\partial t} \right|_{x=X(t)} - \alpha_R \right] dt + q_R(t) \, dW_b(t) \\ p_R(t=T) & = \lambda_R(T) \end{cases}, \tag{45}$$

$$\begin{cases} dp_I(t) & = -\alpha_I \, dt + q_I(t) \, dW_b(t) \\ p_I(t = T) & = \lambda_I(T) \end{cases} .$$ (46)

Determining the roots of the partial derivatives of H_R and H_I with respect to u and v, one gets

$$m\,u(t) = \lambda_R + p_R, \quad m\,v(t) = \lambda_R - p_R$$ (47)

$$m\,u(t) = \lambda_I - p_I, \quad m\,v(t) = -\lambda_I - p_I.$$ (48)

Using the integral representation of the SDEs for the adjoint processes and the relations determined from the maximum principle, one can derive a BSDE for the current velocity, i.e.

$$p_R(t) = \lambda_R(T) - \int_t^T \left[\frac{\partial V}{\partial x}(X(s), s) - \alpha_R \right] ds$$ (49)
$$- \int_t^T q_R(s) dW_b(s)$$

$$= S'(x_0) - \int_0^T \alpha_R \, ds - \int_t^T \left[\frac{\partial V}{\partial x}(X(s), s) - \alpha_R \right] ds$$ (50)
$$- \int_t^T q_R(s) dW_b(s)$$

$$= \underbrace{S'(x_0) - \int_0^t \alpha_R \, ds}_{\lambda_R(t)} - \int_t^T \frac{\partial V}{\partial x}(X(s), s) ds - \int_t^T q_R(s) \, dW_b(s)$$

$$\Rightarrow m\,v(t) = p_R(t) - \lambda_R(t) = - \int_t^T \frac{\partial V}{\partial x}(X(s), s) ds - \int_t^T q_R(s) \, dW_b(s)$$ (51)

Analogously, the osmotic velocity is derived by

$$p_I(t) = \lambda_I(T) + \int_t^T \alpha_I \, ds - \int_t^T q_I(s) dW_b(s)$$

$$= \underbrace{R'(x_0) - \int_0^t \alpha_I \, ds}_{\lambda_I(t)} - \int_t^T q_I(s) \, dW_b(s)$$

$$\Rightarrow m\,u(t) = \lambda_I(t) - p_I(t) = \int_t^T q_I(s) \, dW_b(s),$$ (52)

with the abbreviation

$$\frac{\partial V}{\partial x}(X(s), s) = \left. \frac{\partial V(x, s)}{\partial x} \right|_{x=X(s)}.$$ (53)

Using Itô's formula, the two adjoint processes $q_R(t)$ and $q_I(t)$ are given by

$$\frac{q_R(t)}{m} = \sqrt{\frac{\hbar}{m}} \left.\frac{\partial v(x,t)}{\partial x}\right|_{x=X(t)} \quad \text{and} \quad \frac{q_I(t)}{m} = \sqrt{\frac{\hbar}{m}} \left.\frac{\partial u(x,t)}{\partial x}\right|_{x=X(t)}. \tag{54}$$

Finally, the BSDEs for the current and osmotic velocity read

$$m\,dv(t) = \left.\frac{\partial V(x,t)}{\partial x}\right|_{x=X(t)} dt + \sqrt{\hbar m}\,\partial_x v(x,t)|_{x=X(t)}\,dW_b(t), \tag{55}$$

$$m\,du(t) = \sqrt{\hbar m}\,\partial_x u(x,t)|_{x=X(t)}\,dW_b(t). \tag{56}$$

As a last step, it is possible to derive the Madelung equations using Itô's formula. Suppose the partial derivatives $\partial_t p_R(x,t)$, $\partial_x p_R(x,t)$ and $\partial_{xx} p_R(x,t)$ are continuous. $p_R(t)$ is the adjoint process to the forward motion process and is thus the solution of

$$\frac{\partial p_R}{\partial t} + \big(v(x,t) + u(x,t)\big)\frac{\partial p_R}{\partial x} + \frac{\hbar}{2m}\frac{\partial^2 p_R}{\partial x^2} - \frac{\partial V}{\partial x} + \alpha_R = 0, \tag{57}$$

where $x \in \mathbb{R}$ and $t \geq 0$. Analogously, $\lambda_R(t)$ is the adjoint process to the backward motion process and is given by

$$\frac{\partial \lambda_R}{\partial t} + \big(v(x,t) - u(x,t)\big)\frac{\partial \lambda_R}{\partial x} - \frac{\hbar}{2m}\frac{\partial^2 \lambda_R}{\partial x^2} + \alpha_R = 0, \quad x \in \mathbb{R},\ t \leq 0, \tag{58}$$

which is derived using a so-called backward Itô formula [24]. Taking the difference of Eq. (57) and Eq. (58), results in

$$0 = \frac{\partial v}{\partial t} + v\frac{\partial v}{\partial x} - u\frac{\partial u}{\partial x} - \frac{\hbar}{2m}\frac{\partial^2 u}{\partial x^2} + \frac{1}{m}\frac{\partial V}{\partial x}. \tag{59}$$

Using the relations determined from the entropy production principle $J_I[u,v]$, one can determine a PDE for the osmotic velocity in an analogous way, resulting in

$$0 = \frac{\partial u}{\partial t} + v\frac{\partial u}{\partial x} + u\frac{\partial v}{\partial x} + \frac{\hbar}{2m}\frac{\partial v^2}{\partial x^2}. \tag{60}$$

Notice, that Eq. (59) and Eq. (60) are the Madelung equations derived by Nelson [21]. Hence, the FBSDE system – called quantum Hamilton equations of motion – consisting of the FSDE for the motion process $(X(t))_{t\geq0}$ and the BSDE for the physical momentum $p(t) = m\,[u(t) + v(t)]$, i.e.

$$\begin{cases} dX(t) &= [v(X(t),t) + u(X(t),t)]\,dt + \underline{\sigma}\,dW_f(t), \quad t \geq 0 \\ X(t=0) &= x_0 \in \mathbb{R}^{3N} \end{cases}, \tag{61}$$

$$\mathrm{d}\,m\left[v(t) + u(t)\right] = \frac{\partial V(x(t), t)}{\partial x}\mathrm{d}t + \sqrt{\frac{\hbar}{m}}\frac{\partial m\left[v(x(t), t) + u(x(t), t)\right]}{\partial x}\mathrm{d}W_b(t)\,,$$

$$(62)$$

is equivalent to the Schrödinger equation (for the node-free case, as discussed).

In conclusion, stochastic mechanics based on forward-backward stochastic differential equations leads to a parallel method to describe quantum systems in a more natural way. The evaluation of the optimal control problem introduced by Pavon [24] results in a FBSDE system for the position and the physical momentum of quantum particles, which can be solved without knowledge about the wave function.

4. Applications

Since there are only a few analytic solutions known to uncoupled SDEs, it is unlikely to find a closed solution to most of the quantum systems described by the stationary QHEs. Therefore, a numerical approach is needed. The presented algorithm is based on an iterative scheme. I. e., the numerical approach starts with a guess or fixed function as a first estimate for the osmotic velocity $u^{(0)}(x)$, which is used to calculate a numerical solution to the forward SDE concerning the particle's position $x(t)$. After that, a solution of the backward SDE $u^{(1)}(x)$ is determined. The drift term of the forward SDE is replaced by the new estimate for the osmotic velocity $u^{(1)}(x)$ and the procedure is repeated until the convergence criteria are satisfied. The initial estimate $u^0(x)$ can be chosen as the classical stationary ground state, i.e. $u^0(x) \equiv 0$, since the convergence of the numerical algorithm is (almost) independent of the starting point.

The forward SDE is usually solved by integrating the equations concerning the paths directly. Therefore, the time horizon $[0, T]$ is discretized by $\pi = \{t_i | 0 = t_0 < t_1 < \cdots < t_{n_T} = T\}$. To solve the FSDE, different forward step schemes are available, e.g. the simplest one is given by the Euler-Mayurama scheme

$$x(t_0) = x^\pi(t_0) \tag{63}$$

$$x(t_{i+1}) \approx x^\pi(t_i) + u(x^\pi(t_i))\Delta t + \sigma\Delta W_f(t_i)\,, \tag{64}$$

with constant time step $t_{i+1} - t_i = \Delta t$ and the Wiener increments $\Delta W_f(t_i) = W_f(t_{i+1}) - W_f(t_i)$. The index π is used to clarify that $x^\pi(t_i)$ is the approximated solution regarding the chosen time partition. It should be noted that there exist methods with a higher convergence order, however, a higher computational effort would be necessary without a significantly better result for the osmotic velocity and thus the Euler-Mayurama scheme is chosen in this work.

The direct numerical evaluation of the BSDE is based on a two-step scheme and requires the same partition denoted by π. The Euler-Mayurama method for the backward equation in the j-th iteration reads

$$u^{\pi,j}(x^{\pi,j}(T)) = u^{\pi,j-1}(x^{\pi,j}(T)), \quad q^{\pi,j}(T) = 0,$$

$$mu(x^{\pi,j}(t_i)) \approx mu^{\pi,j}(x^{\pi,j}(t_i))$$
$$= mu^{\pi,j}(x^{\pi,j}(t_{i+1})) - \partial_x V(x^{\pi,j}(t_i))\Delta t - q^{\pi,j}(t_i)\Delta W_b(t_i),$$
$$(65)$$

with $\Delta t > 0$ and the backward Wiener increments $\Delta W_b(t_i) = W_b(t_{i+1}) - W_b(t_i)$. The starting point of the backward equation $u^{\pi,j}(x^{\pi,j}(T))$ is chosen as the solution of the previous estimate of the osmotic velocity at $x^{\pi,j}(T)$ to avoid final costs of the action functional. Because q and u are unknown in each time step, Eq. (65) can not be solved uniquely and the two processes have to be determined by making use of conditional expectations. Suppose $(\mathcal{F}_{t_i})_{t_i \geq t_0}$ is the filtration generated by the forward process $x^{\pi,j}(t_i)$ up to time t_i. Taking the conditional expectation of Eq. (20) w.r.t. $\mathcal{F}_{t_i}$ yields

$$u^{\pi,j}(x^{\pi,j}(t_i)) = \mathrm{E}\left[u^{\pi,j}(x^{\pi,j}(t_{i+1}))\big|x^{\pi,j}(t_i)\right] - \partial_x V(x^{\pi,j}(t_i))\Delta t. \qquad (66)$$

Hence, it is necessary to determine conditional expectations numerically. The employed algorithm is based on a least square Monte Carlo method proposed in [2]. As the name states, it tries to give the best estimate via least-square minimization. In that case, the expectation of the square of the difference of an admissible function and the considered stochastic variable over the set of admissible functions is to be minimized.

The numerical determination of the conditional expectations requires further expansions and approximations. First of all one has to generate a high number of realizations of the forward process per iteration, which leads to an additional index. Thus, for the sake of clarity, we will omit the index for partition and iteration step in the next few formulas. The conditional expectation at time step t_i can be represented by a linear combination of $\mathcal{F}_{t_i}$-measurable functions. In a second step, a functional basis for the numerical estimation has to be chosen. In this work, we chose $u(x)$ to be a multidimensional step-function with L (hyper-)cubes or intervals I_j in space in the range $[a,b] = \bigcup_{i=1}^{d} [\alpha_i, \beta_i] = \bigcup_{j=1}^{L} I_j \subset \mathbb{R}^d$. Therefore, the l-th,

$l = 1, \ldots, N$ generated sample path of the osmotic velocity is given by

$$u^l(t_i) = \mathrm{E}\left[u^l(t_{i+1})|x^l(t_i)\right] - \partial_x V(x^l(t_i))\Delta t$$

$$\approx \frac{1}{\#(x^l(t_{i+1}))\,\Delta t} \sum_{n=1}^{N} \left[\delta(x^l(t_i) - x^n(t_{i+1}))u^n(t_{i+1})\right] - \partial_x V(x^l(t_i))\Delta t.$$

(67)

The function δ is 1, if both positions $x^l(t_i)$, $x^n(t_{i+1})$ are in the same interval and, otherwise, $\delta = 0$. The cardinality $\#(x^l(t_{i+1}))$ returns the number of visits of all N sample paths at step t_i in the interval $I_j \ni x^l(t_i)$. Thus, this estimate is proportional to the average over all products of the velocity from the future time step t_{i+1} with the Wiener increment, where the particle's position at t_{i+1} coincides within a certain interval with the position of particle l at t_i. Keep in mind that all presented stochastic processes are the numerical approximation to the QHEs (Eq. (19) and Eq. (20)) for a certain chosen partition π.

The algorithm is quite heavy to handle, e.g. in comparison to the numerical solution of the one-dimensional Schrödinger equation using, e.g., Numerov's method [5, 15, 22]. This is due to the fact that we need to integrate N sample paths denoted by the index l over n_T steps, store the paths (and increments $\Delta W_f(t_i)$) and then use them to integrate backwards when calculating $u^l(x^l(t_i))$. For that, the estimate of the conditional expectation is required at every step according to Eq. (66) or Eq. (67), respectively, in each iteration. At the end of an iteration step, all $u(x^l(t_i))$ that are in the same hyper-cube w.r.t. $x^l(t_i)$ are averaged for all $l \in \{1, \ldots, N\}$ and $i \in \{0, \ldots, n_T - 1\}$. Finally, the next iteration can be started with the previously calculated step function $u(x)$. The numbers N, L and n_T have to be chosen with caution, also depending on the time partition Δt. For a more detailed overview of the convergence criteria concerning the numerical estimation of the conditional expectation the reader may be referred to [2]. By going to multiple dimensions later, the number of intervals L within this approach has to be increased significantly since the dimension d entails an exponential increase in L. The range of convergence is improved by the use of a randomly distributed starting value $x(0)$ for the forward paths. It should be noted that choosing a random initial value for $x(0)$, the ensemble of these values does not reflect the stationary probability distribution. This is clearly an advantage since then particles are also forced to start and move in regions where it would be not so likely to find them. Thus we get better estimates for the parts from which the particle is repelled.

4.1. One-Dimensional Systems

4.1.1. The Harmonic Oscillator

The one-dimensional harmonic oscillator is an important and didactic model in theoretical physics, and will be analyzed in the following. The harmonic potential $V(x) = m\omega^2 \frac{x^2}{2}$ is transformed into dimensionless variables using the characteristic length $\sqrt{\hbar/m\omega}$ and the characteristic time ω^{-1}, hence, the time-independent Schrödinger equation reads

$$\left[-\frac{1}{2}\frac{\mathrm{d}^2}{\mathrm{d}x^2} + \frac{x^2}{2} \right] \Psi(x) = E\,\Psi(x) \,. \tag{68}$$

The exact solutions are functions of the associated energy eigenvalues $E_n = n + 1/2$, $n \in \mathbb{N}_0$ and are given by

$$\Psi_n(x) = \pi^{-1/4}(2^n\,n!)^{-1/2}\mathrm{e}^{-x^2/2}\,H_n(x)\,, \tag{69}$$

with the Hermite polynomials $H_n(x)$ defined by

$$H_n(x) = \mathrm{e}^{-x^2/2}\left(x - \frac{\mathrm{d}}{\mathrm{d}x} \right)\mathrm{e}^{-x^2/2}\,, \ n \in \mathbb{N}_0\,. \tag{70}$$

The ground-state probability density is hence given by a Gaussian function centred around zero, with standard deviation $\sigma = 1/\sqrt{2}$ and the exact osmotic velocity is $u(x) = -x$. Since the exact solution of the osmotic velocity and the probability density are known, the harmonic oscillator is a good model to evaluate the above introduced algorithm.

The above algorithm is used to find a numerical solution to the quantum Hamilton equations of motion for the harmonic potential. In Fig. 4 (left) the paths of the backward equation and the resulting numerical solution of the osmotic velocity are presented. It is obvious that also the single paths of the osmotic velocity are close to the exact solution given by $u(x) = -x$. The probability density is determined from the osmotic velocity by numerical integration, i.e. by calculating

$$p(x) = |\Psi(x)|^2 = c_0 \exp\left\{ 2\int_{-\infty}^{x} u(x')\mathrm{d}x' \right\}\,, \tag{71}$$

with the normalization constant

$$c_0 = \left[\int_{-\infty}^{\infty} \exp\left\{ 2\int_{-\infty}^{x} u(x')\mathrm{d}x' \right\}\mathrm{d}x \right]^{-1}\,. \tag{72}$$

The numerical solution of $p(x)$ and the exact one are shown in Fig. 4 (right). They are in excellent agreement with each other over a very broad position range, since it is necessary to compute a lot of paths to guarantee the convergence of the numerical algorithm [2]. It is also possible to determine the probability density from the position histogram [17, 18]. However, the accuracy of the numerical integration is much higher than the one of the position histogram and is therefore the preferred method for calculating the related probability density for the motion process. As explained in

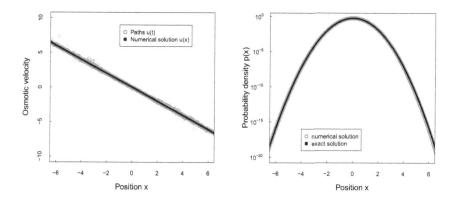

Fig. 4. **Left:** Paths and solution of the osmotic velocity. **Right:** Exact and numerical solution of the probability density for a one-dimensional harmonic oscillator.

Section 3.1.1, the ground state solution of the osmotic velocity can be used to determine the partner potentials, which are necessary to calculate all other solutions to the Schrödinger equation. The first partner potential for the first exited states is given by

$$V_1(x) = V(x) - \hbar \frac{\mathrm{d}u_0(X)}{\mathrm{d}x} = V(x) + \hbar\omega. \qquad (73)$$

Since the drift term of the backward SDE is given by the derivative of the potential, the osmotic velocity of the first exited state equals the ground-state velocity, i.e. the ground-state of the partner Hamiltonian $\mathcal{H}_1$ equals the ground-state probability density. Consequently, no further optimization is needed to calculate all exited states of the harmonic oscillator. The wave function $\Psi_n(x)$ of the n-th state, can thus calculated from the ground-state solution using

$$\Psi_n(x) = \prod_{i=0}^{n-1} \sqrt{\frac{m}{2\left(E_n - E_i\right)}} \left(\omega x - \frac{\hbar}{m}\frac{\mathrm{d}}{\mathrm{d}x}\right)^n \Psi_0(x), \qquad (74)$$

with the energy eigenvalues given by

$$E_n = \mathbb{E}\left[\frac{m}{2}u_0^2(x) + V_n(x)\right] \tag{75}$$

$$= \mathbb{E}\left[\frac{m}{2}u_0^2(x) + V(x) + n\hbar\omega\right] \tag{76}$$

$$= E_0 + n\hbar\omega = \hbar\omega(n + 1/2). \tag{77}$$

Notice that the equation for the energies and the equation for the wave functions are in accordance with the one derived by evaluating the time-independent Schrödinger equation. To conclude, by solving the quantum Hamilton equations of motion it is possible to determine all states of the harmonic potential.

4.1.2. One-dimensional Double-well Potential and Tunneling Phenomenon

The symmetric, quartic, one-dimensional double well problem is characterized by its Hamiltonian

$$\mathcal{H}_0 = -\frac{1}{2}\frac{d^2}{dx^2} + \frac{V_0}{a^4}(x^2 - a^2)^2, \tag{78}$$

where V_0 is the height of the potential barrier and the two minima are located at $\pm a$. This problem is an important model with a lot of applications in physics even though there exists no analytical solution. In [19], Köppe et al. investigated this model system of particles using the quantum Hamilton equations derived in Section 3. The supersymmetric factorization discussed in Section 3 leads to partner Hamiltonians with single-well potentials shown in Fig. 5. As the potential barrier between the wells vanishes, the ground states of the partner potentials become unimodal functions giving rise to node-free ground states. The excited states of the original double well potential are then determined by numerical differentiation. The increasing number of differentiations needed for higher order excited states causes a quickly increasing numerical uncertainty. A longer calculation with more particle paths and a smaller timestep smooths the functions and decreases this problem. In Fig. 6 the first four excited states are shown. From the osmotic velocity determined using the scheme of Section 3, the wave function has been calculated via

$$\psi_i^{(0)}(x) = c_i \exp\left\{\int_{-\infty}^{x} u_i(x')dx'\right\} \tag{79}$$

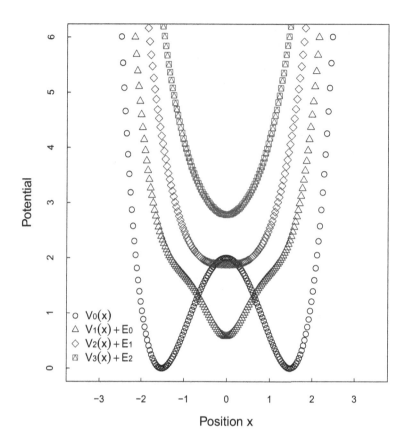

Fig. 5. Potentials for the partner Hamiltonians $\mathcal{H}_i$.

and the energy can be calculated as the expectation value of the stochastic function:

$$E_i^0 = \mathbb{E}\left[\frac{1}{2}u_i^2(x) + V_i(x)\right] = \int_{-\infty}^{\infty} \left(\frac{1}{2}u_i^2(x) + V_i(x)\right)\left|\psi_i^{(0)}(x)\right|^2 dx. \quad (80)$$

The results for the wave function as well as the energies of the eigenstates are in good agreement with those obtained using the Numerov method [22] which is a fourth-order, implicit multistep method for ordinary differential equations. Concerning the numerical efficiency, the Numerov method should be preferred, as it is at least two order of magnitude faster than the

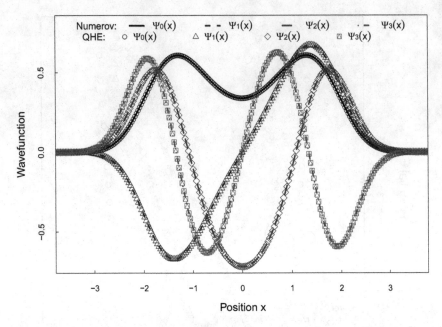

Fig. 6.   Ground state and first excited states of the double well potential. V=2.0. a=1.5.

used Bender-Steiner method, however, the stochastic approach allows for
some deeper insight into the system.

The double-well potential has many applications in physics as it is the
archetypical model for a phenomenon that came as a surprise to the inven-
tors of quantum physics. For sufficiently large barrier, at least the lowest
two energy levels lie below the barrier height. With the classical picture of
a closed physical system in mind, in which the energy is constant at each
time point and not only on average, there would be no way for a particle
located, for instance, in the ground state on the left side of the barrier to
reach the right side. So physicists invented the concept of *tunneling*. The
particle is somehow able to tunnel through the barrier and reach the other
side. This process leads to an energy splitting between the even ground
state and the odd first excited state which is related to the frequency of
this tunneling. In contrast to standard quantum mechanics, the concept
of tunneling is not needed in stochastic mechanics. The noise (added by
a Brownian motion) causes the energy of the particle to fluctuate, eventually leading to enough energy to overcome the potential barrier. Both

pictures allow for a calculation of the energy difference between ground state and first excited state [19]. For the tunneling problem, this consists in a description calculating the contributions of a gas of instanton paths connecting the two minima in the potential. For the Nelson approach we are following here, the frequency of getting from one side to the other clearly has to be related to the mean first passage time of reaching, e.g., the right well starting from the left one. The FSDE for the position process of a quantum particle moving in this potential is given by Eq. (19). For the corresponding Fokker-Planck equation, the mean first passage time to go from a starting point x on the left of the barrier to the exit point of the barrier on the right side, x_t (defined by the point where $V(x) = E_0$) can be given in a closed form

$$\tau_{\mathrm{mfpt}}(x) = 2 \int_x^{x_t} \frac{\mathrm{d}x'}{p_0(x')} \int_{-\infty}^{x'} p_0(x'') \, \mathrm{d}x'' . \tag{81}$$

Here $p_0(x) = \|\psi_0(x)\|^2$ is the probability density in the ground state of the potential. For the double-well potential, there is no exact point in space where the particle starts or stops to interact with the barrier, and it can start at an arbitrary point on the left side. We therefore need to average over the possible starting points in the left well. We choose to calculate

$$\bar{\tau}_{\mathrm{mfpt}} = 2 \int_{-\infty}^{-x_t} \tilde{p}_0(x) dx \int_x^{x_t} \frac{\mathrm{d}x'}{p_0(x')} \int_{-\infty}^{x'} p_0(x'') \, \mathrm{d}x'' , \tag{82}$$

where $\tilde{p}_0(x)$ is renormalized from $p_0(x)$ to be a probability density on $(-\infty, -x_t]$. It turns out that the energy splitting $\Delta E = E_1 - E_0$ is then simply given as

$$\Delta E = \frac{2\hbar}{\bar{\tau}_{\mathrm{mfpt}}} . \tag{83}$$

As one can see in Fig. 7 the best available instanton calculation based on the standard picture of quantum mechanics is in semi-quantitative agreement, while the result based on Nelson's stochastic approach is in perfect agreement with the numerically determined energy splitting. But, of course, this agreement comes at a price: one needs to let go of eighty years old concepts about tunneling and accept that an alternative description, in which tunneling is not needed, is not only numerically accurate but also conceptually much easier.

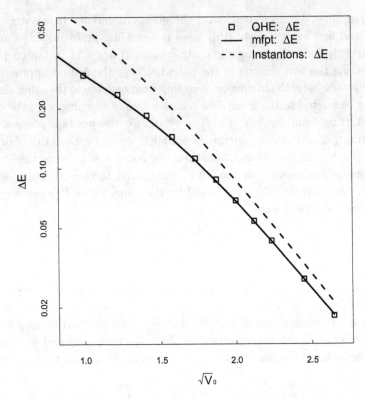

Fig. 7. Energy splitting between ground state and first excited state as a function of barrier height at $a = 1.5$. Symbols are numerical results, the full line is obtained from the mean first passage time and the dashed line is the result of an instanton gas calculation.

4.2. *Higher-Dimensional Systems*

The quantum Hamilton equations including their numerical evaluation for one-dimensional stationary systems were discussed so far. In the next section, we want to exemplify how the equations can be used to analyze systems of higher dimensions as well.

As one of physics basic problems, the two-body problem, exemplified by the hydrogen atom will be discussed next. The system is composed of a bound state of a proton and an electron with masses m_n and m_e, respectively. Both carry the charges e and $-e$, where e is a positive and real constant (not denoting the Euler number but the charge unit). We consider the non-relativistic case, where the interaction between the two particles is described by the Coulomb potential

$$V(|r_{\mathrm{n}} - r_{\mathrm{e}}|) = -\frac{e_0^2}{|r_{\mathrm{n}} - r_{\mathrm{e}}|}, \tag{84}$$

where $e_0^2 = e^2/4\pi\varepsilon_0$ and ε_0 is the permittivity of vacuum. In classical analytical mechanics, the two-body problem is solved with the help of symmetries of Galilean space-time and the resulting conservation laws. E. g., this problem goes along with the conservation of the total momentum due to spatial translation invariance. In terms of stochastic processes like the ones that are considered here, this is transferred to the expectation value of the total momentum. In order to keep the analogy to the classical case, we introduce stochastic processes for the relative coordinate $r(t) \in \mathbb{R}^3$ of the electron w. r. t. the nucleus and the center-of-mass (COM) $R(t) \in \mathbb{R}^3$ of the hydrogen atom, respectively. In [3], it was shown that the motion and dynamics of $r(t)$ and $R(t)$ decouple, where the QHEs for the latter process lead to a simple Brownian motion with diffusion coefficient $\hbar/(m_e + m_n)$. The SDEs for the relative coordinate

$$dr(t) = u(t)\mathrm{d}t + \sqrt{\hbar/\mu}\,\mathrm{d}W_f(t) \tag{85}$$

$$\mu\mathrm{d}u(t) = \frac{e_0^2}{|r(t)|^3}r(t)\mathrm{d}t + q(t)\,\mathrm{d}W_b(t)\,, \tag{86}$$

remain to be solved, with μ being the reduced mass $(1/\mu = 1/m_n + 1/m_e)$ and the stochastic process $q(t) \in \mathbb{R}^3$. Eq. (85) and Eq. (86) are a system of 6 coupled SDEs, where the COM motion has been separated. This can be further generalized to many body problems with dimension d and n particles, resulting in a system of $2 \cdot d \cdot (n-1)$ coupled SDEs. This shows also that analytical or even numerical solutions within the stochastic approach are limited to few-body problems highlighting the close relationship of the QHEs to the classical ones. In classical mechanics, one proceeds to solve the two-body central force problem, i.e., the Kepler problem, exactly by employing the further conservation laws. The considered two-body problem here is a Kepler-like system for the relative coordinate as well. This becomes more obvious when we consider the expectation value

$$\mathrm{d}\mathbb{E}[r(t)] = \mathbb{E}[u(t)]\mathrm{d}t \tag{87}$$

$$\mu\mathrm{d}\mathbb{E}[u(t)] = \mathbb{E}\left[\frac{e_0^2}{|r(t)|^3}r(t)\right]\mathrm{d}t\,, \tag{88}$$

which is a version of Ehrenfest's theorem stating that the time derivative of the expectation values of the position and momentum operators follow the laws of classical mechanics. From the physicist's point of view, it is known from the ground state solution of the Schrödinger equation of this problem that there is no angular momentum in this state. Thus, it is interesting to ask why the particle in the stochastic picture does not get trapped by the proton as it would be the case for the classical Kepler-problem, since there has to be some angular momentum that prevents the particle from falling into the nucleus.

These SDEs for the relative coordinate can be solved numerically via direct evaluation of the SDEs forward and backward in time by discretizing time and spatial axis and a least squares method for conditional expectations as described before. The advantage of the stochastic description is that after determining the velocity field of the osmotic velocity, possible trajectories of the particle can by visualized. In the left plot of Fig. 8, two paths of an electron in the ground state are shown depicting the s-orbital structure of the hydrogen atom. The numerical determined expectation value of the energy (in Rydberg units) is obtained as $E_0 = -0.501 \pm 0.01$ (exact value $-1/2$) and the numerical result of the angular momentum in x direction (e.g.) is $\mathbb{E}[L_x] = 10^{-6} \pm 10^{-7}$ (exact value 0), where the errors are given from 5 independent simulations. Furthermore, the SUSY scheme presented in subsection 3.1.1 allows us to evaluate an excited state within the presented formalism. Remember, it is possible to determine the entire (bound) spectrum from the ground state based on the adjusted scheme for

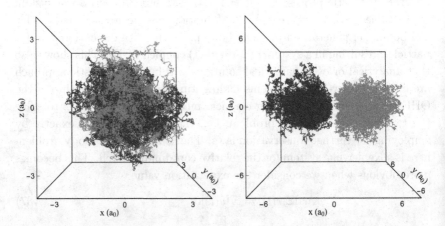

Fig. 8. Example trajectories of electrons in different states of the hydrogen atom generated by the numerically calculated velocity field u are depicted. **Left:** Two possible paths for the ground state are shown. **Right:** Two possible trajectories in the p_x orbital are plotted, where the nodal plane can be seen quite clearly.

stochastic mechanics. This is done by an iterative method by simply adjusting the potential with the help of previously calculated velocities and, finally, by acting with operators on the newly determined osmotic velocities [3]. Two paths of a first excited state are shown on the r. h. s. of Fig. 8, where the motion hints at the typical dumbbell shaped p-orbital. The numerical determined expectation value of the energy (in Rydberg units) is $E_1 = -0.12 \pm 0.08$ (exact $-1/8$) and the expectation value of the angular momentum was calculated as $\mathbb{E}[L_x] = 0.98 \pm 0.08$ (exact 1), where the errors are, again, given by 5 independent simulations.

To summarize, one can see that the numerical result for the two first states are in good agreement with the exact one. However, the problem with the excited states and also the ground state is that a lot of numerical effort is needed due to a $2 \cdot 3$ dimensional problem that is to be solved within a numerical approach including numerical derivatives. Consequently, one ends up with typical problems of many particle physics or high dimensional systems whether it is classical physics where the size of phase space explodes or quantum mechanics where the configuration space explodes with increasing the number of particles. One possible way to get around the bad scaling of dimension is proposed in the appendix. There neural networks are used to solve the coupled forward backward SDEs.

Nevertheless, in the specific case of the hydrogen atom, the considered problem can be further simplified by exploiting its symmetry. We may introduce spherical coordinates $(\rho, \vartheta, \varphi) \in (0, \infty) \times (-\pi/2, \pi/2) \times [0, 2\pi)$. The t-dependence of the processes will no longer be explicitly noted. Using Eq. (85), the new coordinates $\rho(r)$, $\vartheta(r)$ and $\varphi(r)$ obey the following SDEs (using Itô's formula)

$$d\rho = \left(u_\rho + \frac{\hbar}{\mu\rho}\right)dt + \sqrt{\hbar/\mu}\,d\tilde{W}_\rho \qquad \text{with } u_\rho = u \cdot \hat{\rho}, \qquad (89)$$

$$d\vartheta = \left(\frac{u_\vartheta}{\rho^2} + \frac{\hbar}{2\mu\rho^2}\cot\vartheta\right)dt + \frac{\sqrt{\hbar/\mu}}{\rho}\,d\tilde{W}_\vartheta \quad \text{with } u_\vartheta = \rho\,u \cdot \hat{\vartheta}, \qquad (90)$$

$$d\varphi = \frac{u_\varphi}{r^2\sin^2\vartheta}\,dt + \frac{\sqrt{\hbar/\mu}}{r\sin\vartheta}\,d\tilde{W}_\varphi \qquad \text{with } u_\varphi = \rho\sin\vartheta\,u \cdot \hat{\varphi}, \quad (91)$$

with the basis vectors $\hat{\rho}$, $\hat{\vartheta}$ and $\hat{\varphi}$ of unit length. The transformed three dimensional Wiener process $\tilde{W} = SW$ is also a 3 dimensional Wiener process, where S is the spherical transformation matrix. In order to set up the stochastic Hamiltonian $\mathcal{H}_r(r, \vartheta, \varphi, u_\rho, u_\vartheta, u_\varphi)$, we define

$$
\tilde{u}_\pm = \begin{pmatrix} \pm u_\rho + \frac{\hbar}{\mu r} \\ \pm \frac{u_\vartheta}{\rho^2} + \frac{\hbar}{2\mu\rho^2}\cot\vartheta \\ \pm \frac{u_\varphi}{\rho^2 \sin^2\vartheta} \end{pmatrix} \quad \text{and} \quad \tilde{\sigma} = \begin{pmatrix} \sqrt{\hbar/\mu} & 0 & 0 \\ 0 & \frac{\sqrt{\hbar/\mu}}{\rho} & 0 \\ 0 & 0 & \frac{\sqrt{\hbar/\mu}}{\rho\sin\vartheta} \end{pmatrix} \quad (92)
$$

and get

$$
\mathcal{H}_r = \mathcal{L}_r + p \cdot \tilde{u}_+ + \lambda \cdot \tilde{u}_- + \text{Tr}[\tilde{\sigma}q] , \quad (93)
$$

with the Lagrangian

$$
\mathcal{L}_r(r, u) = -\frac{\mu}{2}\left(u_\rho^2 + \frac{u_\vartheta^2}{\rho^2} + \frac{u_\varphi^2}{\rho^2 \sin^2\vartheta} \right) + \frac{e_0^2}{\rho} . \quad (94)
$$

and the three stochastic processes $p(t)$, $\lambda(t) \subset \mathbb{R}^3$ (backward/forward processes adjoint to the drift in the FSDE/BSDE) and $q(t) \subset \mathbb{R}^{3\times3}$. As discussed before, $\lambda(t)$ can be chosen to be 0 $\forall t \in [t_0, T]$ and the remaining stochastic processes obey

$$
dp = -\partial_r \mathcal{H} dt + q d\tilde{W}_b , \quad (95)
$$

where, per definition, we have $\partial_r = \hat{\rho}\partial_\rho + \hat{\vartheta}\partial_\vartheta + \hat{\varphi}\partial_\varphi$. By maximizing $\mathcal{H}$ with respect to u, the components of the process $p = (p_\rho, p_\vartheta, p_\varphi)^T$ can be connected to the osmotic velocity by $(p_\rho, p_\vartheta, p_\varphi) = \mu(u_\rho, u_\vartheta, u_\varphi)$. Finally, one ends up with

$$
\mu du_\rho = \frac{1}{\rho^2}\left(e_0^2 + u_\rho\hbar + \frac{1}{\rho}\left[\mu u_\vartheta^2 + \hbar\cot\vartheta\, u_\vartheta + \frac{\mu}{\sin^2\vartheta}u_\varphi^2 \right] \right.
$$
$$
\left. + \sqrt{\hbar/\mu}\left[q_{\vartheta\vartheta} + \frac{q_{\varphi\varphi}}{\sin\vartheta} \right] \right) dt + (q\, d\tilde{W}_b)_\rho \quad (96)
$$

$$
\mu du_\vartheta = \frac{1}{\rho^2 \sin^2\vartheta}\left(\frac{\hbar}{2}u_\vartheta + \mu\cot\vartheta\, u_\varphi^2 + \rho\sqrt{\hbar/\mu}\cot\vartheta\, q_{\varphi\varphi} \right) dt
$$
$$
+ (q\, d\tilde{W}_b)_\vartheta \quad (97)
$$

$$
\mu du_\varphi = (q\, dW_b)_\varphi , \quad (98)
$$

with the matrix valued stochastic process

$$
q = \begin{pmatrix} q_{\rho\rho} & q_{\rho\vartheta} & q_{\rho\varphi} \\ q_{\vartheta\rho} & q_{\vartheta\vartheta} & q_{\vartheta\varphi} \\ q_{\varphi\rho} & q_{\varphi\vartheta} & q_{\varphi\varphi} \end{pmatrix} \quad \text{and} \quad q d\tilde{W}_b = \begin{pmatrix} (q\, d\tilde{W}_b)_\rho \\ (q\, d\tilde{W}_b)_\vartheta \\ (q\, d\tilde{W}_b)_\varphi \end{pmatrix} . \quad (99)
$$

Since there are no terms with the external potential in the latter two SDEs, we can make the assumption for the ground state of the hydrogen atom that radial probabilities ρ_{nl} depending on n and l

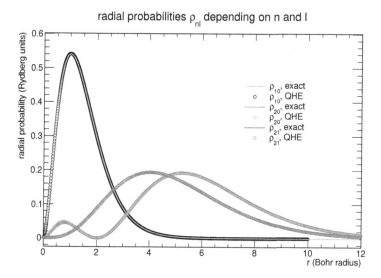

Fig. 9.   Comparison of exact and numerical results of the radial probability density related to the radial osmotic velocity u_ρ for the ground state of the hydrogen atom $n = 1, l = 0$) and the two depicted excited states $n = 2, l = 0$ and $n = 2, l = 1$ with calculated expectation value of the energies $\mathbb{E}[E_{10}] = (-0.50001 \pm 0.00001) \approx E_{\text{Ryd}}$, $\mathbb{E}[E_{20}] = (-0.12502 \pm 0.0001)$ and $\mathbb{E}[E_{21}] = (-0.1253 \pm 0.0001)$.

$u_\vartheta(\rho(t), \vartheta(t), \varphi(t)) = 0$ and $u_\varphi(\rho(t), \vartheta(t), \varphi(t)) = 0$, $\forall t \in [0, T)$ with final values $u_\vartheta(\rho(T), \vartheta(T), \varphi(T)) = u_\varphi(\rho(T), \vartheta(T), \varphi(T)) = 0$. Now recall that the stochastic process q has to ensure that the pair $(p(\cdot), q(\cdot))$ or in this case $(u(\cdot), q(\cdot))$ is a solution to Eq. (98). Therefore, it immediately follows that $q_{ij}(t)$ vanishes for $i \in \{\vartheta, \varphi\}$, $j \in \{\rho, \vartheta, \varphi\}$. Inserting this into Eq. (98), one ends up a set of equations

$$d\rho = \left(u_\rho + \frac{\hbar}{\mu\,\rho}\right) dt + \sqrt{\frac{\hbar}{\mu}} d\tilde{W}_f \qquad (100)$$

$$du_\rho = \frac{1}{\mu\rho^2}\left(e_0^2 + \hbar u_\rho\right) dt + \frac{1}{\mu}(q\,d\tilde{W}_b)_\rho\,, \qquad (101)$$

which has to be solved only for the radial part. Thus, we derived QHEs that allow to solve a reduced one-dimensional problem concerning the distance ρ between the proton and the electron for the ground state of the hydrogen atom. Note that this derivation is valid for any central symmetric potential, not only the Coulomb potential. The differences to the 1D problems so far are the additional drift terms which occur due to the transformation of variables. For example, the term $\hbar/\mu\rho$ in Eq. (100) is a 'probabilistic' force which tries to push the particle away from the center because the volume element $dV = \rho^2 \sin\vartheta d\rho d\vartheta d\varphi$ gets smaller when ρ goes to zero. In the case of the Schrödinger equation treatment for spherical problems, these additional terms occur as a consequence of the transformation of the Laplacian, whereas in the stochastic formulation, they follow from the non-vanishing variance of the stochastic process concerning the position in the mean square limit.

A numerical solution for Eq. (100) and Eq. (101) can be found using the algorithm discussed in [17, 18]. The results for the ground state of the hydrogen atom and two excited states are shown in Fig. 9. We see that the numerical results are in agreement with the exact analytic solutions. Note that the motion of the electron in the ground state is diffusive while being tied to the proton as shown in the left plot of Fig. 8. At first, this may seem odd, since one could expect a perturbed Kepler-like elliptic motion of the electron or something that may be related to the Bohr model of the atom. In the ground state, however, this is not observed due to the lack of an angular velocity when taking expectation values of the angular velocities. The classical Kepler-problem should be recovered in the Bohr correspondence limit which was shown for the Nelson diffusion at least in two dimensions [11]. The radial probabilities for the excited states in Fig. 9 were calculated with the help of an iterative scheme based on the SUSY approach [3]. The two indices correspond to the so called principal quantum number n and the azimuthal quantum number l.

5. Summary and Further Research

We have discussed in this chapter an application of the theory of stochastic optimal control to physics, more precisely to the foundations of quantum mechanics. We have shown that stochastic optimal control is able to close a long-standing gap in the stochastic formulation of quantum mechanics and derive a set of coupled forward-backward stochastic differential equations, which we called quantum Hamilton equations of motion due to the correspondence with their counterpart in classical analytical mechanics. With this derivation, we have a complete quantum analytical mechanics at hand and can begin to explore its possibilities in either solving so far untreatable problems of quantum mechanics or solving problems easier and within a more consistent interpretational framework.

We have shown for a selection of applications how this framework gives new answers to old problems, e.g., the tunneling problem, and provides them numerically more precisely then existing very convoluted treatments. We have also established by an incorporation of a theory for the factorization of Hamiltonians, called supersymmetry in physics, that the complete discrete spectrum of energies of a quantum system including their eigenstates can be derived within our approach. The theory is a natural extension of classical analytical mechanics to the quantum realm, reducing to it in a well-defined limit. This is exemplified by the obvious fulfilment of Ehrenfests theorems as well as the possibility to discuss the hydrogen atom closely paralleling the arguments followed in classical mechanics for the Kepler problem.

The treatment of the hydrogen atom, however, also shows where the theory needs to be extended. Quantum particles possess an internal degree of freedom called the spin, which interacts with their spatial degrees of freedom, which in turn is observable experimentally in shifts of the energy values of the hydrogen atom depending on the spin state of the electron. An extension of the approach of stochastic optimal control to not only treat spatial but also spin degrees of freedom is under way. Another direction of development is an extension away from the Galilean space-time of nonrelativistic quantum mechanics, to which the stochastic approach to quantum mechanics was confined so far, to relativistic physics, i.e., Minkowski space-time and the Dirac equation. As Minkowski space-time is not a Riemannian manifold, this may require to extend the realm of diffusion processes to jump-diffusions, however, stochastic optimal control theory has already been formulated to include those.

A.1. Appendix: Numerics for Higher-Dimensional Problems

The Curse of Dimensionality

The implemented Bender-Steiner method to solve the QHE equations is slow for multidimensional and multi-particle systems. The number of intervals k needed to divide $(\mathbf{a}, \mathbf{b}) \subset \mathbb{R}^n$ for N particles grows exponentially, i.e. $k = s^{n \cdot N}$ (s is the number of intervals in each dimension). Thus, the number of simulated particles and the number of iterations has to be increased to get a smooth osmotic velocity, which is needed for the derivatives. This problem of dynamic programming (DP) was declared as 'the curse of dimensionality' by Richard Bellman in 1957. Furthermore, it is not clear how to solve the time-dependent equations using existing methods.

In this appendix we want to address these problems with a different kind of algorithms using the concepts of machine learning and artificial neural networks. As discussed, the quantum Hamilton principle derived by Pavon can be seen as a stochastic optimal control problem

$$J[\hat{\mathbf{u}}, \hat{\mathbf{v}}] = \max_{\mathbf{u}} \min_{\mathbf{v}} \mathbb{E} \left[\int_{t_0}^{T} (\frac{m}{2}(\mathbf{v} - i\mathbf{u})^2 - V(x,t)dt + \Phi_0(\mathbf{x}_0) \right] \qquad \text{(A.1)}$$

with two controls $\mathbf{v}$ and $\mathbf{u}$ or one complex control $\mathbf{v}_q = \mathbf{v} - i\mathbf{u}$ (see Chapter 3).

Machine Learning for Stochastic Optimal Control

For the following discussion, we adapted some of the concepts in [13] and [14] using a machine learning based approach. Consider a stochastic control problem with finite time horizon T, e.g. equation A.1, defined on a probability space $(\Omega, \mathcal{F}, P)$ with filtration $\mathcal{F}_t$ and $\mathcal{F}_t \subseteq \mathcal{F}_{t+h}$ for all $t, h \geq 0$ and any mentioned stochastic process indexed with t is $\mathcal{F}_t$ measurable. As has been shown, this control problem can be written as a set of stochastic differential equations. In general, these equation have the following form. Let $(W_t, \mathcal{F}_t)$, $t \geq 0$ be a n-dimensional Wiener process and let $b : [0, \infty) \times \mathbb{R}^n \to \mathbb{R}^n$, $g : \mathbb{R}^n \to \mathbb{R}^n$ and $\sigma : [0, \infty) \times \mathbb{R}^n \to \mathbb{R}^{n \times d}$ be measurable functions. The evolution of the systems state X_t is given by the FBSDEs

$$X_t = X_0 + \int_0^t b(s, X_s, Y_s)ds + \int_0^t \sigma(s, X_s) \cdot dW_s^f, \qquad \text{(A.2)}$$

$$Y_t = g(X_T) + \int_t^T f(s, X_s, Y_s, Z_s)ds - \int_t^T Z_s \cdot dW_s^b. \qquad \text{(A.3)}$$

It has been shown [14] that this set of equations, with

$$Y_t = u(t, X_0 + W_t^f) \text{ and } Z_t = \sigma^T J_x(u(t, X_0 + W_t^f)), \qquad \text{(A.4)}$$

is almost surely (using the Ito formula, Feynman-Kac theorem) equivalent to the Hamilton-Jacobi-Bellman equation in u with Jacobian matrix J. It will be useful to have a forward discretization of the backward equation. Note that for all $t_1, t_2 \in [0, T]$ it holds almost surely that

$$Y_{t_2}^j = Y_{t_1}^j - \int_{t_1}^{t_2} f_j(s, X_s, Y_s, J_x(u(t, X_0 + W_t)))ds + \int_t^T J_x(u(t, X_0 + W_t))dW_s^f$$

$$\text{(A.5)}$$

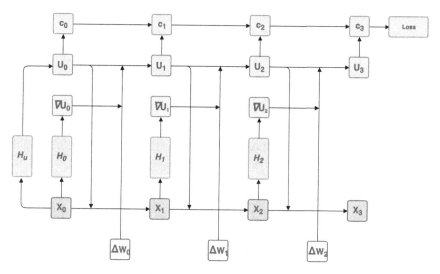

Fig. A.1. Illustration of the neural network for the FBSDE approach for four timesteps and sub-networks $H_\bullet$. The arrows denote functional dependence.

and for a sufficiently large segmentation of this interval it follows [14]

$$Y_{t_{n+1}} \approx Y_{t_n} - f(t_n, X_{t_n}, Y_{t_n}, Z_{t_n})\Delta t_n + Z_s(t_n, X_{t_n}) \cdot \Delta W_{t_n}^f \qquad \text{(A.6)}$$

almost surely. In the stationary case of stochastic mechanics, this can be rewritten as

$$X_{t_{n+1}} \approx X_{t_n} + u(t_n, X_{t_n})\Delta t_n + \Delta W_{t_n}, \qquad \text{(A.7)}$$

$$u_{t_{n+1}} \approx u_{t_n} + \nabla_x V(x)\big|_{x=X_{t_n}} \Delta t_n + q(t_n, X_{t_n}) \cdot \Delta W_{t_n}. \qquad \text{(A.8)}$$

The corresponding infinite horizon cost (or loss) function

$$J[\hat{\mathbf{u}}] = \min_{\mathbf{u}} \lim_{T \to \infty} \frac{1}{T} \mathbb{E}\left[\int_{t_0}^{T} (\frac{1}{2}\mathbf{u}^2 + V(x,t))dt + \Phi_0(\mathbf{x}_0) \right] \qquad \text{(A.9)}$$

can be written in discretized form and approximated as a finite horizon problem with sufficiently large T:

$$J[\hat{\mathbf{u}}] = \min_{\mathbf{u}} \mathbb{E}\left[\sum_{\tau=0}^{T-1} c_\tau(u_\tau, X_\tau) + c_T(X_T)|X_0 \right]. \qquad \text{(A.10)}$$

The optimization task is to approximate the functional dependence of the control on the state. This can be represented by a multilayer feedforward neural network with parameters θ_t,

$$u_t(X_t) \approx u_t(X_t|\theta_t), \qquad \text{(A.11)}$$

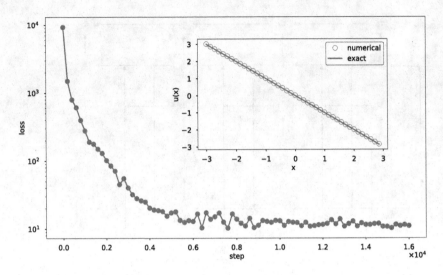

Fig. A.2.   Loss of the ten-dimensional harmonic oscillator and a one-dimensional slice of the averaged control. Three fully-connected layers for each sub-network and 80 timesteps are used with a learning rate of 10^{-6}.

and the optimization problem becomes

$$\min_{\{\boldsymbol{\theta}_\tau\}_{\tau=0}^{T-1}} \mathbb{E}\left[\sum_{\tau=0}^{T-1} c_\tau(u_\tau(X_\tau|\boldsymbol{\theta}_\tau), X_\tau) + c_T(X_T)|X_0\right]. \qquad (A.12)$$

From Eq. (A.7), Eq. (A.8) and Eq. (A.12) a neural network with the architecture shown in Fig. A.1 can be constructed.

The subnetworks $H_\bullet$ are divided in N_{sub} fully-connected layers and the only parameters to be optimized are the weights and biases of every layer. For the optimization, a batch of normal random numbers is generated and for each element of the batch $X_0 \in \mathcal{D} \subset \mathbb{R}^n$ is chosen randomly and fed into the network. Using backpropagation the parameters are adapted for every batch (step). For higher dimensional controls, the gradient is replaced by the Jacobian (see Eq. (A.4)).

The optimal control is learned by the neural network in continuous state space. Consequently, there is so segmentation of this space needed. For example, the harmonic oscillator problem can be easily calculated in ten dimensions with resulting ground state energy $E_0 = 5.001 \pm 0.05$. (the exact value is $E_0 = 5.0$) The network was optimized using the ADAM optimizer [16]. The loss and the solution of the one-dimensional slice of the osmotic velocity are shown in Fig. A.1.

References

[1] A. A. Andrianov, N. V. Borisov, and M. V. Ioffe. Factorization method and Darboux transformation for multidimensional hamiltonians. *Theo. and Math. Phys.*, 61(2):1078–1088, 1984.

[2] C. Bender and J. Steiner. Least-squares Monte Carlo for backward SDEs. In R. A. Carmona, P. Del Moral, P. Hu, and N. Oudjane, editors, *Numerical methods in finance*, volume 12 of *Springer Proceedings in Mathematics*, pages 257–289. Springer Berlin Heidelberg, 2012.

[3] M. Beyer, M. Patzold, W. Grecksch, and W. Paul. Quantum Hamilton equations for multidimensional systems. *Journal of Physics A: Mathematical and Theoretical*, 52(16):165301, 2019.

[4] P. Blanchard, P. Combe, and W. Zheng. Physics Mathematical and Physical Aspects of Stochastic Mechanics, Lecture Notes. *Physics*, 281(8), 1987.

[5] J. M. Blatt. Practical points concerning the solution of the Schrödinger equation. *J. Comput. Phys.*, 1(3):382–396, 1967.

[6] V. G. Boltjansky, R. V. Gamkrelidze, and L. S. Pontrjagin. The theory of optimum processes. *Dokl. Akad. Nauk SSSR*, 110:7, 1956.

[7] E. A. Carlen. Conservative diffusions. *Commun. Math. Phys.*, 94:293–315, 1984.

[8] R. Courant and D. Hilbert. Methoden der mathematischen Physik. *Bull. Amer. Math. Soc.*, 38(1, Part 1):21–22, 1932.

[9] M. Crum, Associated Sturm-Liouville systems. *Q. J. Math.*, 6(1):121–127, 1955.

[10] Gaston Darboux. *Leçons sur la théorie générale des surfaces*. Paris: Gauthier-Villars, 1894.

[11] R. Durran, A. Neate, and A. Truman. The divine clockwork: Bohr's correspondence principle and Nelson's stochastic mechanics for the atomic elliptic state. *J. Math. Phys.*, 49(3):032102, 2008.

[12] F. Guerra and L. Morato. Quantization of dynamical systems and stochastic control theory. *Phys. Rev. D*, 27(8):1774, 1983.

[13] J. Han and W. E. Deep Learning Approximation for Stochastic Control Problems. *Preprint*, 2016.

[14] J. Han, A. Jentzen, and W. E. Overcoming the curse of dimensionality: Solving high-dimensional partial differential equations using deep learning. *Communications in Mathematics and Statistics*, 5(4):349–380, 2017.

[15] B. R. Johnson. New numerical methods applied to solving the one-dimensional eigenvalue problem. *J. Chem. Phys.*, 67(9):4086–4093, 1977.

[16] D. Kingma and J. Ba. Adam: A Method for Stochastic Optimization. *ICLR Conference*, pages 1–15, 2014.

[17] J. Köppe. *Derivation and Application of Quantum Hamilton Equations of Motion*. PhD thesis, Martin-Luther-Universität Halle-Wittenberg, 2018.

[18] J. Köppe, W. Grecksch, and W. Paul. Derivation and application of quantum Hamilton equations of motion. *Ann. Phys. (Berlin)*, 529(3):1600251, 2017.

[19] J. Köppe, M. Patzold, W. Grecksch, and W. Paul. Quantum Hamilton equations of motion for bound states of one-dimensional quantum systems. *J.*

Math. Phys., 59(6):062102, 2018.

[20] J. Ma, P. Protter, and J. Yong. Solving forward-backward stochastic differential equations explicitly – a four step scheme. J. Probab. Th. Rel. Fields, 98(3):339–359, 1994.

[21] E. Nelson. Derivation of the Schrödinger equation from Newtonian mechanics. Phys. Rev., 150(4):1079, 1966.

[22] B. V. Noumerov. A Method of Extrapolation of Perturbations. Monthly notices of Royal Astronomical Society, 84(8):592–602, 1923.

[23] B. Øksendal and A. Sulem. Forward-backward stochastic differential games and stochastic control under model uncertainty. SIAM J. Optim. Theory Appl., 161(1):22–55, 2014.

[24] M. Pavon. Hamilton's principle in stochastic mechanics. J. Math. Phys., 36(12):6774, 1995.

[25] P. C. Sabatier. On multidimensional Darboux transformations. Inverse Probl, 14(2):355, 1998.

[26] C. V. Sukumar. Supersymmetry, factorisation of the Schrödinger equation and a Hamiltonian hierarchy. Journal of Physics A: Mathematical and General, 18(2):L57–L61, 1985.

[27] J. von Neumann. Mathematische Grundlagen der Quantenmechanik. Springer, 1932.

[28] E. Witten. Supersymmetry and Morse theory. J. Differ. Geom., 17(4):661–692, 1982.

[29] K. Yasue. Stochastic calculus of variations. J. Funct. Anal., 41(3):327–340, 1981.